国际材料前沿丛书
International Materials Frontier Series

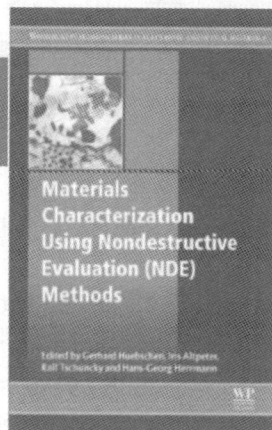

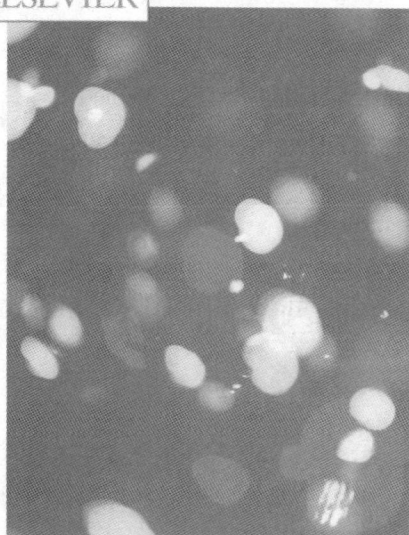

ELSEVIER

Materials
Characterization
Using Nondestructive
Evaluation (NDE)
Methods

Edited by Gerhard Huebschen, Iris Altpeter,
Ralf Tschuncky and Hans-Georg Herrmann

WP

Gerhard Hübschen,
Iris Altpeter,
Ralf Tschuncky,
Hans-Georg Herrmann

材料表征的无损检测方法

Materials Characterization
Using Nondestructive Evaluation (NDE) Methods

影印版

中南大学出版社
www.csupress.com.cn

·长沙·

图字:18 - 2017 - 163 号

内容简介

　　本书概述了能对材料进行长期、短期监测、评估和表征的无损检测方法(NDE)，主要包括材料表征的原子力显微术，扫描电子显微术，透射电子显微术，X射线显微术，X射线衍射技术，微波、毫米波和太赫兹波(MMT)技术，声学显微术，超声波技术，电磁技术以及混合技术。

　　本书内容切合实际，每一章重点介绍了不同的NDE技术，强调了材料的微观结构性质(如相含量和晶粒尺寸)，以及机械性能(如硬度、韧性、屈服强度、织构和残余应力)的测定方法。

　　本书可供土木、结构和机械工程师，材料学家，开发表征技术的物理学家以及汽车、航空航天和发电行业的研发人员使用。同时，本书也可作为高等院校材料、冶金、航空航天等相关专业学生的参考书。

作者简介

Gerhard Hübschen 博士，教授，德国弗劳恩霍夫无损检测研究所(IZFP)研究人员，从事材料的表征、缺陷检测等工作30余年。

Iris Altpeter 博士，教授，德国弗劳恩霍夫无损检测研究所(IZFP)负责人，从事材料的表征、缺陷检测等工作30余年。

Ralf Tschuncky 博士，教授，工作于德国弗劳恩霍夫无损检测研究所(IZFP)，从事材料的表征、缺陷检测等工作15余年。

Hans – Georg Herrmann 博士，德国萨尔州大学教授，德国弗劳恩霍夫无损检测研究所(IZFP)国际区域副主任，主要从事材料性能的检测和生命周期检测等工作。

目　录

Materials Characterization Using Nondestructive Evaluation (NDE) Methods

Related titles

Handbook of terahertz technology for imaging, sensing and communications
(ISBN 978-0-85709-235-9)

Microscopy techniques for materials science
(ISBN 978-1-85573-587-3)

Ultrasonic transducers: Materials and design for sensors, actuators and medical applications
(ISBN 978-1-84569-989-5)

Woodhead Publishing Series in Electronic and Optical Materials: Number 88

Materials Characterization Using Nondestructive Evaluation (NDE) Methods

Edited by

Gerhard Hübschen

Iris Altpeter

Ralf Tschuncky

Hans-Georg Herrmann

AMSTERDAM • BOSTON • CAMBRIDGE • HEIDELBERG
LONDON • NEW YORK • OXFORD • PARIS • SAN DIEGO
SAN FRANCISCO • SINGAPORE • SYDNEY • TOKYO
Woodhead Publishing is an imprint of Elsevier

ELSEVIER

WP
WOODHEAD
PUBLISHING

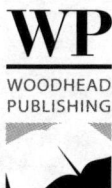

Woodhead Publishing is an imprint of Elsevier
The Officers' Mess Business Centre, Royston Road, Duxford, CB22 4QH, UK
50 Hampshire Street, 5th Floor, Cambridge, MA 02139, USA
The Boulevard, Langford Lane, Kidlington, OX5 1GB, UK

Notices
Knowledge and best practice in this field are constantly changing. As new research and experience broaden our understanding, changes in research methods, professional practices, or medical treatment may become necessary.

Practitioners and researchers must always rely on their own experience and knowledge in evaluating and using any information, methods, compounds, or experiments described herein. In using such information or methods they should be mindful of their own safety and the safety of others, including parties for whom they have a professional responsibility.

To the fullest extent of the law, neither the Publisher nor the authors, contributors, or editors, assume any liability for any injury and/or damage to persons or property as a matter of products liability, negligence or otherwise, or from any use or operation of any methods, products, instructions, or ideas contained in the material herein.

British Library Cataloguing-in-Publication Data
A catalogue record for this book is available from the British Library

Library of Congress Cataloging-in-Publication Data
A catalog record for this book is available from the Library of Congress

ISBN: 978-0-08-100040-3 (print)
ISBN: 978-0-08-100057-1 (online)

For information on all Woodhead Publishing publications
visit our website at https://www.elsevier.com/

Working together
to grow libraries in
developing countries

ELSEVIER Book Aid International

www.elsevier.com • www.bookaid.org

Publisher: Matthew Deans
Acquisition Editor: Kayla Dos Santos
Editorial Project Manager: Heather Cain
Production Project Manager: Debasish Ghosh
Designer: Greg Harris

Typeset by TNQ Books and Journals

Contents

List of contributors

I. Altpeter   Formerly Fraunhofer Institute for Nondestructive Testing (IZFP), Saarbrücken, Germany

J. Epp   Foundation Institute of Materials Science, Bremen, Germany

M.E. Fitzpatrick   Coventry University, Coventry, United Kingdom

T. Fuchs   Fraunhofer Development-Center X-ray Technology EZRT, Fürth/Bay, Germany

R. Hanke   Fraunhofer Institute for Nondestructive Testing (IZFP), Saarbrücken, Germany

G. Hübschen   Formerly Fraunhofer Institute for Nondestrutive Testing (IZFP), Saarbrücken, Germany

B.J. Inkson   The University of Sheffield, Sheffield, United Kingdom

M.K. Khan   Sichuan University, Chengdu, China

R.Gr. Maev   University of Windsor, Windsor, ON, Canada

M. Salamon   Fraunhofer Development-Center X-ray Technology EZRT, Fürth/Bay, Germany

C. Sklarczyk   Fraunhofer Institute for Nondestructive Testing (IZFP), Saarbrücken, Germany

K. Szielasko   Fraunhofer Institute for Nondestructive Testing (IZFP), Saarbrücken, Germany

R. Tschuncky   Fraunhofer Institute for Nondestructive Testing (IZFP), Saarbrücken, Germany

Q.Y. Wang   Sichuan University, Chengdu, China

S. Zabler   Julius-Maximilians-University, Würzburg, Germany

Woodhead Publishing Series in Electronic and Optical Materials

Atomic force microscopy (AFM) for materials characterization

1

M.K. Khan[1], Q.Y. Wang[1], M.E. Fitzpatrick[2]
[1]Sichuan University, Chengdu, China; [2]Coventry University, Coventry, United Kingdom

1.1 Introduction

The use of high-resolution microscopic imaging is widespread in engineering, medical, natural science, and various other fields. The local surface features and defects on the order of nanometers and lower play a crucial role in the functional performance of structures in service life. High-resolution microscopes are used to characterize the surface of materials down to the atomic scale. In many applications the characterization of materials, structures, microstructures, and damage requires spatial resolution of nanometers or lower. This has encouraged further development in capability of microscopes to characterize materials from real-space images.

In materials science and engineering, surface characterization data are usually obtained from a combination of conventional optical and electron microscopes such as scanning (SEM) and transmission (TEM) electron microscopes. However, the complete characterization process of any given material may require the application of various complementary characterization methods to obtain the information at various length scales.

Atomic force microscopy (AFM) is a relatively newly developed technique used for imaging local surface characteristics from submicron to nanometer length scale. It is a powerful nondestructive analytical technique which can be used in air, liquid, or vacuum (Bellitto, 2012; Takaharu et al., 2003; Kageshima et al., 2002; Yaxin and Bharat, 2007). The AFM has capability to generate very high-resolution topographic images of a surface down to atomic resolution. The AFM can be used to obtain the nanoscale chemical, mechanical (modulus, stiffness, viscoelastic, frictional), electrical, and magnetic properties (Bellitto, 2012; Hendrych et al., 2007; Zhang et al., 2004). In materials engineering and its allied fields, AFM can be used for 3D information of surface defects, scribes, scratches, gouges, corrosion pits, etc. The possibility to carry out imaging at such small scales, its small size, and ease in handling, makes AFM one of the very few tools capable of characterizing the surface properties around very small features. Fig. 1.1 shows a typical table-top AFM.

Materials Characterization Using Nondestructive Evaluation (NDE) Methods
http://dx.doi.org/10.1016/B978-0-08-100040-3.00001-8

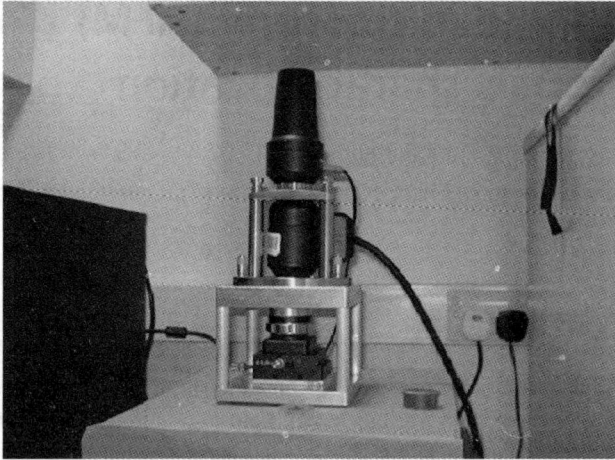

Figure 1.1 A typical atomic force microscope (AFM).

1.2 Comparison of AFM with other microscopy techniques

The best technique for characterization of any material requires knowledge of the properties of the material being analyzed, the length scale of the required information, and the limitations of the technique in use. In material characterization, SEM and TEM are mostly preferred to obtain the submicron or lower scale information. These electron microscopes are very expensive tools. The surface imaging from these microscopes requires samples to be in a vacuum and sometimes preparation to ensure electrical conductivity. The conduction of the electron beam through some samples may damage the surface. The sample chamber of these microscopes is generally very small, which does not allow the observation on large samples. In comparison, AFM has several advantages over SEM and TEM for characterizing materials at very small length scales. The fundamental advantage of AFM is its capability to provide 3D information from the probed surface. 3D information with direct height measurement enables scientists and engineers to make better decisions about the functional performance of the materials. In addition, the AFM can be used in various environments, which make measurement of some sensitive materials possible in their operating environments. Table 1.1 shows the comparison of AFM with conventional SEM and TEM. It can be seen that AFM provides substantial advantages over these techniques. Fig. 1.2 shows the cost comparison of the various surface measurement tools. It can be seen that AFM is much lower in capital cost as compared to other tools. In addition, the operating cost of the AFM is much less than the other techniques.

AFM tips or probes play a key role in measurement of the high-resolution topography. The benefits of sharper tips are numerous, such as smaller contact area and

Table 1.1 Comparison of attributes of various surface characterization tools

Parameter	SEM	TEM	AFM
Measurement environment	Vacuum	Vacuum	Air, water, gas, vacuum, etc.
Surface height	Not possible	Not possible	Possible
Measurement dimension	2D	2D	3D
Size of equipment	Large	Large	Very small
Cost	Expensive	Expensive	Cheap
Usage	Skilled operator required	Skilled operator required	Easy to use
Measurement speed	Fast	Fast	Slow

SEM, Scanning electron microscope; *TEM,* transmission electron microscope; *AFM,* atomic force microscopy.

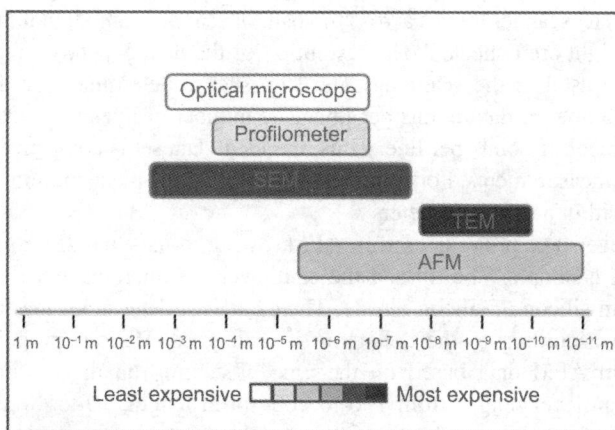

Figure 1.2 Comparison of capital cost of various surface measurement tools.

reduced long-range forces. Most conventional tips are made from silicon nitride and silicon. Now new nanotips produced through whiskers or carbon fiber are capable of providing much higher resolution images (Marcus et al., 1989). The quality of images will keep increasing in the future due to further advancements being made in the tip—surface interaction mechanisms. Combined AFM and STM with tips made up of novel materials are a potential future development which will increase the versatility of AFM data acquisition.

1.3 Principles of AFM technique

In AFM scanning, a cantilever with a sharp tip is used to scan over the surface of the sample. The cantilever probes the surface by sensing the force between surface and tip. The atoms respond to the developed van der Waals force, which can be either short-range repulsive exchange interactions or longer-range attractive, depending on the type of contact. As the tip approaches the surface, the attractive van der Waals force between the cantilever tip and sample deflects the cantilever toward the sample surface. When the cantilever tip is brought in contact with the sample surface, the repulsive van der Waals force develops, which deflects the cantilever in the opposite direction. These deflections, toward or away from the sample surface, are detected by a laser beam. The laser beam strikes the top of the cantilever and reflects back to a position-sensitive four-segment photodetector. The deflection in the cantilever is recorded by the photodetector. The segments of the photodetector are used to track the position of the laser spot on the detector and the angular deflections of the cantilever. While scanning, the AFM tip continuously moves back and forth along the surface features and the resulting deflections are recorded by the detector.

Piezo crystals are ceramic materials that expand or contract in the presence of a voltage gradient and conversely develop an electrical potential in response to mechanical pressure. The scanner moves across the path of scanning and digital image data of the scanner height are collected. The resolution of the data depends on the measurement step size used in the scanning. The step size is determined by the full scan size and the number of data points per line. The number of lines in a data set usually equals the number of points per line. Thus, the ideal data set is comprised of a dense, square grid of measurements. For better results, the Z out-of-plane motion of scanner is restricted to within a few nanometers.

The cantilever-type probe is used in AFM, owing to its suitability to measure the topography of a sample. The tip and the cantilever are micromachined components fabricated from silicon or silicon nitride. The cantilever determines the force applied to the sample according to the deflection it undergoes. Various properties can be measured from AFM data based on the type of coating on the cantilever (Koch, 2005). The cantilever ranges from 100 to 200 µm in length, 10 to 40 µm in width, and 0.3 to 2 µm in thickness. The tip, with continuous measurements, becomes blunt, which reduces the image quality. In contact mode, topography measurement of soft samples sometimes damages the tip or sample itself, owing to the higher contact forces involved in scanning.

The stiffness of the cantilever is very important for measurements in AFM. It depends on its dimension, shape, and the fabricated material. For measurement of soft materials, the AFM is used with a lower stiffness cantilever such as silicon nitride. The cantilever is deflected in the measurement process without deforming the sample surface. The stiffness and resonant frequency are higher for shorter and thicker cantilevers. These types of cantilevers are used in noncontact mode for measurement on slightly harder and rough surfaces.

1.4 Construction and basic components of AFM

The basic components of AFM are a laser diode, a photodetector which works as a scanner, a cantilever, and a sample stage capable of moving in x, y, and z directions. In AFM, a sharp microfabricated tip attached to a cantilever is scanned across the sample. The cantilever is deflected due to the forces developed between the tip and the sample. The deflection is monitored using a laser and photodiode and is used to generate an image of the surface. The probe is placed on the end of a cantilever. The amount of force between the probe and sample depends on the spring constant (stiffness of the cantilever) and the distance between the probe and the sample surface. This force can be described using Hooke's Law as $F = kx$, where F is the force exerted on the cantilever during scanning, k is the spring constant or stiffness of the cantilever, and x is the cantilever deflection. Fig. 1.3 shows the basic schematic representation of AFM construction.

1.5 Working modes of AFM

The AFM is usually equipped to work in three different modes termed as contact, noncontact, and oscillating mode. These modes use different contact forces on the cantilever depending on the interaction between the cantilever tip and sample. In contact mode, the probe experiences repulsive forces. However, in noncontact mode, the tip moves further away from the surface and attractive forces develop. The tapping mode is used on the surfaces where the constant contact of the cantilever tip may damage the surface of the sample. In this mode, the tip is brought into contact with the sample for a very short duration of time to protect the sample as well as the cantilever tip. These modes of working are further explained below.

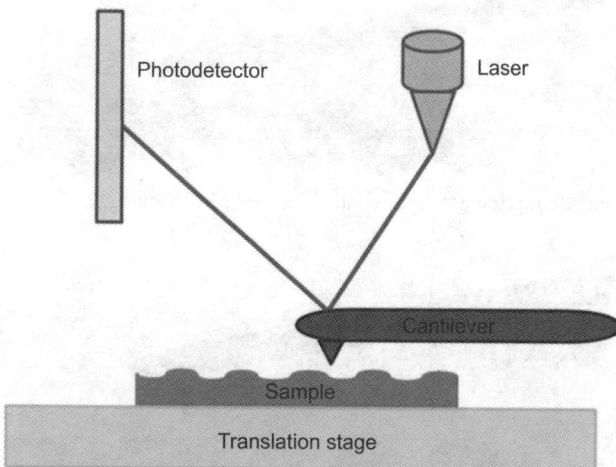

Figure 1.3 Schematic representation of atomic force microscope (AFM).

1.5.1 Contact mode

In contact mode, the AFM probe and sample surface come into contact with separation distance less than 0.5 nm. Contact mode is preferred when the sample surface is not substantially stiffer than the tip. When the cantilever stiffness is less than the measured surface, the contact bends and breaks the cantilever. The cantilever deflects, owing to a constant repulsive van der Waals force exerted on the sample, and an image of the surface is obtained. The system is equipped with a feedback circuit which continuously adjusts the cantilever height to maintain a constant force. A detector monitors the deflection and the resultant force is calculated using Hooke's law. Fig. 1.4 shows the schematic representation of contact mode AFM scanning process.

Contact mode is preferred for surfaces with moderate roughness where relatively quick scanning is required. Owing to the contact with the sample surface, this mode can be used to measure the friction coefficient of the sample surface. The main disadvantage of the contact mode is development of higher contact forces during scanning, which may produce excessive bending, leading to damage of the soft sample or cantilever.

1.5.2 Noncontact mode

In noncontact mode, the AFM cantilever and sample surface do not come into contact and keep a distance of $0.1-10$ nm. The cantilever vibrates very near to the sample surface at the frequency higher than its resonant frequency. When the cantilever moves away from the sample surface, the attractive van der Waals forces decrease the resonant frequency and the amplitude of vibration. Fig. 1.5 shows the schematic representation of noncontact mode AFM scanning process.

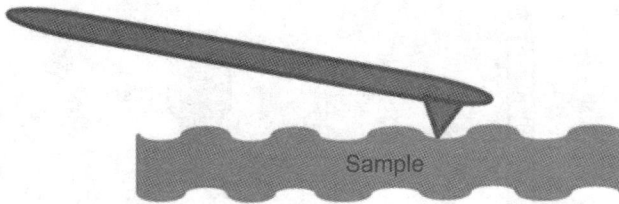

Figure 1.4 Schematic representation of contact mode AFM scanning.

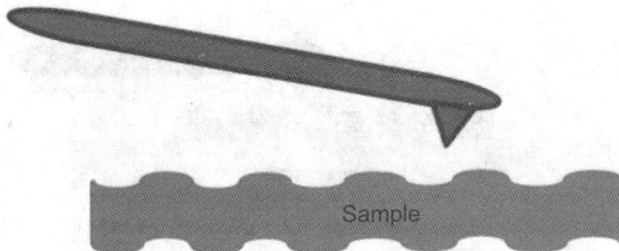

Figure 1.5 Schematic representation of non-contact mode AFM scanning.

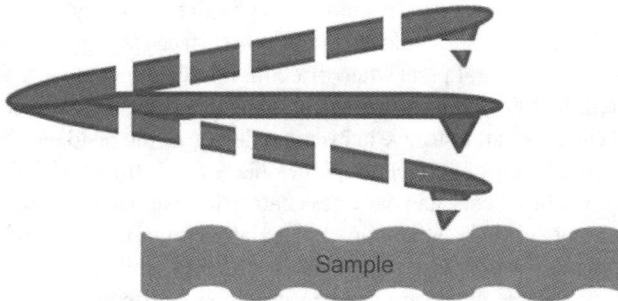

Figure 1.6 Schematic representation of tapping mode AFM scanning.

Some materials are needed to be in their operating environment to determine their properties. The noncontact mode of AFM gives the flexibility to scan the sample surface in various environments. This allows measurements for those surfaces where imaging from electron microscopes was not possible.

In tapping mode scanning in fluids, the native fluid layer of the sample is often significantly thicker than the region where van der Waals forces are significant. So the probe is either out of range of the van der Waals forces it attempts to measure, or becomes trapped in the fluid layer. In these types of environments, the noncontact mode AFM works better under ultrahigh vacuum conditions.

1.5.3 Tapping mode

In tapping mode, the cantilever makes intermittent contact with the surface at a resonant frequency. The intermittent mode is similar to the contact mode. The cantilever vibrates slightly below its resonant frequency at amplitude ranges from 20 to 100 nm. The tip just touches the sample surface during the scanning and contacts the surface at the bottom of its swing. The contact of cantilever is for a very short duration of time in which the lateral force of contact decreases. This mode provides best results for soft materials or films. Fig. 1.6 shows the schematic representation of tapping mode AFM scanning process.

In contact mode, the lateral force exerted by the cantilever on the surface can be quite high. This may damage the surface of the sample or detach loose particles from the surface. For these samples, the tapping mode is used in which the cantilever oscillates, with amplitudes of the order of nanometers during imaging. The tip touches the sample surface for a very short period of time, which avoids high lateral forces and dragging over the sample surface. However, in tapping mode the impact force by which the tip strikes the sample during oscillation may be higher in some cases compared with the contact mode. This may limit the life of the tip.

1.6 Application of AFM for material characterization

The use of AFM in material characterization from micro- to nanoscale level is continuously increasing. Mostly, AFM is used in measurement of properties such as the surface roughness. Surface parameters are important characteristics of the materials which ensure the smooth performance of manufactured product according to the design specifications. Furthermore, the characterization of surface topography is important for development of structural parts where friction, lubrication, and wear are important properties (Thomas, 1999). For measurement of service life characteristics, the observation of local changes in the surface helps in prediction of the residual life of the components. The degradation of surface properties like surface finish, roughness, and the appearance of scratches, dents, and damages control the residual life of the components.

The submicron level characterization of materials is important to completely understand the deformation behavior of materials in different applications. It is now well known that the nanoscale properties can be substantially different than the bulk material properties. In many cases, the nanoscale properties dictate and control the performance of the materials based on the microstructure of the material. AFM can be used to evaluate the 3D information of the local property variation in the materials, which makes it an ideal tool for characterization of complex material systems.

AFM can also be used for force spectroscopy. The AFM force curve is obtained through characterization of tip—sample interactions. The probe tip is brought in contact with the sample and the force is applied to penetrate the tip in the materials surface. The force penetration curve is used to measure the nanomechanical characteristic of the surface such as mechanical, chemical, and adhesion properties (Miyahara et al., 1999).

The quality of the data obtained from AFM depends on the mode of tip—surface interaction. In addition, the sharpness and stiffness of the tip is very important to get the quality data (Malegori and Ferrini, 2010). The sharp and stiffer tip makes a very small contact area with the sample surface. This helps in maintaining the contact with just a few atoms of the surface and atomic resolution data are obtained. The minimum contact area can also be ensured by reducing the tip load on the sample and scanning in a liquid environment. Here, we present applications of different modes of AFM on various materials and discuss the effectiveness of results in material characterization.

1.6.1   Surface properties measurement

In aerospace structural applications, Al 2024 plates are used extensively. These plates are often clad with pure aluminum of thickness ~ 100 μm to prevent the structural areas from corrosion. The extremely thin layer of cladding makes it susceptible to small-scale scratches, dents, and gouges during service life, which may be dangerous for the integrity of aerospace structural parts. Hence, it is important to continuously monitor the surface properties of potential risky areas of structural parts.

Fig. 1.7 shows the 3D surface profile of Al cladding and Al 2024 plate. Surface scanning of both materials was performed using contact mode of the AFM, after surface polishing of samples to a 1 μm finish. It can be seen that the surface roughness of the clad is greater than the surface roughness of the Al 2024-T351. The statistical data of the surfaces for both materials are shown in Table 1.2. The statistical data can be very useful for comparison of the surface data of the materials before and during service life for calculation of the residual life of the components.

AFM measurements can also be performed in liquid or gas media (Horng, 2009). This capability can be very advantageous in applications where better resolution

(a)

(b)

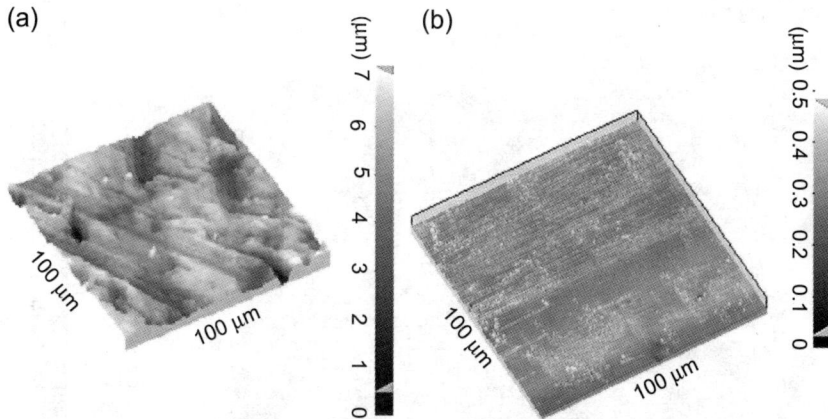

Figure 1.7 Surface scanning of (a) Al-clad and (b) Al 2024-T351.

Table 1.2 **Attributes of the 3D images of Al clad and Al 2024**

S. No.	Parameter	Al clad	Al 2024
1	Surface roughness (nm)	7910	467
2	Image texture aspect ratio	0.35	0.14
3	Skewness	0.82	2.29
4	Kurtosis	4.15	20.9
5	Texture direction	90 °	45 °

data are required, but the stiffness of the sample or surface irregularities does not allow the operation of contact mode. Scanning in fluids reduces the risk of damaging the sample surface or the tip. Fig. 1.8 shows a 3D-scanned AFM image of a collagen. Collagen is the most common protein in the human body found in bone, skin, teeth, tendon, cornea, artery wall, and more. The 3D morphological data of a collagen were obtained by operating the AFM in noncontact mode under ambient conditions, with 35% relative humidity, using a noncontact silicon cantilever. The spring constant of the noncontact probe was 14 N/m. The resonant frequency of the cantilever was 315 kHz. The nominal tip radius was <15 nm. It can be seen that the dimensions and banding pattern of native collagen fibrils were in ranges from micron to nanometer. The tiny structures form into fibrils that consist of many staggered collagen molecules. The collagens form into fibrils because of cross-links between the molecules.

Fig. 1.9 shows the application of Tapping Mode AFM image to the surface morphology of microcapsules at room temperature. The cantilever was operated at a frequency of 1 Hz, and scanning was carried out on a $2 \times 2\ \mu m^2$ area. The AFM sample was prepared by drying a droplet of microcapsule suspension on a substrate. The substrate was attached to a thin glass surface by a double-sided tape.

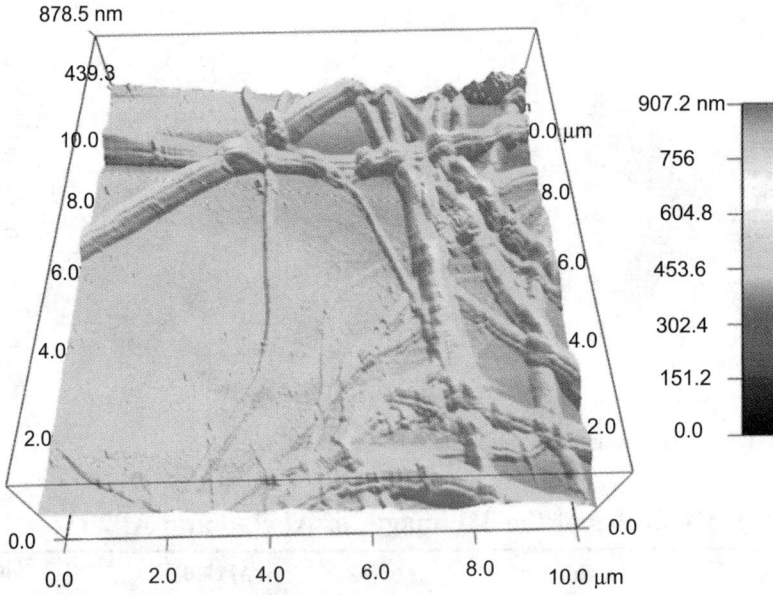

Figure 1.8 3D-scanned noncontact mode image of collagen.

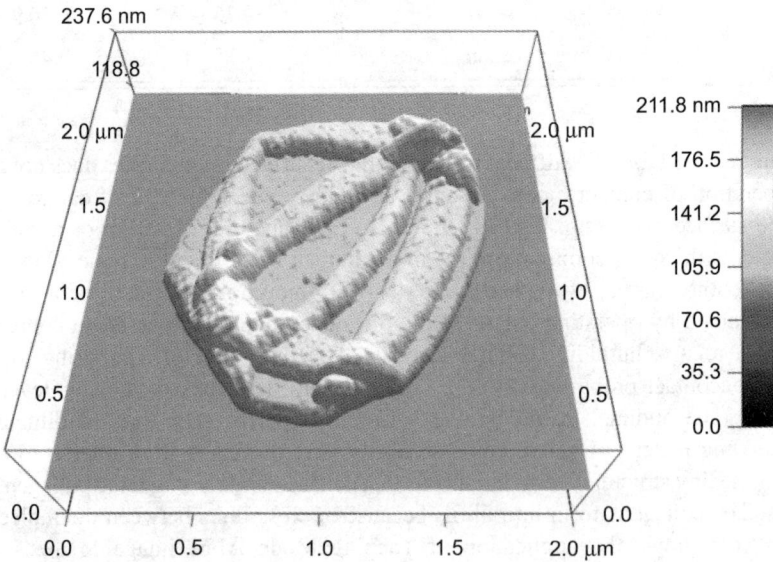

Figure 1.9 3D-scanned tapping mode image of microcapsule.

1.6.2 AFM measurements for hardness and modulus measurements

In some techniques, the measurement of local properties of the order of microns to nanometer range requires additional experimental measurements along with mathematical models. Some equipment has built-in AFMs to be used as an additional tool to obtain the 3D topographic data of surface properties (Saha and Nix, 2001; Kese et al., 2005). The data obtained from these techniques may not be complete without the aid of the AFM data. One such technique is now a well-established standard material characterization known as nanoindentation. Nanoindentation is used for characterization of nanohardness and elastic modulus of materials. In nanoindentation, a sharp Berkovich indenter penetrates into the surface of the material and records the load and penetration depth data. The data can be used for measurement of elastic modulus, hardness, residual stress, and fracture toughness. One particular challenge of testing materials by penetrating at nanometer depths is the measurement of the area of the indented region. In soft materials, while indenting, the plastically deformed material is pushed out of the indented region. This phenomenon is called pileup of the material, and it critically affects nanoindentation hardness and modulus calculations as the actual contact area between the material and indenter tip includes the area contained in the pileup. Most models for analysis of nanoindentation data do not take into account pileup area and therefore underestimate the actual area, which results in an overestimate of the calculated values of modulus and hardness. The measurement of the pileup around the edges of the indenter is usually done by AFM.

Fig. 1.10(a) shows a 3D profile of an indent, which shows pileup around each edge of the indent. Fig. 1.10(b) shows a line profile though a triangular indent and Fig. 1.10(c) shows the AFM height profile from the corner of an indentation through the midpoint of the edge of the indentation. Pileup on the edge can be clearly seen in the AFM trace. Hence the actual contact area includes the area contained in the pileup, and because significant pileup occurred at all the indentations its influence on the determination of the true contact area cannot be ignored. The area of pileup in Fig. 1.10 forms a semiellipse and can be calculated from the method explained in earlier studies (Kese et al., 2005).

1.6.3 AFM measurements for damage characterizations

The surface features that develop due to loading during the service life of components are an important characteristic for structural integrity. Features like scratches, dents, gouges, corrosion pits, etc. eventually lead to the component toward failure. In service life, the surface roughness increases owing to the surface irregularities, which becomes greater than the standard roughness of the material.

In fatigue loading, cracks initiate from the surface of material due to the formation of local plastic deformation in the form of persistent slip markings and bands. The sharp edges of these bands act as local stress concentration and serve as the initiation and driving force for cracks. These slip markings and bands appear on the surface of the material in the form of extrusions and intrusions.

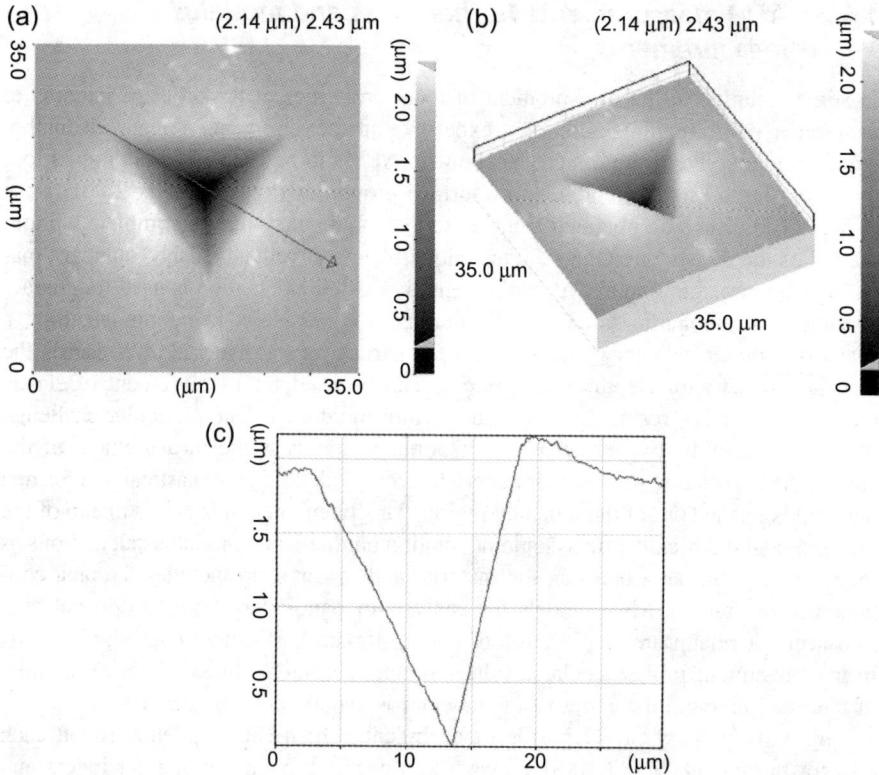

Figure 1.10 (a) Contact mode AFM scanned image of an indent in Al 2024; (b) 3D image of the indent shown in part (a) of the figure; (c) the scan across the indent from which the data in part (a) were obtained.

The access to slip bands for measurement of surface features near the fracture is usually challenging but critical for better understanding of the crack initiation mechanism. The plasticity developed on the surface of the material is localized, and AFM measured surface profiles are very useful as they provide 3D information of the slip bands and markings (Man et al., 2004; Polak et al., 2003). The topography of surface slip bands and markings changes significantly in various materials owing to the difference in the crystallographic planes, preferred orientation of the grains, and their size (Mughrabi et al., 1981). Hence, case-by-case investigation of the developed local plasticity in different materials is required for the better understanding of crack initiation mechanism. Fig. 1.11(a) shows the AFM scanned image of an area 90×90 μm near to the fracture region in AISI 310 stainless steel alloy after fatigue loading of 3×10^8 cycles at stress ratio $R = -1$. Fig. 1.11(b) and (c) shows the cross-section and 3D image of the area, respectively. The AFM height profile from the top corner of the area through the slip bands to the other corner of the area showed multiple crests and troughs. This showed that persistent slip bands were present in arrays of intrusions and extrusions around the whole periphery of the specimen, especially below the crack

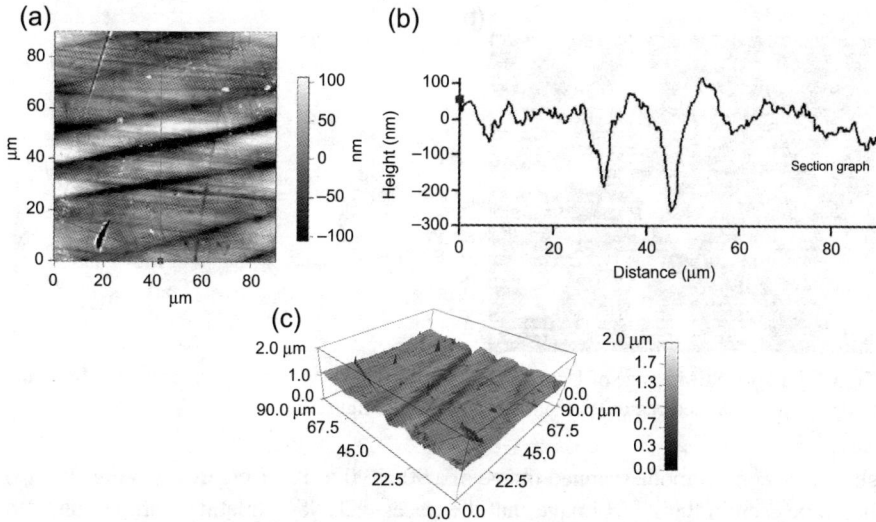

Figure 1.11 (a) Contact mode AFM scanned image of an area near to crack initiation for the specimen failed at 3×10^8 fatigue cycles; (b) height profile through the area (*from one corner to the other corner*) showing significant dislocations; (c) 3D surface topography of the region.

initiation region. The highest depth of the intrusion reached 300 nm. The bands and markings with higher intrusion depth serve as potential fatigue crack initiation and propagation regions.

1.6.4 AFM measurements for characterizations of surface treatment effects

The use of mechanical work to transfer energy into alloys for improvement of the surface properties and fatigue life is now very common. Techniques like laser shock peening, shot peening, ultrasonic shot peening, and ultrasonic nanocrystal surface modification (UNSM) have shown substantial effectiveness in improvement of the fatigue life. These techniques plastically deform the material surface and produce compressive residual stresses up to a significant depth.

In one such technique, known as UNSM, the alloy surface is targeted by striking ultrasonically with a certain force through a sharp carbide tip. The process decomposes the microstructure into nanocrystals and produces a compressive residual stress in the material, which may go as deep as 500 µm in the surface of material. The formation of a nanostructured layer on the material surface in the surface treatment is very effective in delaying the fatigue crack initiation. The process also improves the surface roughness and wear rate of the material.

Fig. 1.12(a) shows the SEM image of the surface of titanium alloy Ti-6Al-4V after UNSM treatment. The UNSM processing marks can be seen on the specimen surface. The measurement of the accurate width and height of these marks is very important for prediction of accurate wear rate and fatigue behavior of the materials. Fig. 1.12(b)

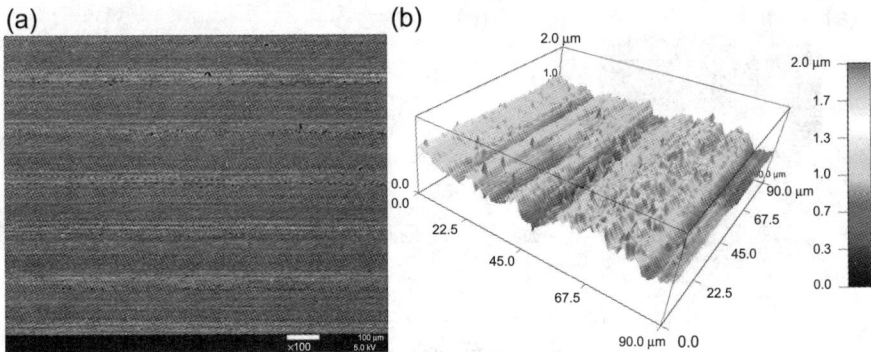

Figure 1.12 (a) SEM image of UNSM treatment on Ti6Al4V; (b) contact mode AFM scanned image of the encircled region shown in part (a) of the figure.

shows the contact mode scanned image of a 90×90 μm area encircled in Fig. 1.12(a). It can be seen in the AFM image that the traces of UNSM treatment are not smooth, and there are crust and troughs on the material surface. The height and depth of these traces can be reduced to improve the wear rate of the surface by lowering the impact force used in the UNSM treatment. However, this may also reduce the strength and depth of the compressive residual stress, which may affect the fatigue performance of the material. Hence, a balance between the surface features and corresponding impact force is usually required for optimal use of process parameters. A multiple combinations of UNSM parameters are used before the optimum process parameters are obtained, which gives the best combination of fatigue life and wear rate. In this process, the AFM data in each iteration are very useful in characterization of the material.

1.7 Conclusions

Microscopic imaging through high-resolution microscopy is of fundamental importance in characterization of materials. The study of submicron scale surface features is now mostly done by AFM. Besides its main advantage of providing 3D topographic information of surfaces at atomic scale, AFM data can be obtained with minimal surface preparation and without any environment restriction, which is not matched by any other techniques. It can record topographic images and can be used to obtain the information on nanoscale chemical, mechanical (modulus, stiffness, viscoelastic, frictional), electrical, and magnetic properties. In materials engineering and its allied fields, AFM can be used in generation of 3D information of surface defects, scribes, scratches, gouges, etc.

The AFM probes play a key role in realizing high resolution topography. The benefits of sharper tips are numerous, such as smaller contact area and reduced long-range forces. Most conventional tips are made from silicon nitride and silicon. Recent developments in producing nanotips through whiskers or carbon fiber may find potential application in AFM for high-resolution images. The quality of images will keep increasing in future due to further advancements being made in the tip—surface interaction mechanisms. In addition, the automation of AFM scanning, multiprobe AFM, better control on fast speed scanning, and increase in the robustness in the scanning process are possible future developments. These developments will enhance the capability of engineers and scientists to acquire the throughput of AFM from surfaces.

AFM is a relatively newly developed technique. It is a powerful surface analytical technique which can be used in air, liquid, or vacuum. Despite recent developments in AFM, instrumentation for precise control of tip movement, it is still highly desirable to confirm the reliability of AFM topography with complementary techniques such as SEM or TEM.

Acknowledgments

M.E. Fitzpatrick is grateful for funding from the Lloyd's Register Foundation, a charitable foundation helping to protect life and property by supporting engineering-related education, public engagement, and the application of research.

References

Bellitto, V., 2012. Atomic Force Microscopy-imaging, Measuring and Manipulating Surfaces at the Atomic Scale. TeOp Publishing, USA.

Hendrych, A., Kubínek, R., Zhukov, A.V., 2007. The magnetic force microscopy and its capability for nanomagnetic studies- the short compendium. In: Méndez-Vilas, A., Díaz, J. (Eds.), Modern Research and Educational Topics in Microscopy, vol. 2. Formatex, Badajoz, Spain, pp. 805—811. ISBN: 13: 978-84-611-9420-9.

Horng, T.L., 2009. Analyses of vibration responses on nanoscale processing in a liquid using tapping-mode atomic force microscopy. Appl Surf Sci 256, 311—317.

Kageshima, M., Jensenius, H., Dienwiebel, M., Nakayama, Y., Tokumoto, H., Jarvis, S.P., et al., 2002. Noncontact atomic force microscopy in liquid environment with quartz tuning fork and carbon nanotube probe. Appl Surf Sci 188, 440—444.

Kese, K.O., Li, Z.C., Bergman, B., 2005. Method to account for true contact area in soda lime glass during nanoindentation with the Berkovich tip. Mater Sci Eng A 404, 1.

Koch, S.A., 2005. Functionality and dynamics of deposited metal nanoclusters [Ph.D. thesis]. Groningen University Press, Groningen, The Netherlands. ISBN: 90-367-2289-6.

Malegori, G., Ferrini, G., 2010. Tip-sample interactions on graphite studied in the thermal oscillation regime. J Vac Sci Technol B 28, C4B18.

Man, J., Petrenec, M., Obrtlik, K., Polak, J., 2004. AFM and TEM study of cyclic slip localization in fatigued ferritic X10CrAl24 stainless steel. Acta Mater 52, 51—61.

Marcus, R.B., Ravi, T.S., Gmitter, T., Chin, K., Liu, D., Orvis, W.J., et al., 1989. Formation of atomically sharp silicon needles, 01631918, Washington, DC, USA.

Miyahara, K., Nagashima, N., Ohmura, T., Matsuoka, S., 1999. Evaluation of mechanical properties in nanometer scale using AFM-based nanoindentation tester. Nanostruct Mater 12, 1049—1052.

Mughrabi, H., Herz, K., Stark, X., 1981. Cyclic deformation and fatigue behaviour of alpha-iron mono- and polycrystals. Int J Fract 17, 193−220.

Polak, J., Man, J., Obrtlik, K., 2003. AFM evidence of surface relief formation and models of fatigue crack nucleation. Int J Fatigue 25, 1027−1036.

Saha, R., Nix, W.D., 2001. Soft films on hard substrates- nanoindentation of tungsten films on sapphire substrates. Mater Sci Eng A 319−321, 898.

Takaharu, O., Hiroshi, S., Hideo, A., Atsushi, I., 2003. Self oscillation technique for AFM in liquids. Appl Surf Sci 210, 68−72.

Thomas, T.R., 1999. Rough Surfaces, second ed. Imperial College Press, London, ISBN 978-1-86094-100-9.

Yaxin, S., Bharat, B., 2007. Finite-element vibration analysis of tapping-mode atomic force microscopy in liquid. Ultramicroscopy 107, 1095−1104.

Zhang, F.M., Liu, X.C., Gao, J., Wu, X.S., Du, Y.W., Zhu, H., et al., 2004. Investigation on the magnetic and electrical properties of crystalline Mn0.05Si0.95 films. App Phy Let 85 (5), 786−788.

Scanning electron microscopy (SEM) and transmission electron microscopy (TEM) for materials characterization

2

B.J. Inkson
The University of Sheffield, Sheffield, United Kingdom

2.1 Introduction

Electron microscopy has been a revolutionary imaging technology for scientists and engineers over the past 80 years, opening up the world of nanoscale materials and enabling characterization of their unique properties. The power of electron microscopes to image submicron-sized objects, even down to single atomic positions, has led to the development of completely new nanotechnologies, and also enabled remarkable developments to take place by nanoscale engineering of macrosized components. Many everyday objects such as mobile phones, plasma screen televisions, and materials for car and aeroplanes have required extensive use of electron microscopy to facilitate their development.

Electron microscopy has become a key technology for materials characterization across a wide range of industries. Historically, access to electron microscopy methods has largely been restricted to large institutions able to invest in expensive instrumentation. However, perhaps the most exciting development in electron microscopy is the desktop electron microscope, which enables imaging of objects down to <15 nm at a fraction of the price of a traditional instrument. It is now economic to make electron microscopes accessible to small businesses, university students, and even schools. This step change in accessibility will enable the embedding of electron microscopy into all levels of materials development, optimization, failure analysis, and education.

Electron microscopy is a complex, mature technology that has evolved over the past 80 years to have many strands and methodologies. In this chapter, we will focus on the main themes relevant for engineers wishing to carry out noninvasive imaging of component surfaces and key microstructural features. We will look at the two main classes of microscopy, scanning electron microscopy (SEM) and transmission electron microscopy (TEM), and how to decide which technique is most appropriate for characterizing common material properties including surface quality, grain size, and local chemistry.

Materials Characterization Using Nondestructive Evaluation (NDE) Methods
http://dx.doi.org/10.1016/B978-0-08-100040-3.00002-X

2.2 Why electron microscopy?

The most common microscope in the world today is a light microscope. Light microscopes are cheap, robust, and typically noninvasive, leaving the imaged object unchanged by the imaging process. There are many variants of light microscopy, but there is typically a limit to their ability to distinguish objects and features smaller than about 0.1 μm (100 nm) (Bradbury and Bracegirdle, 1998; Mertz, 2009) (Fig. 2.1). The resolution of a microscope is the ability to separate two features a given distance apart as individual objects in the image. For a light microscope, the limit of resolution is an intrinsic property due to the wavelength λ of visible light radiation, which ranges from 400 nm (blue) to 700 nm (red).

Shorter wavelengths of radiation interact more strongly with nanoscale materials, and can produce higher resolution images (Reimer and Kohl, 2008; Williams and Carter, 2009; Goodhew et al., 2001; Buseck et al., 1988; Spence, 2013). Considering the electromagnetic spectrum, types of radiation with wavelengths smaller than visible light include Ultraviolet (100−400 nm) and X-rays (10 pm−10 nm). X-rays are widely used to characterize materials with an image spatial resolution of around 1 μm (see chapters: X-ray microtomography for materials characterization and X-ray diffraction (XRD) techniques for materials characterization).

Particles can also behave as waves, known as wave-particle duality. The wavelengths of electrons are dependent on their momentum, which can be changed by accelerating them through a range of voltages. Higher accelerating voltages produce high-energy electrons with smaller wavelengths. Electrons with wavelengths of 40 pm down to 1 pm can be readily achieved using accelerating voltages of 1−300 kV (Reimer and Kohl, 2008; Williams and Carter, 2009). A microscope designed to use 200−300 keV electrons thus offers a potential step change in image spatial resolution

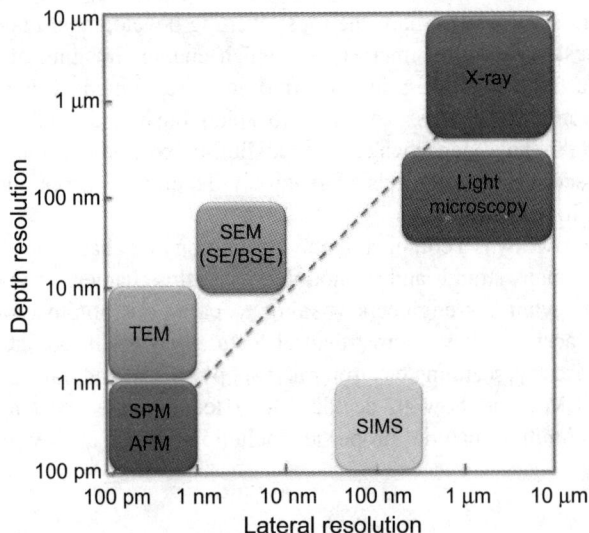

Figure 2.1 Comparison of the typical spatial resolution of key materials characterization imaging methods SPM, AFM, TEM, SEM, SIMS, X-ray imaging, and light microscopy.

over optical microscopy, down to less than 0.1 nm (100 pm) (Buseck et al., 1988; Spence, 2013). This atomic level resolution is comparable to that of the nondestructive surface imaging techniques scanning probe microscopy (SPM) and atomic force microscopy (AFM) (see chapter: Atomic force microscopy (AFM) for materials characterization), and to that of destructive secondary ion mass spectrometry (SIMS) (Fig. 2.1).

2.2.1 Key advantages of imaging with electrons

High resolution: The main advantage of electron microscopy is the high image resolution compared to light microscopy. State-of-the-art transmission electron microscopes can analyze the position and chemistry of individual atoms (Reimer and Kohl, 2008; Williams and Carter, 2009; Goodhew et al., 2001; Buseck et al., 1988; Spence, 2013), although we will see in this chapter that in practice the resolution of data depends on the type of electron microscopy technique chosen and the microscope operating conditions.

Magnification range: Electron microscopes offer a very high magnification range, typically in the range 10–500,000 times for SEM, and 2000 to 1 million times for TEM. This enables characterization of microstructures at many different length scales, from micro- to nanoscale, within an imaging session. However, it is important to understand the difference between magnification and resolution. The magnification of an image is the size of a feature in the image relative to the actual size of the same feature on the sample. Magnification is making the image bigger, whereas resolution is the ability to distinguish the features of the object in the image. High magnification is useless if the microscope operating conditions have not been optimized to achieve high resolution.

Electron generated signals: A major advantage of using electrons rather than light is that because electrons have negative charge they interact very strongly with atoms. Electron interactions with the specimen lead to a wide range of phenomena which generate emission of signals from the specimen. These signals, summarized in Section 2.4, can be detected and used to form structural and chemical images of specific areas of the specimen. EM is thus multifunctional, being both a nanoscale imaging and spectroscopy tool.

2.2.2 Key disadvantages of imaging with electrons

Vacuum: The strong interactivity of electrons with matter can lead to difficulties in methodology. Electrons interact with air molecules, so electron microscopes require some degree of vacuum to ensure that sufficient electrons in the electron beam are not scattered on their journey from the electron source through the electron optics to the specimen. Typical pressures used in electron microscopes range from 0.1 to 10^{-4} Pa for SEM (low vacuum), and from 10^{-4} to 10^{-7} Pa for TEM (high vacuum).

The requirement for a vacuum environment can be in conflict with the nature of the material to be imaged. Samples with a high vapor pressure such as biological materials and liquids will change by loss of water and other volatiles in vacuum. To combat this problem, special variable pressure and environmental electron microscopes have been developed that can operate at reduced vacuum (up to 4 kPa) in the close vicinity of the specimen to keep the specimen as unchanged as possible (Stokes, 2008).

High electron energy: The interaction of electrons with materials is a function of the kinetic energy of the incident electrons. Typical electron microscopes operate with 1−300 keV electrons, dependent on the microscope used. Transfer of energy into the imaged material can lead to changes in the material, a scenario which must be risk assessed by the scientist for every electron microscopy investigation. In fact, some material changes are inevitable at the atomic level, but may range in a spectrum from negligible to significant. The point at which the electron beam induced damage becomes unacceptable will be dependent on the material being quantified and the methodology being used.

2.3   Types of microscopes

Two generic classes of electron microscopy exist, SEM and TEM, depending on the type of microscope used. Fig. 2.2 shows schematically the fundamental design principles of SEM and TEM microscopes.

SEM and TEM microscopes all generate highly focused beam of electrons, which impact the specimen inside a vacuum chamber. However SEM microscopes are designed primarily to examine material surfaces (like reflection light microscopes), whereas TEM microscopes are primarily designed to examine the internal structure of specimens (like transmission light microscopes).

SEM: In SEM the electron beam is focused to a spot, and is scanned sequentially across the specimen (Fig. 2.2(a)) (Stokes, 2008; Reimer, 1998; Egerton, 2011; Goldstein et al., 2007; Kuo, 2014; Pawley and Schatten, 2014). At each location, signals are emitted from the specimen and collected by detectors. The detector signal is synchronized with known location of the beam on the specimen, and the signal intensity is used to modulate the corresponding image pixel. The signals collected in series are combined to form an image whose dimensions/pixel distribution depends on the scan pattern chosen. Typical electron energies are 1−30 keV.

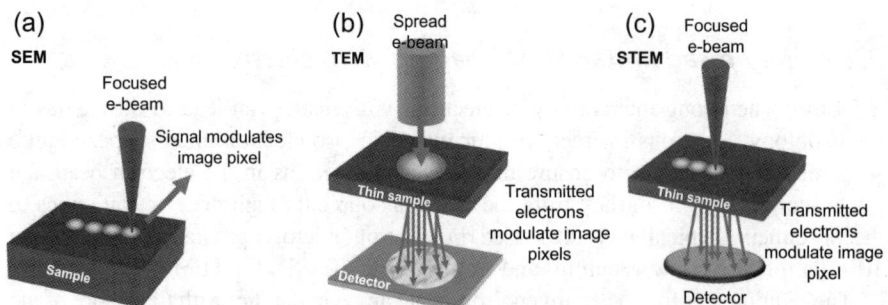

Figure 2.2 Schematic of SEM, TEM, and STEM imaging methodology. (a) Serial collection of data points in SEM. (b) Parallel image acquisition in TEM. (c) Serial collection of transmitted electrons in STEM.

TEM: In TEM the electron beam is incident onto a defined area of the specimen (Fig. 2.2(b)). Electrons transmitted through the specimen are focused by lenses and collected by a parallel detector to form an image. Electron energies in TEM are much higher than SEM, typically 80−300 keV, to enable them to penetrate through material (Reimer and Kohl, 2008; Williams and Carter, 2009; Goodhew et al., 2001; Buseck et al., 1988; Spence, 2013).

Hybrid versions of SEM and TEM are also possible in modern machines. SEM microscopes can be fitted with detectors to collect electrons transmitted through thin specimens (Fig. 2.2(c)). This scanning mode methodology with transmitted signal collection is usually called scanning transmission electron microscopy (STEM) (Egerton, 2011). Modern TEMs fitted with extra scan instrumentation can also scan a highly focused beam, enabling STEM implementation at higher electron energies (Keyso et al., 1998; Pennycook and Nellist, 2011).

The choice of which microscope to use is not always a simple choice. A plethora of microscopy instrumentation and techniques are now available, meaning that the same sample may be subjected to a wide range of surface and internal microstructure investigations. This chapter aims to give a basic guide to opportunities available, which can be individually followed up with dedicated texts on individual techniques.

2.4 Interaction of electrons with materials

There is a range of mechanisms by which electrons interact with matter, which are exploited in electron microscopy to generate a range of signals which are collected and used to form an image of the specimen.

We distinguish the primary electrons which impact the specimen, from the secondary electrons (SEs) that originate from inside the specimen:

Primary electrons: The electrons which are targeted at the specimen.
Secondary electrons: Electrons with kinetic energy produced by primary electron−matter interactions. SE emission occurs when they escape from the sample.

Primary electrons incident on the specimen each follow an individual trajectory through the specimen. Interactions of the primary electrons with atoms and electrons cause scattering, resulting in a change in direction of the electrons from their original path (Williams and Carter, 2009; Goodhew et al., 2001). The probability that an individual electron will experience a scattering interaction depends on the complexities of the electron path through the sample, including atom arrangement, types of element encountered, and it's path length.

The spectrum of electron−specimen interactions can be divided into elastic interactions and inelastic interactions.

2.4.1 Elastic versus inelastic electron scattering

Elastic scattering occurs when there is no loss of energy of the incident primary electron. Elastically scattered electrons can change direction but do not change their wavelength. Coherent elastic scattering produces the effect of electron diffraction, which is used to analyze crystal structure (Spence and Zuo, 1992; Fuller et al., 2013).

Inelastic scattering occurs when there is an interaction that causes loss of energy of the incident primary electron. Inelastically scattered electrons have a longer wavelength. Inelastic scattering occurs by many mechanisms, and the energy is transferred to the specimen generating a range of useful signals that are exploited to characterize the material (Fig. 2.3).

Specimen thickness: The thicker the specimen, the longer the electron path, and the more likely it is that an electron will experience multiple scattering events. In thick specimens it becomes unlikely that an electron will experience only elastic scattering, so the ratio of elastic to inelastic scattered electrons decreases with increasing specimen thickness. Electrons lose kinetic energy with each inelastic scattering event. An electron that loses all its energy to the sample is absorbed.

Interaction volume: The volume of electron interaction within the specimen, determined by the 3D distribution of many different electron trajectories, is called the interaction volume (Fig. 2.3(b)). The maximum depth (average) that a given energy of electrons can penetrate to in a given specimen before being absorbed is called the electron range.

Electron transparency: Specimens that are thin enough for some of the primary electrons to pass right though the specimen are called electron transparent. This depends on the kinetic energy of the incident electrons (determined by the microscope accelerating voltage) since higher energy electrons can withstand more energy loss events. Electron transparency also depends on the chemical composition

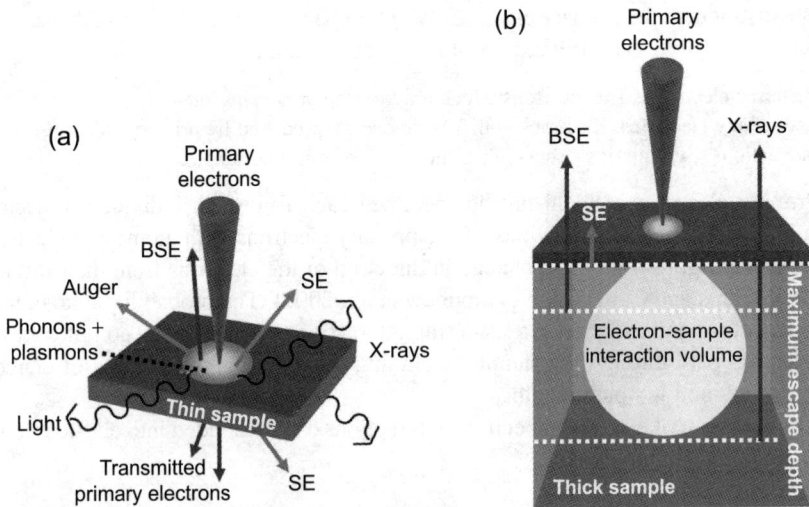

Figure 2.3 The interaction of incoming primary electrons with a sample. (a) Useful signals generated by electron−matter interactions in a thin sample. (b) Absorption of SE, BSE, and X-rays in thick samples, by inelastic scattering within the interaction volume, limits the sample depth from which they can escape.

of the sample. For a given specimen thickness, materials containing heavier elements (higher atomic number Z) have stronger electron interactions and absorb electrons faster. To be electron transparent a sample has to be thinner than the electron range in the specimen.

2.4.2 Signals from the specimen

When a beam of electrons hit a sample, the key signals emitted from the specimen are scattered primary electrons, SEs, and X-rays (Fig. 2.3) (Reimer, 1998; Egerton, 2011; Goldstein et al., 2007; Kuo, 2014; Pawley and Schatten, 2014). These signals are abundant, easy to collect, and form the basis of most microstructural characterization studies by SEM. The transmitted primary electron signal, used in TEM, will only be abundant if the sample is ultrathin.

Primary electrons, which are scattered through $90°-180°$, and emerge back out of the specimen surface are called backscattered electrons (BSEs). Transmitted primary electrons and BSEs may have experienced elastic or inelastic scattering.

SEs are generated by several inelastic scattering mechanisms. The most common are slow SEs, which are ejected loosely bound outer shell or valence electrons with very low $0-50$ eV energies. Fast SEs are knocked out of inner atomic shells and can have up to 50% of incident primary energy. Low energy Auger electrons with specific energies from electron transitions between energy levels also occur.

X-rays are generated by two main inelastic scattering mechanisms (Egerton, 2011; Goldstein et al., 2007; Garratt-Reed and Bell, 2007). Bremsstrahlung X-rays are caused by the slowing down of primary electrons as they pass through atoms and have a continuous range of energies up to the primary electron incident energy. Characteristic X-rays are produced when primary electrons knock an electron out of an atom, and the subsequent transition of a second electron between energy states causes an X-ray of specific energy (the difference in energy levels) to be generated (Fig. 2.4(a)). These X-rays of specific energy can be directly attributed

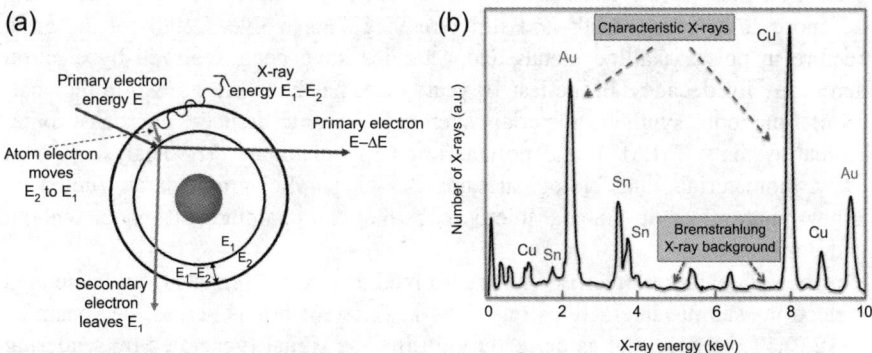

Figure 2.4 Electron-induced X-ray emission. (a) Characteristic X-ray generation. An X-ray is generated when an electron moves to fill the empty electron energy level vacated by the secondary electron. (b) Schematic of an energy dispersive X-ray (EDX) spectrum exhibiting large numbers of characteristic X-rays at specific energies.

to the specific elements in the sample which generated them, and are thus used for chemical analysis (Fig. 2.4(b)).

Cathodoluminescence (CL) is emission of a visible or ultraviolet light signal from semiconducting samples. CL originates from the excitation of electron—hole pairs by incident primary electrons. When the excited conduction band electrons recombine with the holes, light is emitted.

Some electron—matter interactions do not generate emitted signals, but may alter the specimen internally. The most important of these are the generation of phonons (heat) and plasmons (collective oscillations of electrons). Specimen modification is further discussed in Section 2.12.2.

2.5 What material features can we analyze using electron microscopy?

Modern materials are increasingly being engineered simultaneously at the nano-, micro-, and macrolevels. Hierarchical structures have evolved in nature over millennia, but novel manufacturing and fabrication methods mean that many synthetic materials are now complex composites, with their structure, chemistry, and morphology varying over different length scales. Electron microscopy is a versatile tool with a range of methodologies to characterize the microstructural features of a sample from 100 pm to 100 μm length scales.

Fig. 2.5 illustrates some of the types of microstructural features that can be analyzed using electron microscopy. At the highest resolution, the arrangement of atoms into different types of crystal structures can be determined. The analysis of the aggregation of atoms to form clusters, nanoparticles, nanowires, and thin films has been a key contribution of electron microscopy to nanotechnology development. Materials are rarely fabricated how you would want them to be, with defects such as dislocations, cracks, and surface contamination layers all detectable by electron microscopy characterization. As structures start to reach >100 nm in size, more complex materials and structures begin to evolve. Details of the grain structure in polycrystalline metals and ceramics have been analyzed by electron microscopy for decades. In the last 10 years there has been big growth in the analysis of functional synthetic materials such as electronic devices, microelectromechanical systems (MEMS), and polymeric energy materials. The analysis of soft matter, biomaterials, and biological samples is a parallel growth area, fueled by the new range of environmental microscopes that can examine soft matter without rapid destruction.

Images in electron microscopy are formed by collecting a signal generated by electron—sample interactions (eg, SEs) at different points across the specimen (Fig. 2.2). If the specimen is perfectly uniform, the signal (generated by scattering of many incident electrons) collected at each point would be, on average, constant so the whole image would be of constant intensity. A constant intensity image is like a single color artwork: expensive to collect, but not at all informative.

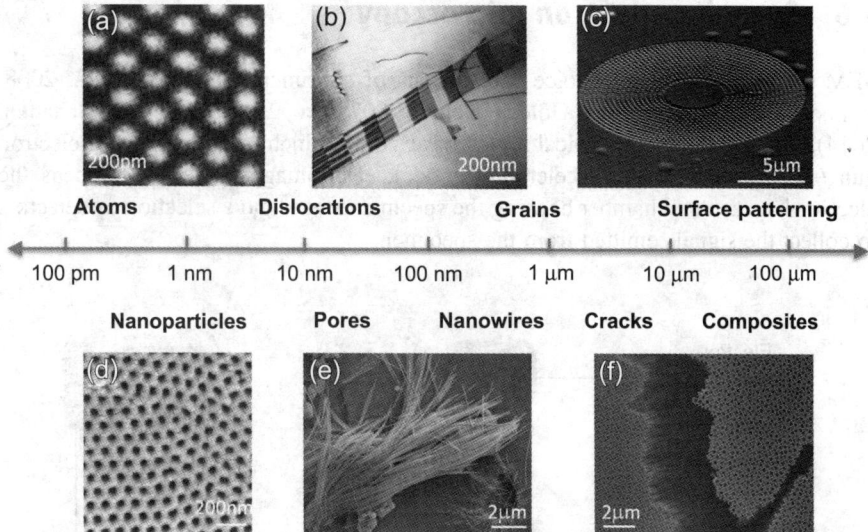

Figure 2.5 The nano-to-micro imaging range of SEM and TEM enables multiscale quantification of microstructural features from atoms to complex microstructures. (a) Atoms in FeAl. (b) Crystal defects in TiAl. (c) Surface patterning of Al. (d) Columnar pores in alumina. (e) 50 nm diameter CoPt nanowires. (f) Fracture of an alumina thin film.

To obtain useful information we want the collected signal to be variable across the specimen, giving us modulation in the image intensity that we call contrast.

Image contrast: *The range in image intensity generated by the change in collected signal intensity at different locations on the specimen.*

Measured signal intensity will vary across the specimen due to changes in the local microstructure, shape, and orientation of the specimen, which alter the mechanisms of signal formation and collection. The example electron microscopy images given in Fig. 2.5 illustrate the complexity of microstructures that can be imaged by electron microscopes, and the complex image contrast across length scales from atomic to macroscopic that results. The detailed analysis of contrast in electron microscopy images can be used to precisely interpret the microstructure of the specimen being analyzed, however it can take time, patience, and sometimes quite a lot brainpower to understand the data.

2.5.1 Practical electron microscopy

It has never been easier to use electron microscopes, with computerized machines offering a high degree of automated set-up, ease of use, and excellent reliability. New users can quickly learn the basics which underpin all machines. More advanced instrumentation may include the addition of a variety of different specimen holders, signal detectors, and spectrometers. The detailed nuances of advanced techniques may take years, sometimes decades, to master.

2.6   Scanning electron microscopy

SEM is the technique of choice for analysis of specimen surfaces (Stokes, 2008; Reimer, 1998; Egerton, 2011; Goldstein et al., 2007; Kuo, 2014; Pawley and Schatten, 2014). Fig. 2.6 shows the typical layout of an SEM, which encompasses the electron gun (electron source and accelerating anode), electromagnetic lenses to focus the electrons, a vacuum chamber housing the specimen stage, and a selection of detectors to collect the signals emitted from the specimen.

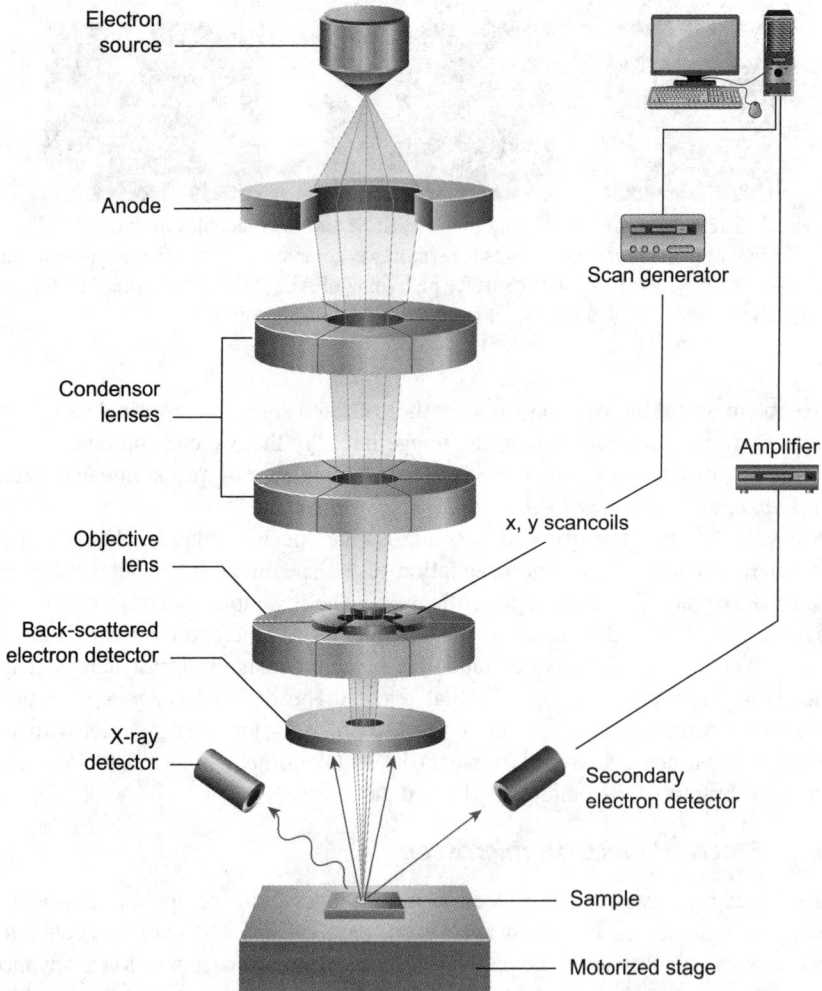

Figure 2.6 Schematic diagram of the core components of an SEM microscope.

2.6.1 Key features of the SEM microscope

Electrons: The electron gun in an SEM typically accelerates electrons through $1-30$ kV accelerating voltage. $15-30$ keV electrons are typically used for routine imaging. A $1-5$ keV low voltage SEM (LVSEM) operating mode can be used to reduce electron penetration and achieve higher resolution SE imaging (Egerton, 2011; Pawley and Schatten, 2014).

Vacuum: The pressure inside the SEM chamber is usually low vacuum $0.1-10^{-4}$ Pa. Specialist variable pressure SEM (VPSEM) and environmental SEMs (ESEM) can operate in reduced vacuum (up to 4 kPa) to inhibit evaporation of volatile components of the specimen (Stokes, 2008).

Specimen: SEM can take specimens of up to $3-20$ cm in diameter depending on the specimen stage installed in the chamber. The stage is usually motorized, and sometimes computer controlled, with $3-5$ degrees-of-freedom movement. Linear translation (x, y, z), tilt, and rotation modes can all be used to change the position of the specimen with respect to the incoming electron beam. Larger or rough specimens may be limited in their range of tilt and motion upwards (z-axis), since no stage motion should lead to collision with the lowest part of the final electron lens (the objective lens pole piece). Stage maneuverability is important to enable images of different areas of the specimen to be taken, and at different tilt angles, so that the collected signals can be optimized.

Attachment of the SEM specimen to the stage is flexible with a wide range of specimen holders available, frequently in the form of flat metal disks called specimen stubs (Echlin, 2009). Important criteria for specimen mounting are secure fixing (particularly for high tilt examination), good electronic contact with the specimen holder and stage (to inhibit electrostatic charge accumulation), and vertical positioning to ensure the optimum focal range of the electron beam can be reached.

Control of the electron beam: Electromagnetic lenses are used to focus the electrons into a beam, adjust beam astigmatism, move the beam across the specimen, and to scan the beam to generate images (Reimer, 1998). In modern machines, a range of different scan patterns are preprogrammed for the user, including adjustable scan dimensions, scan speed/spot dwell time, and pattern repetition. These parameters afford the user a high degree of control of the imaging process, and an expert user can achieve images with resolution of a few nm.

2.6.2 Specimen preparation

Careful specimen preparation is key to good electron microscopy, and a few extra hours spent optimizing specimen preparation and cleanliness can save days of wasted effort in microscopy time and image analysis (Echlin, 2009). In essence, dirty specimens lead to SEM images of the surface dirt. This is especially true for SE imaging, where the SEs are emitted from the surface top <20 nm. Specimens which have accumulated layers of contamination, such as hydrocarbons from prolonged exposure to air, or grease from fingers, will have SE images strongly affected by these contamination layers.

There is no one fail-safe way of preparing specimens, but good practice includes keeping specimens in clean environments such as sealed boxes, desiccators, and vacuum packs. Water/humidity exposure can be reduced using desiccating media or storage in vacuum chambers. Specimens from dirty environments such as oil, aqueous solutions, or air can have their surfaces systematically cleaned by methods including soaking in solvents and ultrasonic agitation. It is important that all cleaned samples are subsequently handled using gloves.

Once clean, nonconducting or poorly conducting samples usually require covering with an ultrathin coating, which serves to conduct away accumulated surface electrostatic charge. Suitable coating materials commonly used include carbon, gold, and platinum, for which a range of coating instrumentation is available. Electrostatic charging occurs due to a difference in the incoming current (negatively charged primary electrons) and outgoing current (number of SEs, BSEs, and transmitted primary electrons). Charging causes uncalibrated deflection of the incoming electron beam, and must be minimized. Charging can be close to zero at $1-4$ kV accelerating voltages, so LVSEM is sometimes used to reduce charging effects (Egerton, 2011; Goldstein et al., 2007; Kuo, 2014; Pawley and Schatten, 2014).

This chapter focuses on nondestructive materials characterization, which may limit options for objects too large to fit into the SEM chamber. If the sample can undergo some specimen preparation of a destructive manner, including being cut-up, then a much wider variety of surface techniques become available including slicing and dicing, acid etching, sputtering by bombardment with energetic ions, and abrasion/surface polishing with harder materials (Echlin, 2009; Ayache et al., 2010). Optimizing specimen preparation can be an iterative process, and first SEM images should be carefully examined for specimen artifacts, which are unwanted changes to the specimen caused by poor specimen preparation methodology.

2.6.3   SEM detectors

The detectors define the imaging modes and application range of a given SEM. It is worth investing in the best quality detector you can afford, and if possible several detectors on the same instrument (Fig. 2.6). This enables the collection of simultaneous and complementary data sets analyzing different physical information (eg, chemistry and morphology) of the same microstructural features.

All modern SEMs have SE detectors, with many having BSE detectors and X-ray spectrometers.

SE detection: SEs are abundant, and SE imaging is the basic imaging mode used to image specimen shape. The most frequent type of SE detector is the Everhart−Thornley (ET) detector, which comprises a biased grid to attract the electrons, a scintillator to convert the SE signal to light, and a photomultiplier tube to amplify the signal (Goodhew et al., 2001). ET detectors are usually fitted on one side of the vacuum chamber (Fig. 2.6), and SE signal intensity can often be enhanced by tilting the specimen surface toward the detector.

BSE detection: BSEs are less abundant than SEs but much more energetic. BSEs can be detected by ET detectors by altering the grid bias voltage to selectively repel low energy SEs. However, most frequently BSE detectors are installed directly above the specimen where primary electrons that have been deviated close to 180° can be intercepted (Fig. 2.6) (Goodhew et al., 2001). The BSE detector may be located on the bottom of the electron column, or actually a short distance inside the column (in-lens detector). BSE detector types are typically based on semiconductor or scintillator plates, arranged in two to four segments so different solid angles can be sampled to enhance chemical or topographic contrast.

BSEs are also used for electron backscatter diffraction (EBSD) methodology, which analyzes BSEs that undergo coherent electron diffraction by crystals in the specimen and are subsequently emitted with special angular distributions (Schwartz et al., 2009). EBSD requires use of a special holder to tilt the specimen 70° to the electron beam, an EBSD BSE detector which comprises a phosphor screen to collect the diffracted BSEs over a large solid angle, and a digital camera.

X-ray spectrometers: X-rays are abundant and emitted all directions from the specimen. Since X-rays are hazardous to human health, the specimen chamber must absorb X-rays, usually by having a thick lead lining. The X-ray detector is located in line-of-sight of the specimen inside the chamber (Fig. 2.6). X-rays are counted according to their energy in energy dispersive X-ray analysis (EDX) (Egerton, 2011; Goldstein et al., 2007; Garratt-Reed and Bell, 2007). EDX detectors are based on semiconductor chips that convert individual X-rays to electron—hole pairs, which then form an electronic current. The electronic signal generated is proportional to the energy of the incoming X-ray. In this way during a collection time of a few minutes, the detector can count the number of X-rays detected of given energy at a chosen location on the specimen. This forms an energy dispersive X-ray spectrum (Fig. 2.4(b)).

2.7　Key microstructural features analyzed by SEM

The SEM is the tool of choice for many industries due to its ability to image materials and structures with submicron resolution. Here we overview the SEM methods implemented to analyze morphology, chemistry, and crystallography; however, there are a great number of further methodologies available to advanced users.

2.7.1　Specimen shape

The accurate measurement of the surface topography of a specimen with spatial resolution <1 nm is an extremely important function of an SEM. High resolution topographic imaging is achieved using SEs. SEs have <50 eV energy and are generated throughout the interaction volume; however, they are rapidly absorbed by inelastic scattering with the material. The only SEs that can escape are those <20 nm from specimen surface with sufficient energy to overcome the specimen work function (Fig. 2.3(b)).

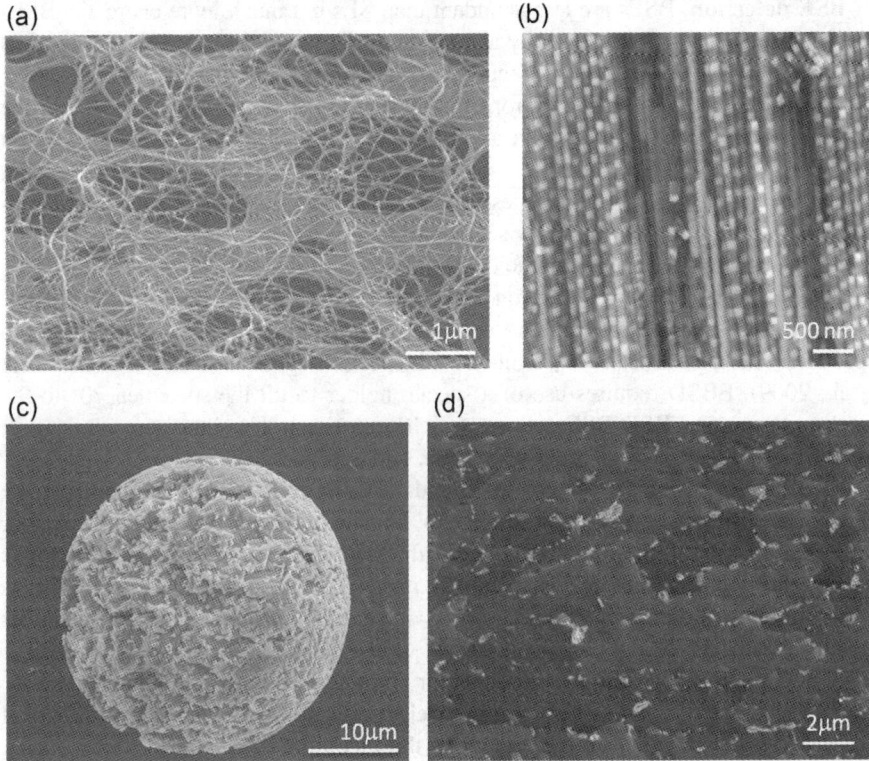

Figure 2.7 Morphology characterization using secondary electron contrast. (a) Carbon nanotubes on a holey carbon film. (b) CoPt multilayer nanowires, with alternating Pt (high SE emission) and Co (lower SE emission) layers. (c) Enhanced SE emission at surface irregularities on a Sn-microparticle. (d) Sn grains (gray contrast, low SE emission) with Cu-based particles (high SE emission) at the grain boundaries.

The number of SEs emitted from a sample depends on the angle of incidence of the electron beam to the surface, and on the local surface shape. As the surface becomes more parallel to the beam (increased tilt), there is a larger intersection of the electron interaction volume with the surface, and therefore enhanced SE escape. Angled regions of surface such as slopes, ridges, points, and edges exhibit increased SE emission and become bright in an SE image. This so-called edge effect means that sharp changes in surface relief are delineated by bright lines in SE images, and SE images therefore often appear to have a 3D quality (Figs. 2.3 and 2.7) (Goodhew et al., 2001; Reimer, 1998). Rough surfaces can exhibit light and dark zones according to local changes in the surface orientation (Fig. 2.7(c)). Additionally, because the SE detector is on one side of the chamber, shadowing of surface features can occur if the SEs do not have a clear flight-path to the detector. It is therefore often useful to image the sample at several tilt values when analyzing surface topography, and it is possible to generate 3D images from multidirectional SE imaging.

SEM has been used extensively for the quantitative analysis of surface topography. Typical applications include measuring microstructural feature sizes and their distribution/density, quantifying the 3D morphology of objects, measuring particles (size, number, and shape), examining porosity (size, distribution, and tortuosity), cracks (size, shape, length, and interconnection), and quantifying failure sites such as fracture surfaces and tribological wear surfaces.

2.7.2 Specimen composition

The chemical composition of materials and the distribution of different phases can be key to their structural and functional behavior. Quantification of chemical composition can be used to understand and control key features such as oxidation, surface reactivity, mechanical strength, and conductivity. Chemical imaging in SEM can be carried out by both BSEs and X-ray analysis (Egerton, 2011; Goldstein et al., 2007; Garratt-Reed and Bell, 2007).

BSE imaging: Primary electrons are backscattered by electrostatic interaction between the positive atomic nuclei and negative primary electrons (Rutherford backscattering). The repulsive power of atomic nuclei is proportional to their positive charge, determined by the number of protons (element Z-number). Heavier elements (high Z) generate more BSEs than light elements (low Z). The BSE signal will be proportional to the average Z-number in the sampled part of the interaction volume (the BSE escape depth being dependent on incident electron energy). Therefore in a BSE image, the contrast will vary both due to topographic contrast and chemical contrast. For a flat specimen (minimized topographic contrast), regions of higher average Z will emit more BSEs (be brighter) than regions of lower average Z. For scientists, this is very rapid method to look for distributions of two or more phases which have sufficiently different average Z-number to cause chemical contrast modulation in the BSE image—for example, in multiphase aerospace alloys and composites.

X-ray spectroscopy: All materials generate Bremstrahlung and Characteristic X-rays under bombardment with electrons. The Characteristic X-rays are generated by electron transitions within atoms in the sample (Fig. 2.4(a)), and can occur with energies up to the energy of the incident electrons. X-ray spectra are usually collected by focusing the electron beam onto the specimen, and counting the number of X-rays reaching the detector with different energies. This is called an EDX spot spectrum (Fig. 2.4(b)). The Characteristic X-ray peaks are matched to specific elements in the sample by comparing their energy with elemental standards. X-rays can escape from throughout the electron interaction volume, which can be 1 µm deep into the specimen (Fig. 2.3(b)). Therefore, even for a highly focused electron beam, the spatial resolution of X-ray analysis in SEM will be limited to ≈ 1 μm^3 in thick specimens.

Quantitative analysis of specimen composition by X-ray spectroscopy is very complicated because the X-rays collected will be a function of factors including the incident electron energy (determining the range of possible X-rays energies), the probability of the electronic transition which generates the X-ray (depends on the element and local bonding), the X-ray absorption after generation (depends on the X-ray energy, material, and specimen thickness), and the collection efficiency of the detector system (Williams and Carter, 2009; Garratt-Reed and Bell, 2007).

Despite the complexity of accurate quantitative X-ray analysis, the determination of the average chemical composition of a material to an accuracy of approximately 1% weight can be routinely achieved, and qualitative analysis of relative changes in chemistry across a given sample is an extremely useful methodology. EDX spectroscopy in SEM has found widespread use in nondestructive chemical analysis of materials. With advances in computerized instrumentation, X-ray spot spectra can now be obtained from many locations across a sample enabling elemental distributions along line scans and specific areas of surface (X-ray maps) to be obtained.

2.7.3 Surface crystallography

Electron backscatter diffraction: EBSD is a technique that can determine the local crystal structure and crystal orientation at the surface of a specimen. The methodology collects elastically scattered BSEs which have undergone coherent Bragg scattering as they leave the specimen. A dedicated EBSD detector collects the scattered BSEs over a large solid angle, which forms electron backscatter diffraction patterns made up of Kikuchi bands (Schwartz et al., 2009). The EBSD patterns can be analyzed to give the crystalline structure and orientation of the crystal that the BSEs were scattered by. EBSD analysis in SEM is now very automated, and has found widespread application in the analysis of crystallography of metallic alloys. EBSD can be used to quantify which crystal structures are present and their orientation, sizes and morphology of individual grains, the collective texture of alloys, and crystallographic relationships between phases.

2.8 Transmission electron microscopy

TEM is the technique of choice for analysis of specimen internal microstructure, evaluation of nanostructures such as particles, fibers, and thin films, and imaging of atoms (Reimer and Kohl, 2008; Williams and Carter, 2009; Goodhew et al., 2001; Buseck et al., 1988; Spence, 2013). Fig. 2.8 shows the key components of a TEM microscope, which comprises the electron gun, electrostatic lenses to focus the electrons before and after the specimen, and a transmitted electron detection system.

2.8.1 Key features of the TEM microscope

Electrons: The electron gun in a TEM typically accelerates electrons through 80−300 kV accelerating voltage to give them sufficient energy to pass through up to 1 μm of material (Reimer and Kohl, 2008). 200−300 keV electrons are typically used for routine imaging, with lower energy <100 keV electrons used for analysis of very light elements such as carbon to reduce specimen damage. Very specialist

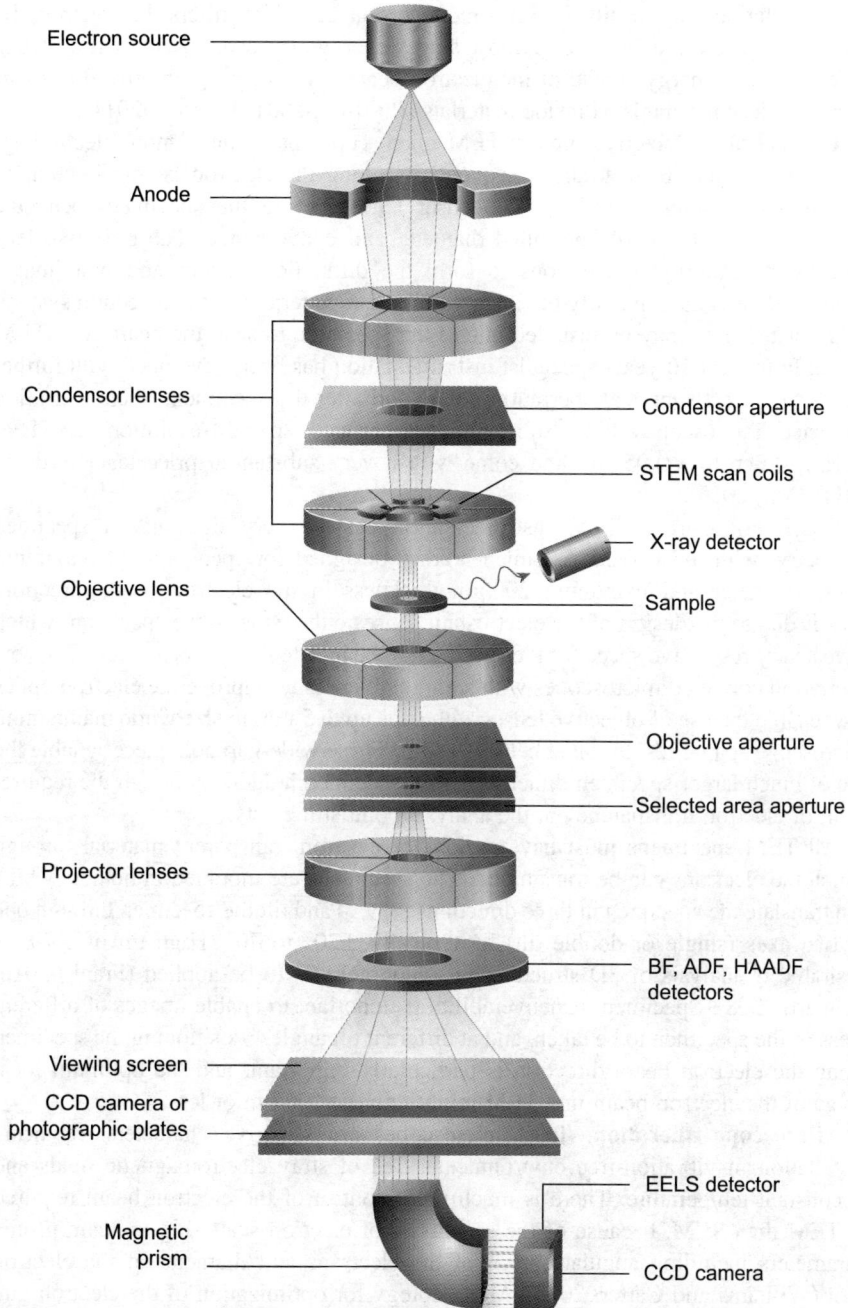

Electron source

Anode

Condensor lenses

Condensor aperture

STEM scan coils

X-ray detector

Objective lens

Sample

Objective aperture

Selected area aperture

Projector lenses

BF, ADF, HAADF detectors

Viewing screen

CCD camera or photographic plates

EELS detector

Magnetic prism

CCD camera

Figure 2.8 Schematic of core components of a TEM microscope.

instrumentation may be fitted with a monochromator, which filters the electrons by energy to generate a beam of almost single energy (Brydson, 2011; Erni, 2015). A reduction in energy spread of the electron beam is required for improved analysis of energy loss mechanisms inside materials (Brydson, 2001; Egerton, 2011).

Control of the electron beam: TEM microscopes have many more electromagnetic lenses than SEM, arranged sequentially along the electron beam direction as an electron column (Fig. 2.8). Condenser lenses before the specimen focus the electrons into a beam of controlled diameter and convergence. The objective lens focuses the transmitted electrons to form the diffraction pattern and first image. Projector lenses then magnify the image/diffraction pattern onto the detection system. Additional lenses may be installed before the specimen to scan the beam for STEM mode. In the last 10 years specialist instrumentation has been developed with further lenses to correct spherical aberration before and after the specimen. Such aberration corrected microscopes have substantially improved spatial resolution of TEM/STEM, down to <0.05 nm, and come with a very substantial price tag (Brydson, 2011; Erni, 2015).

Specimen: Current TEM instrumentation can be very limiting on specimen geometry, with most TEM specimen holders designed for specimens of maximum 3 mm diameter and maximum 200 μm thickness in the electron beam direction. This is due to the design of the electrostatic lenses either side of the specimen, which have a very restrictive space for the specimen to be inserted and maneuvered. The new aberration corrected microscopes with additional lenses to improve the electron optics now enable the use of objective lenses with gaps up to 5 mm in size while maintaining microscope spatial resolution at below 0.1 nm. These wide-gap pole pieces enable the use of much larger specimen dimensions and specimen holders, although the requirement of electron transparency at the analysis point still exists.

All TEM specimens must have regions of electron transparent material through which the electrons can be transmitted. The specimens are mounted in holders which can translate the specimen in three directions (x, y, z) and tilt the specimen through one or two axes (single or double tilt) by typically $\pm20°$ to $40°$. High tilt of $\pm70°$ is desirable if analysis of 3D structure by tomography is to be applied (Frank, 2010; Banhart, 2008). Specimen maneuverability is important to enable images of different areas of the specimen to be taken, and at different tilt angles. Positioning the specimen along the electron beam direction is particularly important, and the optimum focal range of the electron beam in a TEM might only be 100 μm or less.

Microscope operation: TEM microscopes are sensitive characters, requiring installation in vibration-free environments, free of stray electromagnetic fields and at constant temperature. There is much more control of the electron beam required in TEM than SEM, because of the sensitivity of electron scattering to beam profile parameters including angular spread of the electrons and diameter of the electron spot (Williams and Carter, 2009). Methodology for optimization of the electron gun parameters, alignment of multiple lenses, and alignment of apertures which partially block the electron beam at different points, must be determined for each microscope. Alignment of the instrument may take upwards of 30 min at the beginning of the working day, and may need to be periodically adjusted.

Detection of electrons: The transmitted electrons pass through the specimen and are focused by the postspecimen lenses to form an image. The image can be monitored live on a phosphor screen, or a wide-angle camera. Images are recorded by a parallel recording device, with an array of pixels that should be as large as possible. All modern TEMs are fitted with digital electron detection systems, the most common being charged coupled devices (CCD), which convert incoming electrons into an electronic pulse per pixel. Older microscopes may still operate with photographic film, which is being phased out. In STEM mode, a TEM needs to be fitted with additional axial bright-field (BF) and annular dark-field (ADF) detectors (Pennycook and Nellist, 2011; Brydson, 2011).

2.8.2 TEM specimen preparation

TEM specimen preparation is a varied and often complex job. Decades of work have generated a variety of recipes for creation of thin material from thick (Ayache et al., 2010). Materials made of heavy elements (high Z-number) may need to be 100 nm or thinner to transmit electrons. True nondestructive characterization by TEM is therefore only possible for specimens such as nanoparticles, nanofibers, and light materials such as molecules and nanocarbons, which are electron transparent in their natural state. If some degree of destructive specimen preparation is acceptable, then as with SEM, a wide variety of thinning techniques can be applied such as electrochemical dissolution, chemical etching, ion sputtering, mechanical abrasion, and controlled fracture (Echlin, 2009; Ayache et al., 2010). Specimen cleanliness and specimen handling protocols are even more important for TEM than SEM, especially if the goal is sub-nm spatial resolution.

2.9 TEM imaging modes

The contrast in TEM images is determined by the geometry of the electron illumination, the electron scattering within the specimen, and the path of the transmitted electrons through the postspecimen lenses, apertures, and detection system (Reimer and Kohl, 2008; Williams and Carter, 2009; Goodhew et al., 2001; Buseck et al., 1988; Spence, 2013). There are a great number of different TEM imaging modes developed to obtain the maximum amount of information from the specimen, and the most common are summarized here.

Bright-field (BF) imaging: Electrons which are scattered as they pass through the sample, either elastically (no energy loss) or inelastically (energy loss), move at an angle to the axis of the electron beam. Scattered electrons can be blocked by using an aperture (objective aperture) positioned in the back focal plane of the objective lens (Fig. 2.8). By doing this, a BF image is formed from only unscattered electrons. In a BF image, areas of the specimen which are actively scattering have fewer electrons, and therefore darker contrast. There are many microstructural features which locally increase the electron scattering, particularly inelastic scattering, and so can be

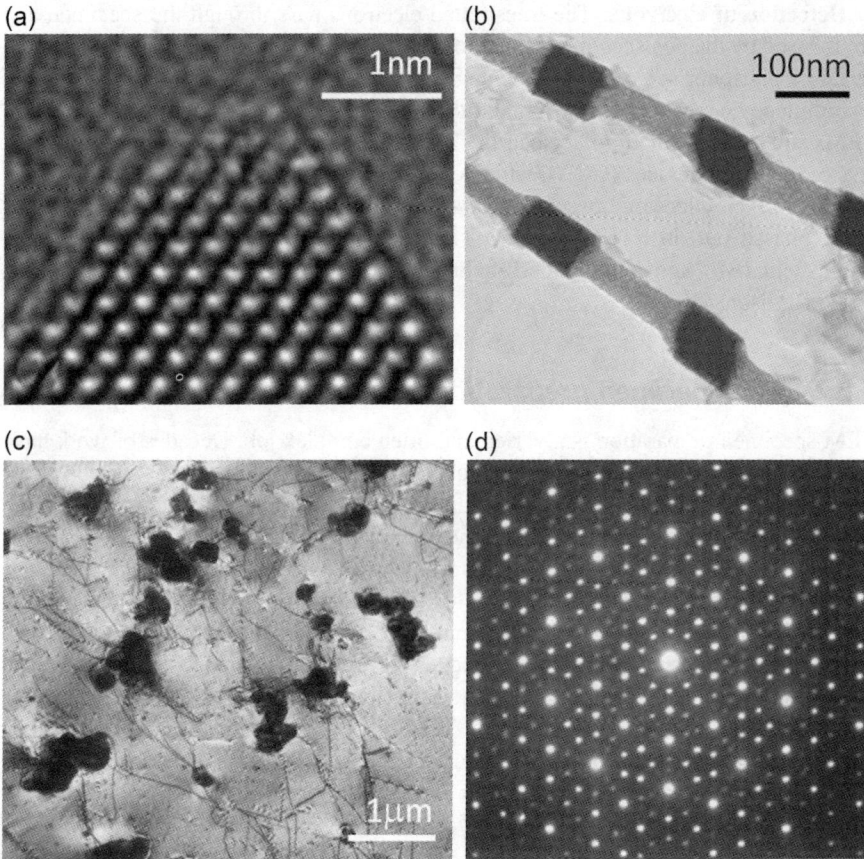

Figure 2.9 Materials characterization using TEM imaging modes. (a) Atomic structure of a CeO_2 nanocrystal, [110] HRTEM, aberration corrected. (b) Bright-field image of CoPt nanowires, with Pt layers causing significant scattering (dark contrast). (c) Bright field image of an FeAl reinforced with Y_2O_3 particles which are scattering electrons. The dark lines are dislocations in the FeAl matrix. (d) Diffraction pattern formed by elastically scattered electrons ($Al_{64}Cu_{20}Fe_{11}Co_5$ alloy).
Image (a) courtesy Günter Möbus and image (b) courtesy Yong Peng.

located as darker contrast in a BF image (Fig. 2.9) (Williams and Carter, 2009; Goodhew et al., 2001). These include regions of increased thickness, increased mass (high Z-number/density), grain boundaries, and dislocations. Areas with strong elastic Bragg scattering will also appear dark in a BF TEM image, which can be used to detect changes in crystal orientation such as grain boundaries.

The objective aperture can also be placed in the back focal plane to block unscattered electrons and choose a selection of scattered electrons to form a dark-field (DF) image. DF images are used to map regions of sample that generate specific electron scattering (those that pass through the aperture), which appear bright in the DF image.

Electron diffraction: Elastically scattered electrons can change direction but do not change their wavelength. Constructive and destructive interference of coherent elastically scattered electrons can generate strong beams of transmitted electrons at specific angles to the incoming electrons determined by the crystal structure and orientation of the material. This phenomenon, called electron diffraction or Bragg scattering, is analogous to X-ray diffraction (Spence and Zuo, 1992; Fuller et al., 2013; Hammond, 2009; Giacovazzo et al., 2011). When the orientation of a crystal is in a favorable orientation with respect to the incoming electron beam, elastically scattered electrons can be focused by the postspecimen lenses to form an electron diffraction pattern (Fig. 2.9(d)).

Electron diffraction is one of the most frequently used analysis modes of TEM (Williams and Carter, 2009; Fuller et al., 2013). The diffraction pattern is formed in the back focal plane of the objective lens and magnified by the projector lenses onto the recording device. Detailed analysis of diffraction patterns can determine the crystallography of the sample volume generating the diffraction pattern including lattice type, point group, lattice parameters, local crystal orientation, existence of different phases, and phase orientation relationships. The structural analysis by electron diffraction can be directly correlated with in situ BF imaging of the specimen, and potentially with chemical information from EDX or electron energy loss spectroscopy (EELS) analyses. The nanoscale spatial resolution of electron diffraction is an advantage over X-ray diffraction; however, it can only be applied to electron transparent specimens.

High-resolution TEM (HRTEM): If specimens are ultrathin, of the order of 100 nm thickness or less, the elastic scattering dominates over inelastic scattering. The electron waves interacting with the crystal lattice diffract and form complex interference patterns visible at magnifications of 400 k or more (Buseck et al., 1988; Spence, 2013). Under some imaging conditions, the patterns correspond to atom positions, and this is called high-resolution TEM imaging (Fig. 2.9(a)).

Scanning TEM (STEM): Some TEM instruments are fitted with scan coils, which can scan a focused electron beam across the specimen (Keyso et al., 1998; Pennycook and Nellist, 2011). This STEM mode is extremely useful for carrying out sequential chemical analysis across areas of the specimen. Advanced instruments, especially those fitted with extra lenses to correct spherical aberration in the electron beam (STEM probe correction), can obtain focused spots of less than 0.1 nm diameter (Pennycook and Nellist, 2011; Brydson, 2011). This coupled with ultrastable scan electronics means that the electron beam spot can be sequentially positioned with accuracy down to different atom columns.

High angle annular dark field (HAADF): In STEM mode, unscattered electrons are collected to form BF images, and scattered electrons are collected to form ADF images. Electrons incoherently scattered through very high angles can be collected using a high angle annular dark field (HAADF) detector (Pennycook and Nellist, 2011). HAADF images show very strong contrast changes due to local changes in atomic number of the specimen (Z-contrast), and can be used to analyze chemistry at the atomic scale.

2.10   TEM spectroscopy

The inelastic scattering of electrons transfers energy to the sample, resulting in the emission of SE, BSE, X-rays, and transmitted electrons with reduced energy. Due to the lack of space in the objective lens, TEM microscopes are not usually fitted with SE or BSE detectors. However, EDX detectors are often mounted, together with EELS spectrometers after the specimen to collect and analyze the energy of the transmitted electrons (Fig. 2.8). These methodologies enable exceptionally high spatial resolution chemical analysis to be carried out on TEM specimens.

2.10.1   X-ray analysis in TEM (EDX)

The spatial resolution of X-ray analysis in TEM is much higher than in SEM, because the X-rays can only originate from the very restricted thickness of the electron transparent TEM specimen (usually less than 200 nm thickness) rather than the large interaction volume in SEM. The local chemical composition of a material can be determined to an accuracy of approximately 0.1% weight, 10 times better than SEM (Williams and Carter, 2009; Egerton, 2011; Goldstein et al., 2007; Garratt-Reed and Bell, 2007). If available, the use of highly focused STEM probes <1 nm diameter can further increase the spatial resolution of the X-ray spectrum. Advanced X-ray techniques in TEM include X-ray analysis along lines (line scans), using 2D scan areas to obtain X-ray maps of different elements (Fig. 2.10(a)), or tilting the specimen to obtain X-ray distribution in 3D volumes (EDX tomography). All of these techniques enable nondestructive measurement of elemental profiles across nanoscale microstructural features.

2.10.2   Electron energy loss spectrometry

Inelastic scattering inside the specimen causes the primary electrons to lose energy. This can be detected after the specimen by passing the transmitted electrons through

Figure 2.10 Advanced materials characterization methods in SEM and TEM. (a) TEM 2D STEM-EDX map of Ti-Kα (green), Cu-Kα (blue), and Al-Kα (red) X-rays emitted from Cu-Al-Cu-Ti multilayer nanopillars. (b) Moving a carbon nanotube using a microgripper inside a SEM. (c) Electron tomography of a CeO$_2$ nanoparticle. Energy filtered TEM images in multiple directions are analyzed to give a surface rendered 3D reconstruction. Images (a) and (c) courtesy Günter Möbus.

energy filters to separate the electrons according to their energy loss. The usual methodology is to use a postcolumn EELS spectrometer after the normal electron optics, which uses a magnetic prism to deflect the electrons according to their energy (Fig. 2.8) (Brydson, 2001; Egerton, 2011). The electrons can then be counted according to their energy to form an EELS spectrum, with typical energy resolution of 0.5−1 eV. 2D maps of electron energy loss across the specimen can be formed in STEM mode by taking many sequential EELS spectra (STEM spectrum imaging) (Pennycook and Nellist, 2011). Alternatively EELS spectrometers with imaging optics can be used to generate images of the specimen at selected electron energy loss, called energy filtered TEM (EFTEM). Improved quality EELS spectra with energy resolution <0.1 eV can be achieved by the use of monochromators, which ensure that the electrons incident on the sample have as close as possible all the same energy (same wavelength) (Brydson, 2011).

EELS evaluates the distribution of scattered electrons with a given energy loss. Specific electron energy loss values can be linked to a range of inelastic scattering mechanisms including plasmon scattering, phonon scattering, SE emission, and fine detail of atomic bonding in different crystal structures (Brydson, 2001; Egerton, 2011). EELS analysis is also extremely useful to determine the distribution of lighter chemical elements including *B*, *C*, *N*, and *O*, whose X-rays are of such low energy they are significantly absorbed by specimens and detectors, and thus difficult to analyze. Indeed EDX and EELS analyses are complementary methodologies and can be carried out on the same area of specimen.

2.11 Key applications of TEM

The joy of using TEM is that structural and chemical analyses can be carried out on the same sample across lengths scales ranging from atoms up to 100 μm.

At maximum magnification, HRTEM and HAADF STEM with atomic resolution are used to evaluate the structure of crystals, grain boundaries, and interfaces between different phases. Electron diffraction studies determine detailed quantitative information about crystal structures and lattice parameters. The evaluation of imperfect structures is as important as perfect ones, with HRTEM revealing the structure of dislocation cores and planar defects such as stacking faults. The rapid developments in high spatial resolution TEM/STEM EDX and EELS can directly relate chemical modulations to structural inhomogeneities, such as precipitates, amorphous layers at grain boundaries, and impurity segregation to cracks and surfaces.

TEM is contributing significantly to the evaluation of nanostructures including nanoparticles, carbon nanotubes, graphene, and thin films. 3D morphology and chemistry of nanoparticles can be determined, with important applications including the evaluation of catalysts and nanoparticles being developed for drug delivery. TEM has also played a key role in the determination of the structure and function of nanocarbons including carbon nanotubes, fullerenes and graphene.

At the micro- and macroscales, the use of TEM has been integral to the development of electronics, systematically evaluating the structure of semiconducting and polymer devices, and determining how structure and chemistry relate to their functional performance. The current drive in many technologies, including energy materials and biomaterials, is to fabricate objects out of multiple materials with geometries defined in a hierarchical way, even down to the nanometer level. This inevitably leads to the existence of fabrication defects such as inhomogeneous layers, dislocations, pores, cracks, and strain. TEM is an excellent tool to quantify the nature of defects across the nm to 100 μm imaging range (Fig. 2.5).

2.12 Is electron microscopy a nondestructive technique?

For SEM and TEM techniques to be nondestructive, they should leave no permanent change to the material under investigation. Under some circumstances electron microscopy can alter a specimen, so we should consider what the changes might be, and if the magnitude of the changes are acceptable or not.

2.12.1 Specimen preparation

A specimen has to fit inside the vacuum chamber of an electron microscope to be analyzed. The size of specimen therefore depends on the electron microscope being used, which can vary from large semiconductor wafers (SEM) down to 3 mm disks of <200 μm thickness (TEM) (Echlin, 2009; Ayache et al., 2010). Large components, such as aircraft alloys, may have to be cut up in order to get the area to be analyzed to fit inside the microscope. This may not be an easy or acceptable process, particularly for biological materials. Coping strategies may include preparing model specimens, or removing small samples from noninvasive locations.

2.12.2 Specimen changes during imaging

Specimens are subjected to high-energy electrons in large numbers (the electron dose) during the SEM and TEM imaging process (Reimer and Kohl, 2008). Electrons that are elastically scattered within the specimen transfer no energy and so do not change the specimen. However, inelastic scattering transfers energy to the specimen causing changes which may be reversible or permanent. Permanent changes to the specimen, often called *radiation damage*, depend on many factors including the inelastic scattering mechanism, the specimen, the electron energy, and the local electron dose. Common changes to the specimen resulting from inelastic scattering are discussed below.

Electrostatic charging: When electrons are absorbed into the specimen and different numbers of electrons are absorbed than ejected, there will be a net buildup of charge. If the specimen is conductive and part of a circuit to the specimen stage, electrons can flow to balance charge. Charging of poorly conducting specimens can result in distortion of the primary beam and loss of image resolution.

Atom displacement: If primary electrons transfer significant energy to single atoms, particularly in >200 keV TEM, atoms can be completely displaced from their location. Atoms at surfaces can be sputtered laterally or into the vacuum. In crystalline materials atom displacement can generate crystal defects including interstitial atoms, lattice vacancies, dislocations, and planar faults, which must be distinguished from intrinsic material defects.

Covalent bond breaking: In molecular and polymeric materials, key bonds may be broken by the ejection of SEs (Kuo, 2014; Pawley and Schatten, 2014). This can lead to the breakdown of polymer chains and molecules, and at surfaces the loss of the smaller components into the vacuum.

Heating: The generation of phonons can heat the specimen. If there is poor thermal conduction to the surrounding material (heat dissipation), local heating can result in breaking of bonds in polymeric and biological molecules, atom diffusion, and loss of volatile materials such as dehydration (Kuo, 2014; Pawley and Schatten, 2014). In extreme cases, holes can be "drilled" into specimens by ejection of energized atoms off the surface.

2.12.3   Strategies for minimizing specimen damage

There have been many strategies developed to keep SEM and TEM as nondestructive as possible. To reduce the damage, and damage rate, during your investigations you can do the following:

- **Reduce the electron energy** (lower the accelerating voltage), so each electron has less energy to transfer to the specimen.
- **Reduce the number of electrons** per area (local intensity) by spreading the electron beam over a larger area, for example, by reducing the magnification. Part of the beam can be blocked with an aperture.
- **Reduce the imaging time** to reduce the total electron dose.
- **Cool the specimen** to remove heat. This is a highly effective method for biological samples, and cooling to cryogenic temperatures (usually by liquid nitrogen using special specimen stages) is called Cryo-electron microscopy (Cryo-EM).

2.13   Outlook for SEM and TEM

Since their inception, SEM and TEM have proven to be incredibly useful, highly versatile techniques for materials characterization. In this chapter we have covered the basic principles and core methodologies available to scientists and engineers at the onset of using electron microscopy for materials analysis. The high spatial resolution of SEM and TEM, in both imaging and chemical characterization modes, is highly complementary to other nondestructive materials characterization techniques including X-ray imaging and crystallography, light microscopies, and scanning probe microscopies (Fig. 2.1).

Learning to effectively use electron microscopes is a progressive task, and to the advanced user there are a wide range of further techniques which may be of benefit. Electron microscopy has evolved to be a highly complex discipline. However, despite its apparent maturity, there are some key trends in instrumentation, which are significantly advancing capability and thereby range of applications. The first is to view the SEM and TEM as a nanolaboratory rather than simply as an imaging device. A wide range of SEM and TEM attachments enable the application of stimuli to the specimen to investigate physical behavior in situ (Banhart, 2008). These include heating and cooling the specimen, applying controlled electrical fields, liquid/gaseous environments, mechanical manipulators, and mechanical testing (Fig. 2.10(b)). Each of these methodologies, some at minimal cost, expands the EM toolbox from noninvasive imaging to capability in functional testing.

The second revolution in SEM and TEM has been to move from 2D imaging methodologies to 3D imaging methodologies, called tomography (Wang, 2015). Nondestructive tomography can be carried out in SEM and TEM by imaging the sample along multiple directions (achieved by tilting/rotating the sample), and reconstructing the 2D image sets into 3D reconstructions of the sample microstructure (Fig. 2.10(c)) (Frank, 2010; Banhart, 2008). Because electron microscopes are both structural and chemical imaging tools, combined high-resolution structural, morphological, and chemical tomography is possible including chemical tomography using EDX, EELS, and HAADF signals (Banhart, 2008). 3D materials characterization by SEM and TEM tomography is highly complementary to the large-volume, lower resolution nondestructive X-ray tomography characterization techniques (see chapter: X-ray microtomography for materials characterization).

These advancements in SEM and TEM capability have come at a fortuitous time, with the advancement in demand for multiscale quantification of microstructural features from atoms to complex microstructures. Complex new materials, in particular inhomogeneous composites, mesostructures, and inorganic−organic hybrid materials, can be all characterized by using SEM and TEM techniques. Furthermore, the spread of desktop SEM instrumentation into smaller institutions and schools will encourage our next generation of engineers and scientists to get involved in developing enabling technologies for the planet.

References

Ayache, J., Beaunier, L., Boumendil, J., Ehret, G., Laub, D., 2010. Sample Preparation Handbook for Transmission Electron Microscopy. Springer Handbook. ISBN: 978-0387981826.

Banhart, J. (Ed.), 2008. Advanced Tomographic Methods in Materials Research and Engineering. Oxford University Press.

Banhart, F. (Ed.), 2008. In-situ Electron Microscopy at High Resolution. World Scientific.

Bradbury, S., Bracegirdle, B., 1998. Introduction to Light Microscopy, second ed. BIOS.

Brydson, R., 2001. Electron Energy Loss Spectroscopy. CRC Press, Oxford. ISBN: 13: 978-1859961346.

Brydson, R. (Ed.), 2011. Aberration-corrected Analytical Electron Microscopy. Wiley. ISBN: 978-0-470-51851-9.

Buseck, P.R., Cowley, J.M., Eyring, L. (Eds.), 1988. High-resolution Transmission Electron Microscopy: and Associated Techniques. Oxford University Press. ISBN: 13: 978-0195042757.

Echlin, P., 2009. Handbook of Sample Preparation for Scanning Electron Microscopy and X-ray Microanalysis. Springer Handbook. ISBN: 978-0-387-85731-2.

Egerton, R.F., 2011. Physical Principles of Electron Microscopy: An Introduction to TEM, SEM, and AEM, second ed. Springer.

Egerton, R., 2011. Electron Energy-loss Spectroscopy in the Electron Microscope, second ed. Springer Handbooks. ISBN: 978-1-4419-9583-4.

Erni, R., 2015. Aberration-corrected Imaging in Transmission Electron Microscopy: An Introduction, second ed. Imperial College Press.

Frank, J., 2010. Electron Tomography: Methods for Three-dimensional Visualization of Structures in the Cell, second ed. Springer.

Fuller, B., Howe, J., Fultz, B., 2013. Transmission Electron Microscopy and Diffractometry of Materials, fourth ed. Springer. ISBN: 13: 978-3642297601.

Garratt-Reed, A.J., Bell, D.C., 2007. Energy Dispersive X-ray Analysis in the Electron Microscope (Microscopy Handbooks), Kindle ed. CSC Press.

Giacovazzo, C., Monaco, H.L., Artioli, G., Viterbo, D., Milanesio, M., Gilli, G., et al., 2011. Fundamentals of Crystallography. Oxford University Press. ISBN: 13: 978-0199573660.

Goldstein, J., Newbury, D.E., Joy, D.C., Lyman, C.E., Echlin, P., Lifshin, E., et al., 2007. Scanning Electron Microscopy and X-ray Microanalysis, third ed. Springer.

Goodhew, P.J., Humphreys, J., Beanland, R., 2001. Electron Microscopy and Analysis, third ed. CRC Press. ISBN: 13: 978-0748409686.

Hammond, C., 2009. The Basics of Crystallography and Diffraction, third ed. OUP.

Keyso, R.J., Garratt-Reed, A.J., Goodhew, P.J., Lorimer, G.W., 1998. Introduction to Scanning Transmission Electron Microscopy. CRC Press. ISBN: 13: 978-1859960660.

Kuo, J. (Ed.), 2014. Electron Microscopy: Methods and Protocols (Methods in Molecular Biology), third ed. Humana Press. ISBN: 13: 978-1627037754.

Mertz, J., 2009. Introduction to Optical Microscopy. Roberts and Company Publishers. ISBN: 13: 978-0981519487.

Pawley, J., Schatten, H. (Eds.), 2014. Biological Low-voltage Scanning Electron Microscopy. Springer. ISBN: 13: 978-1489995841.

Pennycook, S.J., Nellist, P.D. (Eds.), 2011. Scanning Transmission Electron Microscopy: Imaging and Analysis. Springer.

Reimer, L., Kohl, H., 2008. Transmission Electron Microscopy: Physics of Image Formation, fifth ed. Springer. ISBN: 13: 978-0387400938.

Reimer, L., 1998. Scanning Electron Microscopy, second ed. Springer, Heidelberg. ISBN: 3-540-63976-4.

Schwartz, A.J., Kumar, M., Adams, B.L., Field, D., 2009. Electron Backscatter Diffraction in Materials Science. Springer Handbooks.

Spence, J., Zuo, J.M., 1992. Electron Microdiffraction. Springer.

Spence, J.C.H., 2013. High-resolution Electron Microscopy, fourth ed. Oxford University Press. ISBN: 13: 978-0199552757.

Stokes, D., 2008. Principles and Practice of Variable Pressure: Environmental Scanning Electron Microscopy. Wiley-Blackwell. ISBN: 13: 978-0470065402.

Wang, M. (Ed.), 2015. Woodhead publishing series in electronic and optical materials. Industrial Tomography: Systems and Applications. ISBN: 13: 978-1782421184.

Williams, D.B., Carter, C.B., 2009. Transmission Electron Microscopy. Springer.

X-ray microtomography for materials characterization

3

R. Hanke[1], T. Fuchs[2], M. Salamon[2], S. Zabler[3]
[1]Fraunhofer Institute for Nondestructive Testing (IZFP), Saarbrücken, Germany; [2]Fraunhofer Development-Center X-ray Technology EZRT, Fürth/Bay, Germany; [3]Julius-Maximilians-University, Würzburg, Germany

3.1 Introduction

Since the 1970s, X-ray computed tomography (CT) has become a well-established and routinely used modality in modern diagnostic radiology (Hounsfield, 1973). Beyond that, since the late 1980s, X-ray CT has emerged as a very important and widespread tool in industrial inspection as well as in material sciences (Hanke and Fuchs, 2008).

Outside medical radiology, today X-ray CT is used in three major fields of applications:

- Nondestructive testing (NDT) and evaluation for safety purposes of, eg, critical parts of airplanes, cars, engines, turbines, or power generators. Typically, in safety inspection the task is to search and detect cracks, delaminations, blow holes, and similar defects. Thus, these applications deal with high contrast details and require high spatial resolution.
- Metrology measurement based on three-dimensional (3-D) volume representations of the samples, in order to perform nominal-actual value comparison of a part's geometry in terms of size and shape. In this field, X-ray CT is clearly superior to alternative metrology tools like tactile or visual systems, since CT allows for a measurement of inner surfaces and hidden structures in 3-D.
- Quantitative materials analysis of geological or biological samples, compound materials like carbon- or glass-fiber reinforced plastics (CFRP resp. GFRP), advanced ceramics, composites, plastics, and alloys. Here, the accuracy of reconstructed mass densities or attenuation coefficients is crucial (Nachtrab et al., 2011). Thus, for quantitative purposes, the focus lies on CT image quality in terms of homogeneity, absolute contrast, and noise, as well as on the reduction of artifacts.

The last of these groups of application includes the field of materials characterization, which shall be discussed in the following sections.

3.2 Imaging physics

The basic principle of CT data acquisition is depicted in Fig. 3.1. The object is positioned between a suitable X-ray source and a detection device, as for instance a two-dimensional flat panel detector. While the object is rotating, projective images

Materials Characterization Using Nondestructive Evaluation (NDE) Methods
http://dx.doi.org/10.1016/B978-0-08-100040-3.00003-1

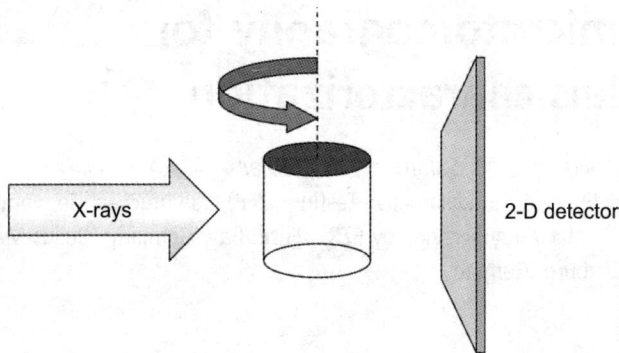

Figure 3.1 Schematic illustration of a standard micro-CT setup. The sample is positioned on a rotational stage between the X-ray source and a two-dimensional (2-D) X-ray imaging device.

are recorded from a certain number of angular positions. The particular issues related to each of the components, like X-ray source, imaging device, mechanical setup, number, and features of the projection images will be discussed in the next chapters.

As indicated in the introduction, the essential requirement for an X-ray CT system for materials characterization is high spatial resolution. The process of materials characterization is less dominated by the need for fast data acquisition like it is the case for, eg, production integrated testing. In addition, for characterization of new sophisticated materials like fiber-reinforced plastics or metal ceramic composites, the main challenge rises more from image contrast, material discrimination, and spatial resolution rather than scanning time (Zabler et al., 2012).

Thus, the preferred X-ray sources for materials characterization provide very small focal spot sizes at correspondingly low X-ray flux, but with additionally high stability in intensity and position of the X-ray tube's electron beam on the target.

3.2.1 X-ray microfocus tubes

The development of microfocus X-ray technology started in the late 1970s. Today, for modern microfocus X-ray tubes, the progress in development results at focal spot sizes down to 0.4 µm and thus enables high resolution direct magnification X-ray microscopy by the use of such microfocus tubes as a source of radiation.

In this context, it is important to note that in fact radioscopic image resolution is directly related to the tube's focal spot size and is of about the same order of magnitude, whereas CT resolution in practice is in the best case the same as the spatial resolution of a single projection. Usually, it is slightly worse than that due to the additional limitations encountered with CT measurements, like long-term reliability of focal spot size and position, inaccuracies of the mechanical manipulation system, and stability of the imaging detector's calibration.

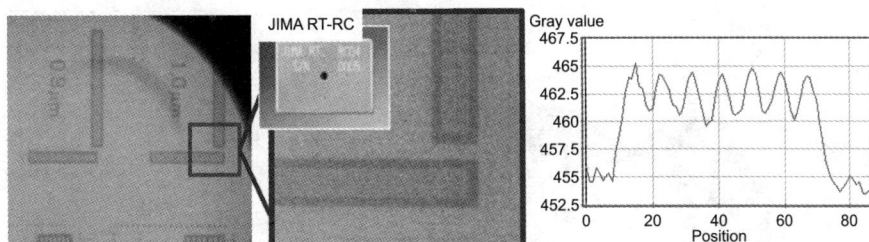

Figure 3.2 Focal spot determination using a JIMA "RT-RC 02" resolution test pattern. Highly magnified radiograph of pattern (left), enlarged view on the 1 μm pattern (middle), extracted profile from the 1 μm structure (right).

The characterization of microfocus X-ray tubes mainly is based on spot size measurement. Even though significant progress has been made in decreasing spot sizes below 1 μm, there are still no standardized methods for determination of focal spot sizes smaller than 5 μm. The currently applied measurement procedures in the sub-micron regime are based on test patterns, eg, JIMA "RT-RC 02" (cf. Fig. 3.2) or the Siemens star-pattern used by companies like Carl Zeiss (formerly XRadia), providing finest periodic structures of discrete or continuously decreasing size. Some representative experimental results on this topic will be discussed in more detail in Section 3.7 of this chapter.

Further details on the different determination methods can be found at (Salamon et al., 2008a,b). The latest developments in standard microfocus X-ray technology are related to higher stability of the focal spot position based on internal cooling of the focusing device reducing the thermal elongation due to a temperature gradient of approximately 60°C between standby and focused scan mode. Detailed information about state of the art (design, function and characteristics like focal spot size, energy range, power, etc.) can be read, eg, in Salamon et al. (2008a,b, 2009).

Advanced developments toward lab-based real nanometer-focus X-ray sources currently follow two approaches of new X-ray target conceptions. On the one hand, there are new liquid metal jet targets (LMJ), generating high brilliance in the low energy region between 7 and 20 keV, with the gallium's K_α-line at 9.24 keV (cf. Fig. 3.3, right (Otendal, 2006; Vogel, 2015; Hemberg et al., 2003)). On the other hand, new needle or pin targets realize focal spot sizes down to 50 nm in combination with electron focusing optics of a latest generation's electron microscope and thus enabling magnification factors up to $M = 1000$ (Stahlhut et al., 2013, 2014).

LMJ-based CT setups, in combination with advanced detector technology (high speed or high resolution), can be used for materials characterization both in terms of dynamic material analysis as well as structure characterization, respectively (Fig. 3.4). Moreover, the high brilliance in combination with stabilized manipulation systems allows even phase contrast imaging or the application of focusing

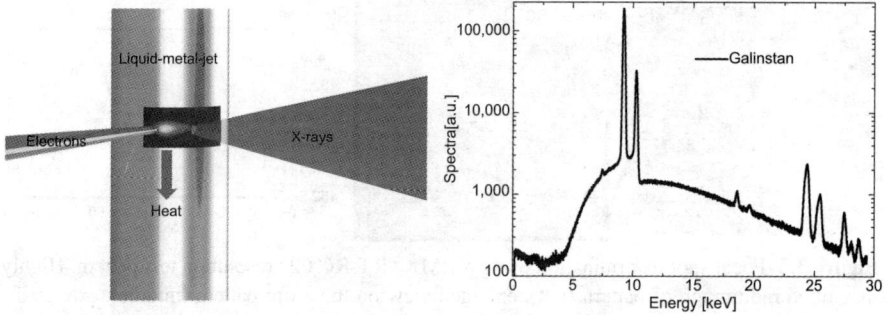

Figure 3.3 Left: illustration of the gallium–indium metal jet, which allows for a variable focal spot size and very high brilliance at the same time *(Courtesy Excillum)*. Application areas are essentially plastic materials and biological samples. Currently, a spot size of around 80 μm by 20 μm can be achieved. Right: typical X-ray spectrum resulting from an alloy of 69% gallium, 22% indium, 9% tin with 70 keV energy of the incident electrons. The maximum tube voltage available is 160 kV in this case.

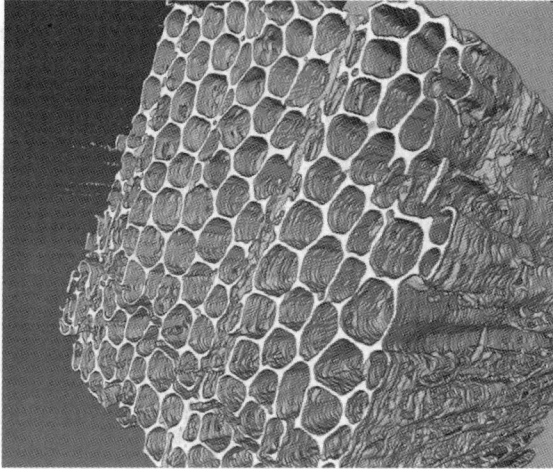

Figure 3.4 Wood cell structure as an example of a CT with a liquid metal jet X-ray source. Courtesy Balles, A. University of Würzburg, Germany.

devices (refractive lenses, zone plates, etc.) in laboratory environments within reasonable measuring time (Zabler et al., 2012).

To further reduce the focal spot size toward the nanometer-focus range—compared to state-of-the-art microfocus X-ray tubes, which are based on transmission targets—a new pin target concept is realized inside an electron microscope (cf. Figs. 3.5 and 3.6). Especially produced needles, made of molybdenum or tungsten (Stahlhut, 2012) with tips in the range of about 50 nm will be placed directly into the electron beam and thus produce X-rays with a focal spot size, limited only by the diameter of the metal tip.

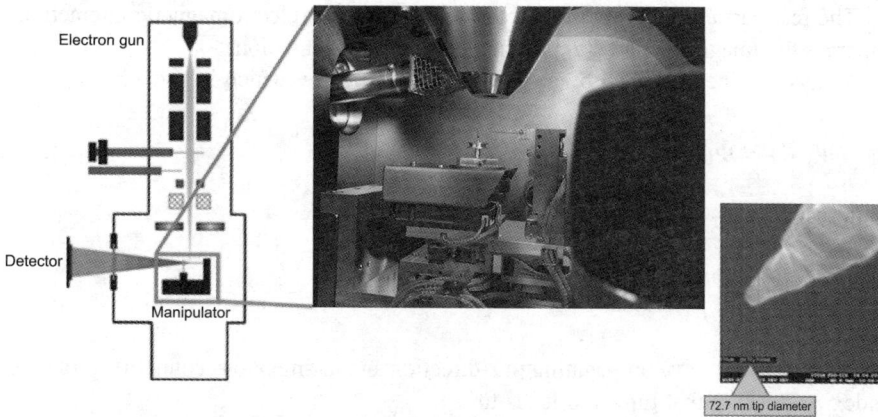

Figure 3.5 Upgrade of an electron microscope: sketch (left). Sample and X-ray source are inside of the microscope's vacuum vessel (center). The extremely small needle-targets are produced by electrochemical etching, tip size below 100 nm (right).

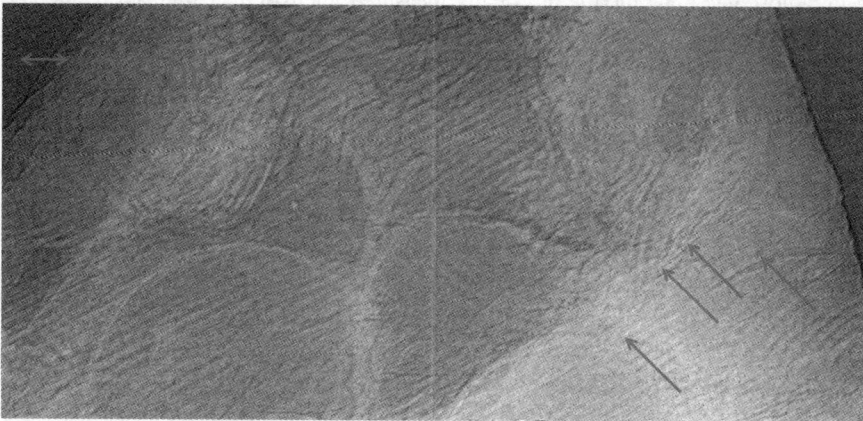

Figure 3.6 Highest resolution achieved by nanometer imaging: the radiographic resolution achieved is less than 100 nm by phase contrast. Thereby the sample size must not be larger than 100 μm. Projection radiography were taken to prepare 3-D imaging of an AlGeSi alloy; blue arrow: 500 nm wave-like surface details; red arrows: hairline cracks down to 100 nm; Acquisition parameters 120 min exposure, 30 kV tungsten X-ray spectrum, directly converting CdTe sensor with 1000 μm sensitive layer and 55 μm pixel size.
Courtesy Stahlhut, P. University of Würzburg, Germany.

3.2.2 Interaction of hard X-rays with materials

X-ray interaction with condensed matter can be discussed both from the aspect as particle interaction (electron with photon) on the one hand or as electromagnetic wave interaction, on the other hand. These two perspectives are described by the complex refraction index $n(x, y, z)$ that is given by

$$n(x, y, z) = 1 - \delta(x, y, z) + i\beta(x, y, z) \qquad [3.1]$$

The real part δ represents the refractive contribution (electromagnetic interaction) whereas the imaginary part β represents the photon attenuation.

X-rays may be described by an electromagnetic wave, which is given by

$$\Psi(\vec{r}) = \Psi_0 \exp(i\vec{k}\,\vec{r}) \qquad [3.2]$$

with wave-vector

$$k = |\vec{k}| = \frac{n\omega}{c} \qquad [3.3]$$

Assuming the wave propagating in x-direction and taking into account the refractive index $n = n(x)$ in that direction leads to

$$\Psi(x) = \Psi_0 \exp\left(i\frac{\omega}{c}n\cdot x\right) = \Psi_0 \exp\left(i\frac{\omega}{c}(1-\delta)\cdot x\right)\exp\left(-\frac{\omega}{c}\beta\cdot x\right) \qquad [3.4]$$

In radioscopic imaging, the physically measured intensity $I(x)$ after X-ray attenuation of electromagnetic waves by some material with a thickness x is given by the square of absolute value of the complex wave function

$$I(x) = |\Psi(x)|^2 = I_0 \exp\left(-2\frac{\omega}{c}\beta\cdot x\right) \qquad [3.5]$$

with $\beta = \frac{\mu}{2\omega}c$ where μ represents the linear attenuation coefficient, this equation turns out to be the well-known absorption law of Lambert Beer.

The next two sections will cover photon attenuation and phase contrast imaging, based on these results.

3.2.2.1 X-ray attenuation

The application of X-ray micro-CT to materials characterization basically means to learn about the spatial distribution of the X-ray sensitive attenuation coefficient, which describes the material under investigation. Thus, it is crucial to understand the two dominating processes of X-ray interaction with condensed matter, given by photon absorption and incoherent Compton scattering within the energy regime between 10 keV up to about 300 keV.

The behavior of matter is described by the sum of contributions of the single attenuation processes:

$$\frac{\mu(Z,E)}{\rho} = \frac{\tau}{\rho} + \frac{\sigma_{Rayleigh}}{\rho} + \frac{\sigma_{Compton}}{\rho} + \frac{\pi}{\rho} \qquad [3.6]$$

The mass-attenuation coefficient as defined by Eq. [3.6] is varying with energy E of the photon and atomic number Z of the absorber.

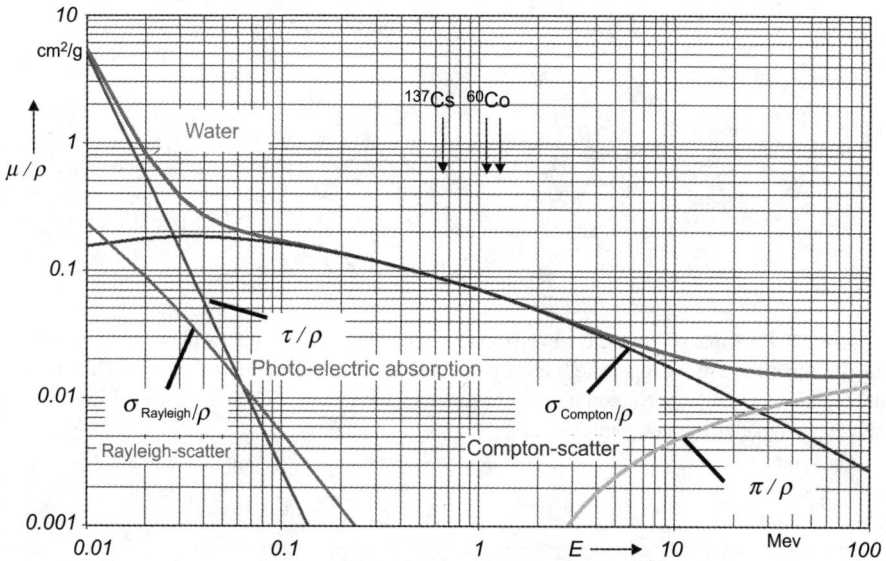

Figure 3.7 The mass-attenuation coefficient as a function of photon energy in case of water. Based on the tables in Hubbell, J.H., Seltzer, S.M., 1989. Tables of X-Ray Mass Attenuation Coefficients and Mass Energy-Absorption Coefficients (Online). National Institute of Standards and Technology. http://www.nist.gov/pml/data/xraycoef/.

Values of those coefficients for all known chemical elements can be taken from tables (Hubbell and Seltzer, 1989, 1982) as functions of energy. Fig. 3.7 shows the functional behavior of the different contributions in the case of water.

Coherent scattering (Rayleigh) and pair production are not relevant for the discussed X-ray regime and thus are not treated here.

Photon absorption

The process of photon absorption is shown in Fig. 3.8. Photons are mainly absorbed by inner and relatively strong bound core electrons. During interaction, the whole energy of the photon is transferred to the orbital electron, which is emitted with energy E_{kin} while another characteristic fluorescence photon is emitted or converted to an Auger-electron.

The attenuation coefficient for photon absorption is given by

$$\mu_{photon} \propto \frac{Z^n}{E_\gamma^{3.5}} \qquad [3.7]$$

The dependency of the atomic number is proportional to Z^n with $3 < n < 4$.

Compton scattering

Attenuation by Compton scattering is the second most important absorption process and becomes dominant with increasing atomic numbers Z compared to photon absorption, as was shown in Fig. 3.7.

$$E_{kin,e^-} = E_\gamma - E_b$$

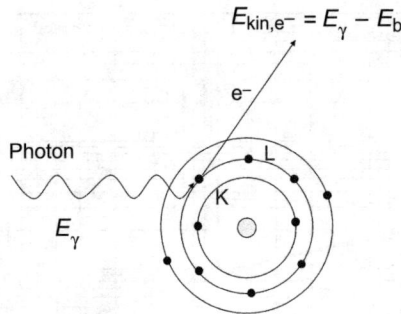

Figure 3.8 Principle of photoelectric process responsible for photon absorption. Most probably an electron on one of the inner shells close to the nucleus is knocked out of the atom. Thereby the photon's energy E_γ is spent completely for the ionization energy E_b and additional kinetic energy E_{kin} of the charged particle.

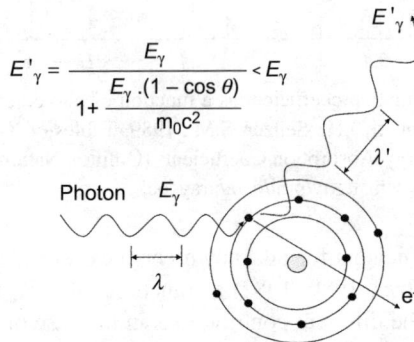

$$E'_\gamma = \frac{E_\gamma}{1 + \dfrac{E_\gamma \cdot (1 - \cos\theta)}{m_0 c^2}} < E_\gamma$$

Figure 3.9 Principle of X-ray attenuation by Compton scattering (bottom right) and the theoretical relation between the scattered photon's energy and the scattering angle.

The Compton process is shown schematically in Fig. 3.9, including the relationship between the incoming and the scattered photon's energy. The interaction probability and the mass-attenuation coefficient resulting therefrom are given by the Klein-Nishina cross-section and can be approximated by

$$\frac{\mu_{KN}}{\rho} \propto \left(\frac{Z}{A}\right) \cdot f(E) \qquad\qquad [3.8]$$

where $f(E)$ is a slowly decreasing function with photon energy E. Thereby, A denotes the atomic mass of the element number Z; thus Z/A is approximately 0.5 or smaller for all elements with $Z > 1$ (see also Hanke and Boebel, 1992).

3.2.2.2   Phase contrast imaging

Phase contrast imaging techniques both in 2-D and 3-D have found their way into X-ray materials characterization on a large scale, first at synchrotron facilities, then in the laboratory over the last 20 years (Mayo, 2002). In spite of their common physical origin—optical interferences through wave-propagation—three distinct imaging techniques are being used in quite different setups and therefore have to be addressed separately:

1. inline phase contrast imaging
2. Talbot interferometry or grating-based phase contrast
3. Zernike phase contrast

Since the latter (Zernike phase contrast) is exclusively used for transmission X-ray microscopy experiments on synchrotron beamlines, we refrain from a more detailed description here. The interested reader may refer to (Zernike, 1935; Holzner et al., 2010).

The formation of X-ray images (just as visible light) is described by electromagnetic waves, propagating through a medium described by the complex refraction index n. The solution of the Helmholtz wave-equation phrases by propagation of a wave was already given above by Eq. [3.2] The wave-vector $\vec{k}$ describes the direction of propagation and has the amplitude $\left|\vec{k}\right| = 2\pi/\lambda$.

In reality, plane waves cannot be observed. Every kind of radiation originates from a source of finite size and has a finite spatial and temporal coherence. For the formation of X-ray inline phase contrast images, it is of particular importance to consider the partial coherence of X-ray beams. If the latter is fit then we are able to observe constructive and destructive interferences in X-ray images over some tens of micrometers length. Von Ardenne (von Ardenne, 1939), in 1956, was the first to infer from the edge-enhancement. He observed in high-resolution X-ray shadow images that Fresnel fringes could be used to highlight structural details.

In textbooks (eg, Goodman, 1996) Fresnel wave propagation is introduced with a small aperture of size b (Fig. 3.10). An incoming plane wave is diffracted by the aperture and propagates along the z-axis over a distance d until it reaches the detector plane. Geometrically, the coordinates of the point of observation P are related as:

$$\cos \theta = d/|\vec{r}| \tag{3.9}$$

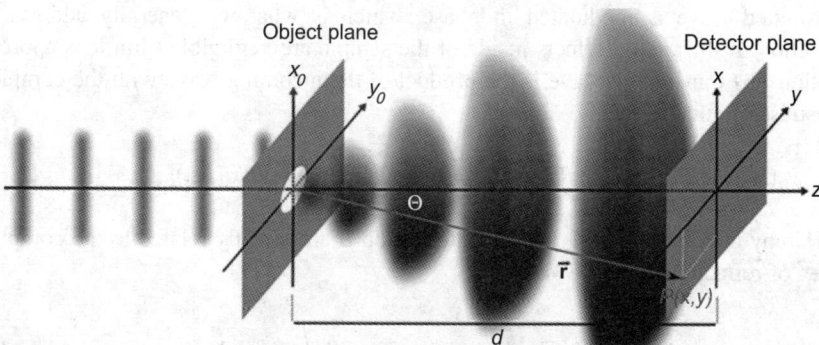

Figure 3.10 Schematic representation of Fresnel-Kirchhoff diffraction.

with

$$\overrightarrow{\mathbf{r}} = \begin{pmatrix} x - x_0 \\ y - y_0 \\ d \end{pmatrix} \qquad\qquad [3.10]$$

The complex wave amplitude at point $P(x,y)$ is described by the Fresnel-Kirchhoff diffraction integral:

$$\psi_d(x,y) = \frac{1}{i\lambda} \iint_{\sigma} \psi_0(x_0, y_0) \frac{\exp\left(i\overrightarrow{\mathbf{k}}\,\overrightarrow{\mathbf{r}}\right)}{|\overrightarrow{\mathbf{r}}|} \cos\Theta \, dx_0 dy_0 \qquad [3.11]$$

Considering only near-field diffraction and thus replacing the spherical wave kernel in Eq. [3.11] by parabolic waves yields the Fresnel diffraction, which is a simple convolution of ψ with the kernel p_d.

$$\psi_d(x,y) = (\psi_0 * p_d)_{x,y}, \quad p_d(x,y) = \frac{e^{ikd}}{i\lambda d} \exp\left\{i\frac{k}{2d}\left[x^2 + y^2\right]\right\} \qquad [3.12]$$

p_d is called the Fresnel propagator, and according to Fourier analysis, Eq. [3.13] can be rewritten in reciprocal space as a multiplication (operator formulation):

$$\Psi_d(u_x, u_y) = \Im_{2D}[\psi_d(x,y)] = \Psi_0 \cdot P_d, \quad P_d \sim \exp\left[-i\pi\lambda d\left(u_x^2 + u_y^2\right)\right] \qquad [3.13]$$

where $\Im_{2D}$ designs the 2-D Fourier transform and (u_x, u_y) the spatial frequencies, reciprocal variables to the spatial coordinates (x,y). Here it is important to highlight differences between the optics textbook and real X-ray imaging physics. The X-ray transmission through a real sample has little in common with the ideal aperture in Fig. 3.10. X-rays are only fractionally attenuated (cf. Eq. [3.5]). In addition, the transmitted wave is modulated in phase, which is what we generally address as refraction. If diffraction effects inside of the sample are negligible (thin lens approximation) the transmitted wave is the product of the incoming wave with the complex transmission function T:

$$\psi_0(x_0, y_0) = \psi_{\text{in}} \cdot T, \quad T(x_0, y_0) = \exp[i\Phi(x_0, y_0) - B(x_0, y_0)] \qquad [3.14]$$

Hereby the phase-shift Φ and the attenuation B are directly related to the complex index of refraction $n = 1 - \delta + i\beta$

$$\Phi(x_0, y_0) = k\int_{L}[1 - \delta(\overrightarrow{\mathbf{r}})]dz, \quad B(x_0, y_0) = k\int_{L}\beta(\overrightarrow{\mathbf{r}})dz \qquad [3.15]$$

where L designs the line-integral along the direction of propagation. The parameters delta and beta are energy- and material-dependent (via the given mass density and atomic number). The interest in measuring Φ stems from the difference in magnitude of the two effects: delta is generally one to three orders of magnitude larger than beta. Fresnel diffraction beholds a huge contrast gain.

While Ψ_{in} can be considered a quasi−plane wave at synchrotrons, rewriting Eq. [3.12] for laboratory setups (finite source-to-sample distance s) requires a coordinate transform for the propagation distance d and the detector coordinates x,y which are scaled with the magnification M

$$d \rightarrow d_{eff} = d\,\frac{s}{s+d} = \frac{d}{M}, \quad x,y \rightarrow x' = x\,\frac{s+d}{s} = x \cdot M, \quad y' = y \cdot M \qquad [3.16]$$

Fig. 3.11 shows some examples of high-resolution radiographs which display the edge-enhancement effect by Fresnel-type inline phase contrast.

Series of inline phase contrast projection images can in turn be used to build approximate 3-D volume images of $\delta(x,y,z)$ (holotomography). This approach, however, involves the numerical solution to the inverse problem of phase retrieval since the phase is obscured when intensity is recorded, ie, by taking the squared modulus of $\psi(x,y)$ in Eq. [3.12]. The most common method for calculating projection images

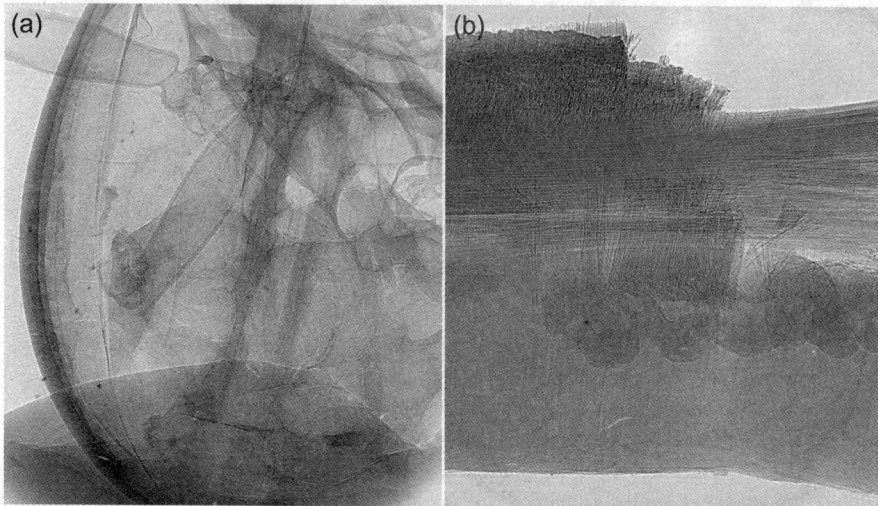

Figure 3.11 High resolution radiographs recorded at the liquid metal jet anode at the chair of X-ray microscopy showing (a) part of the head of a bee (the facetted eye is on the left side) and (b) a laser weld CFRP-polyethylene sample with porosity. Structures are highlighted by inline phase contrast due to Fresnel propagation, the X-ray spectrum is peaked gallium-K_α-emission (9.24 keV).
Courtesy Fella, C., Balles, A. University of Würzburg, Germany.

$t(x,y)$ from phase contrast radiographs is by applying the Paganin-filter prior to tomographic back projection.

$$t(x, y) = -\frac{1}{\mu}\log\left(\Im_{2D}^{-1}\left\{ \frac{\Im_{2D}\{[I(x, y, z = d)]\}/I_{in}}{4\pi^2 d(\delta/\mu)\left(u_x^2 + u_y^2 \right) + 1} \right\} \right) \qquad [3.17]$$

Fig. 3.12 shows the application of the Paganin-type phase retrieval to a high-resolution scan of graphite particles from a Li-ion battery.

The Talbot interferometer is the answer to the question, "How can one record the refraction through a sample?" For X-rays which pass through an oblique interface, the angle of refraction is extremely small; hence an instrument with extremely high angular precision and a 2-D pixel matrix are needed.

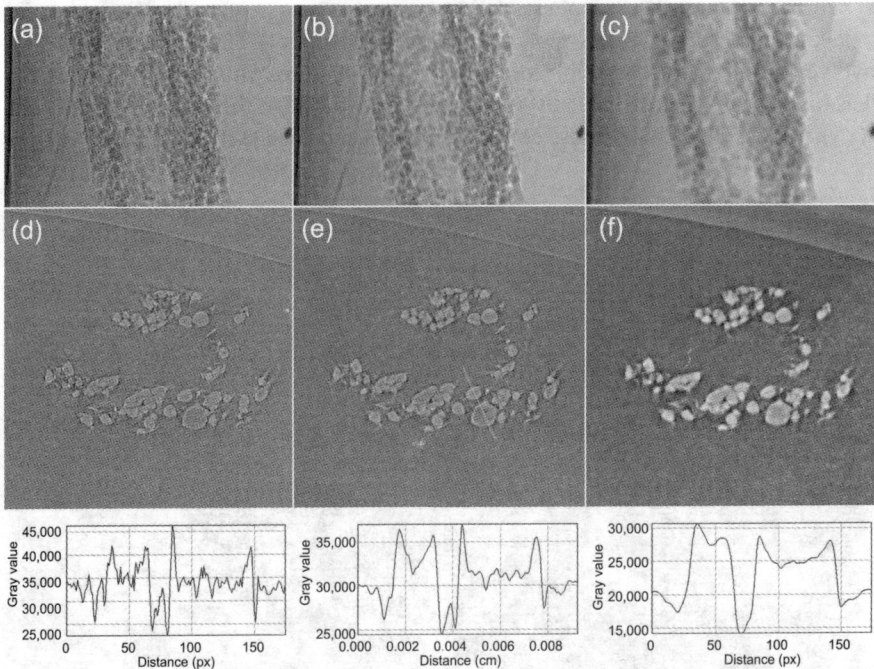

Figure 3.12 High-resolution radiographs and CT slices showing graphite particles from a Lithium-ion battery *(Courtesy Gold, L. Fraunhofer ISC Würzburg, Germany)*. The projection images (0.67 μm pixel sampling, *Courtesy Balles, A., Fella, C. University of Würzburg, Germany*) make use of inline phase contrast to highlight the particles (a and d). Images are then converted to phase projection maps by applying the Paganin phase retrieval filter of short (b and e) then of larger filter radius (c and f). The line profiles (*yellow line* in e) across two particles illustrate how the filter works: intensities from edge-enhanced structures are converted into pseudoattenuation values while also smoothing the image contrast (the filter has low-pass properties).

First experiments were done by rocking-curve imaging for which Bragg mirrors provided the angular resolution. Using the Talbot effect—the self-imaging of periodic structures (gratings) in near-field diffraction images—came in handy to replace this first and relatively difficult setup. In short, when Eq. [3.13] is applied for a single spatial frequency which is $1/p^2$ with p the pitch or period of the grating, the Fresnel propagator P_d becomes unity for multiples of the so-called Talbot distance $d_T = 2p^2/\lambda$. Hence the transmission image of the grating is reproduced. If then the incoming wave front is disturbed by a sample, the reproduced image of the grating at d_T is altered by a local displacement of the grating lines, which are proportional to the first derivative of the object phase $\Phi'(x,y)$.

The remaining problem of measuring this small displacement is solved by superimposing a second (analyzer) grating G2 at d_T and reading the intensities at different positions (phase steps) of G2 as depicted in Fig. 3.13. The high angular resolution is now given by the ratio between grating period and Talbot distance. Another advantage of the Talbot interferometer is that the sampling of the X-ray camera which measures the intensity does not have to match the periodicity of the grating. Rather, it can be much larger than p thus integrating the phase-shift over many periods of G2. When this technique was introduced to the laboratory for the first time in 2006, it became clear that $\Phi'(x,y)$ could be measured for objects that were as big as the grating's area which is reaching 20 cm by today's available technology.

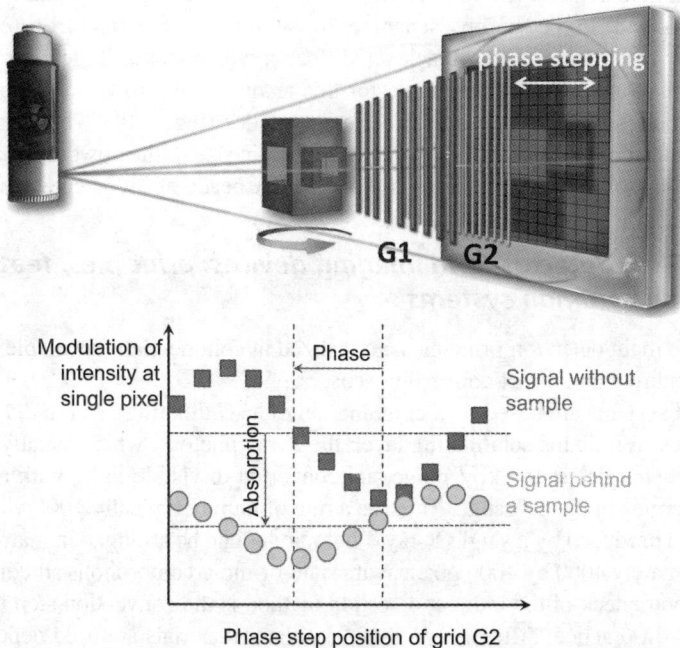

Figure 3.13 Typical laboratory setup of a Talbot interferometer using a microfocus anode (top) and evaluation of the amplitude and phase from the Talbot imaging data (bottom) (Fraunhofer IIS).

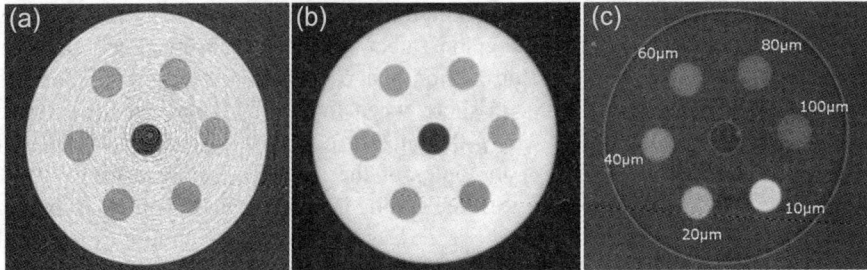

Figure 3.14 Three CT slices showing a Plexiglas cylinder with six compartments of Polymethylmethacrylat (PMMA) powder of different grain sizes, ranging from 10 to 100 μm *(Microbeads Norway)*: (a) attenuation contrast CT, (b) phase CT and (c) dark field CT. The latter shows a clear sensitivity to the mean grain size whereby smaller grain yield a stronger signal. Courtesy Revol, V. CSEM Switzerland.

The applicability of phase contrast imaging to laboratory systems makes this technique favorable to both applications in medicine and component inspection (mostly fiber composites and porous materials). Two main improvements were introduced after 2006: first, the use of a third (source) grating G0 which combines the technique with high-power rotating anodes, thus shortening the measurement times significantly (so-called Talbot–Lau interferometers); and second, the extraction of the visibility, ie, the dark field contrast from the grating-based phase contrast image data. The latter quickly showed to be highly sensitive to subpixel microstructures (eg, grain-boundaries, precipitates, pores, cracks, and fibers), which makes it ideal for the study of defects in composites. An example for this technique is shown for a powder test sample which was scanned in the three image modes (Fig. 3.14). The sensitivity of the dark field contrast increases with decreasing grain size of the powder, which varies from 100 μm down to 10 μm (cross-linked PMMA beads of 1% accurate size).

3.2.3 X-ray detectors and imaging devices: principles, features, and common systems

Today, two main detection principles are realized in commercially available flat panel systems: indirect and direct converting sensors.

Indirect systems are based on a combination of a scintillating layer and a matrix of photodiodes. Within the scintillating layer, the X-ray photons, which usually are associated with energies in the keV range, are converted to visible light, with associated photon energies in the eV range. Thus, as a rule of thumb, typically 1000 visible light photons are produced by a single X-ray. Photodiodes can be arranged in matrices of up to approximately 4000 by 4000 pixels, with state-of-the-art amorphous silicon technology. The bottleneck of the indirect detection method is the conversion step within the scintillator. In practice, different types of scintillator materials are used depending on applied X-ray energy range, spatial resolution and other scan characteristics like X-ray flux or scan speed. Due the fact that micro-CT has to cope with relatively low X-ray flux at relatively high X-ray energies between 50 and 200 keV, predominantly the thickness of the scintillator layer is affecting the detection efficiency. On the other hand, the resolution is decreased with rising thickness of the layer requiring an optimized combination of scintillator and photodiode. For this reason, structured scintillation materials like CsI, fabricated as crystalline, needle-type structures are used to avoid lateral signal degradation along the scintillation path.

The respective size of the pixel can be as small as 50 μm, thereby defining the limit for the sampling distance of the X-ray image at the imaging plane. For laboratory purposes, where a low energy is applicable and the sample sizes are small, also pixel sizes of 6.5 μm are available with good efficiency and adequate resolution.

3.3   Principles of microcomputed tomography

The difficulties and even barriers in transferring and applying medical CT equipment and knowledge on industrial applications and systems may be discussed from two major natural aspects.

First, the human body is made almost completely out of lightweight elements like hydrogen, carbon, nitrogen, and oxygen. Only our bones contain Calcium as an element with a higher atomic number in nonnegligible amounts. Technical objects, on the other hand, are typically made out of metals, like aluminum, titanium, steel, copper, and others. One exemption is carbon fiber—reinforced polymer, but materials like PVC or Teflon contain a high amount of halogens, ie, elements with a high atomic number. Since the physics of X-ray absorption and scatter depend strongly on the atomic number of the material, the X-ray approach to technical objects is in principle different to humans. Most important, the X-ray energies applied to industrial applications range between 50 and 600 keV using X-ray tubes or up to 10 MeV in case of a linear accelerator.

Second, in medical X-ray CT the size of the "sample" never varies significantly from examination to examination. The diameter of grownups lies in between 300 mm (head) and 600 mm (shoulder) maximum, the length between 1 and 2 m, and the required spatial resolution is somewhere around 300 μm (head) and 600 μm (thorax), respectively, which equal a typical number of 1000 resolution elements in each direction within a single 2-D cross-section through the body. In consequence, the distance of the focal spot of the X-ray source and the imaging device is fixed in medical CT systems. On the contrary, industrial systems have to cope with a range of three orders of magnitude in size. From small material samples with diameters of a few millimeters up to huge objects like cars or shipping containers with dimensions of several meters, every size is possible.

In addition, the required spatial resolution varies between less than one micron and several millimeters for the largest samples, respectively. Thus, a typical industrial system is equipped with a set of linear axes, in order to adjust the magnification factor from case to case. Except for the most heavy and largest objects, in industrial CT the detector and the X-ray source remain fixed on their position during the measurement, while the sample is rotated within the X-ray beam.

All applications mentioned so far have become feasible in industrial routine because of the dramatic progress that the digital detector technology has made in the past 20 years.

Today, so-called flat panel detectors are available with up to 4096 by 4096 pixels, covering an X-ray sensitive area of 40 by 40 cm. Most often a pixel spacing of 50, 100 or 200 μm is available; thus, for instance, a spatial resolution of 5, 10, or 20 μm can be achieved by ten-fold magnification.

At the same time, a new generation of high-power microfocus X-ray tubes has emerged, which allows for focal spot sizes of less than one micron. Today, X-ray tubes are commercially available that provide focal spot sizes down to already half a micron.

Using these X-ray tubes in combination with large area digital detector matrices, recently the first submicrometer CT systems have been developed with spatial resolution of less the 500 nm and laboratory CT setups with focal spot sizes down to even 50 nm are in research progress.

Fig. 3.15 depicts the mechanical setup of a typical CT machine. The design differs from a medical CT system significantly. Typically, there are between 4 and 6 degrees of freedom, eg, for the positioning of the sample between tube and detector, for lateral movement of the sample and detector, and for a vertical shift of the sample and detector. Since these mechanical translations have to be done with very high accuracy, the whole system is mounted on a rigid and air-cushioned foundation.

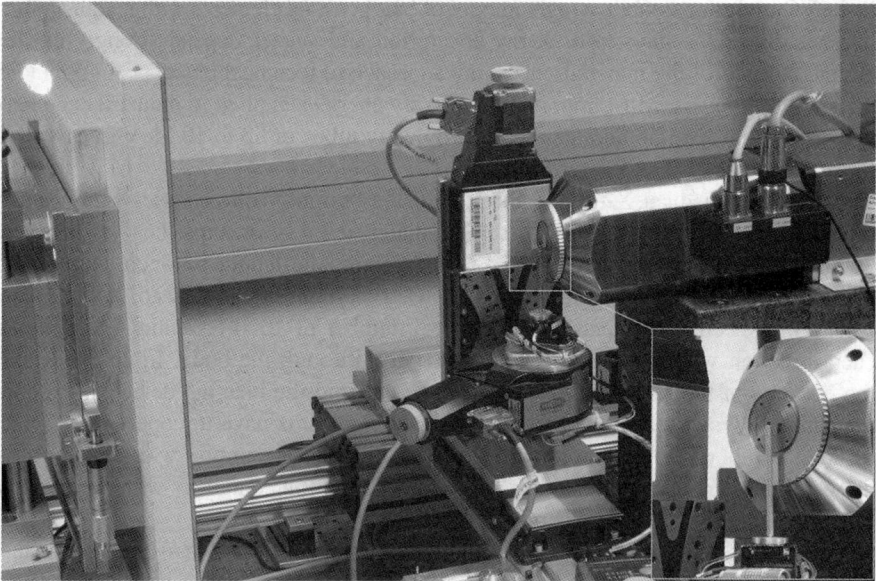

Figure 3.15 View into a CT cabinet: the μ-focus X-ray tube (right), the flat panel detector (left) and the rotational table on a vertical lifting stage (middle). The lab-system is located in a concrete cabinet for versatile applications. It can also be stored in a compact cabinet completely shielded with lead. Note the small object size shown in the right image corner and the small Focus Object Distance (FOD) corresponding to a high magnification.

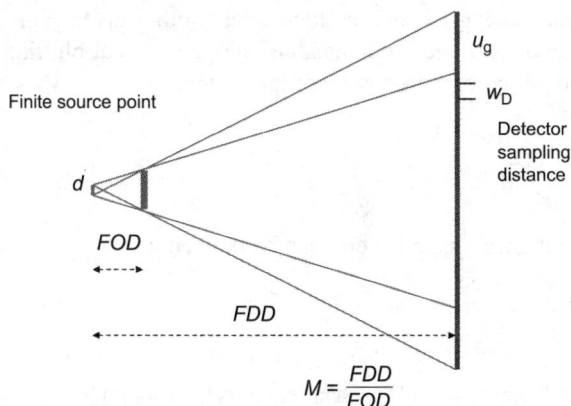

$$M = \frac{FDD}{FOD}$$

Figure 3.16 Schematic view of the geometrical conditions in projection imaging.

In order to achieve high resolution in X-ray imaging, a geometrical magnification is often the means of choice. Due to the physical properties of hard X-rays, ie, electromagnetic waves with wavelengths below 1 Å (Angstrom), until today there are not really commercial and robust techniques available to build lenses or more sophisticated optics for this kind of radiation.

There are two specific distances which play a key role in the determination of the magnification: the focal spot detector distance (FDD) and the distance between the focal spot and the sample (the "object") under examination (FOD) (cf. Fig. 3.16).

Usually the latter distance is assumed to be equal to the distance of the focal spot to the virtual axis of the rotational stage of the CT system. The rotational axis usually defines one of the coordinate axes, oriented orthogonal to the cross-sectional planes which are known as tomographic slices, or in short, tomograms.

3.4 Geometrical considerations and data acquisition

In order to achieve high spatial resolution in the cross-sectional tomographic images, sampling has to be sufficient and adequate in two domains: first, the projection pixel spacing must be small enough to resolve small details in the projection image of the transmitted radiation, and second, the angular increment while sampling the series of projection images dominates the spatial resolution depending on the distance of a reconstructed pixel from the axis of rotation.

The effective spatial resolution in projection imaging can be measured by the blurring resulting from the combined influence of focal spot size width, detector pixel spacing (which is usually equal to the pixel size), and the object's motion during the data acquisition. The latter effect plays an important role in the range of 1 μm detail size and below since random displacement of the mechanical axes and the drift of the focal spot's position are of the same magnitude during total data acquisition times of several minutes and more.

With the pixel spacing w_D and the focal spot width d given (Fig. 3.16), the ideal magnification factor is found by demanding the geometrical blurring to be equal to the effective sampling, ie, at the center of the system:

$$\frac{w_D}{M} = d \cdot \left(1 - \frac{1}{M}\right) \qquad [3.18]$$

There from, the ideal magnification can be derived as:

$$M = 1 + \frac{w_D}{d} \qquad [3.19]$$

A fundamental feature of a CT measurement is the rotation of the object. Thereby a series of projection images is recorded, each from a slightly different view angle. The tomographic reconstruction algorithm provides the pixel (in case of 2-D slices) or voxel (in case of 3-D volumes) data sets from these input data by applying an inverse Radon transform.

Precisely, these input data, which are also referred as projection or sinogram data, result from an inversion of Beer's law, ie,

$$p(\theta, t) = -\ln\frac{I(\theta, t)}{I_0} \cong \int_{L(\theta,t)} d\eta \, \mu(x, y) \qquad [3.20]$$

L denotes all existing paths through the object within the x-y-plane.

This is, of course, a simplified formulation of the projection data acquisition, with a spatial parameter t, denoting the offset of each ray from the center of rotation and the angular direction θ. Second order effects as for instance X-ray scatter and beam hardening are neglected. The projection data obtained as defined in Eq. [3.19] require the primary intensity I_0 to be known, which is incident on the object. From the X-ray tube's point of view, this means that the radiation output must be very constant over time or has to be monitored while the object is rotating inside the X-ray beam. One basic approach to derive the normalization factor is using the unattenuated output of the X-ray source given as average gray value within a region of interest (Fig. 3.17).

Another important issue to be accounted for is the least distance the sample can be moved toward the exit window of an X-ray tube. This smallest distance limits the magnification which can be achieved, without the object touching the tube during rotation (Fig. 3.18).

This is the reason why in micro-CT systems transmission targets are preferred in most cases.

A further approach to keep a large distance between source and the object is to apply detectors with small pixels (ie, 6 µm) allowing for a magnitude lower magnification at same high resolution but also limiting the versatility of the CT scanner to smallest object sizes.

I_0 As average grey value within ROI!

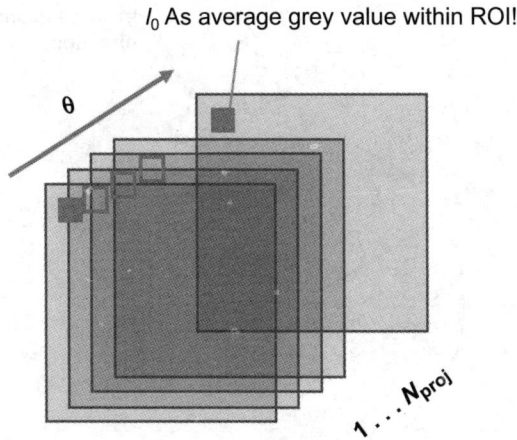

Figure 3.17 Simple approach to normalize the primarily measured intensities at each detector pixel. The unattenuated output of the X-ray source is determined by averaging the gray values within a region of interest, which has to lie outside of the object's shadow for all projection angles.

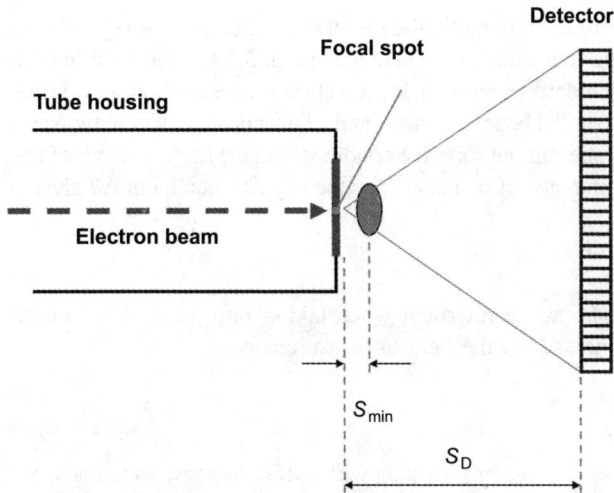

Figure 3.18 Limitation of achievable magnification with a transmission target tube. The minimum distance S_{min} of the rotational axis to the focal spot is given by the half of the sample's diameter.

As mentioned above, spatial resolution in reconstructed tomographic slices depends on an additional parameter, the angular sampling distance. In general, the operator of a CT system can choose this parameter freely since there is no physical restriction demanding a certain number of angular positions. Nevertheless, there exists a reasonable number of projections which are recommended to be acquired, in order not to

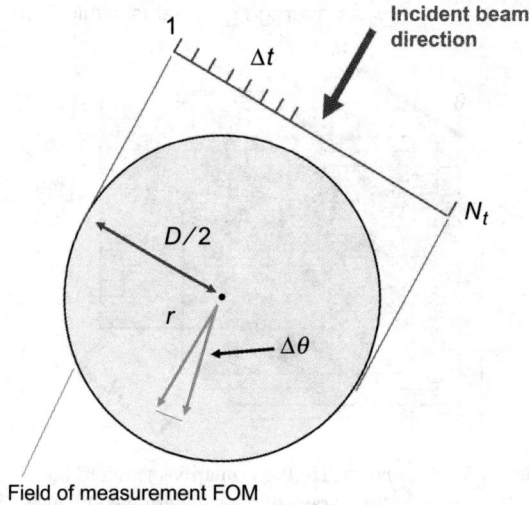

Figure 3.19 Sketch of the field of measurement. Radial and angular sampling is illustrated within the cross-sectional tomographic plane.

lose spatial resolution, in particular out of the center of the image. The following "back of the envelope" consideration, sketched in Fig. 3.19 delivers a rule of thumb, which is valid for all standard filtered back-projection reconstruction algorithms.

Assuming parallel beam geometry and a linear detector sampling Δt in a tangential direction given, one can put forth the condition that the same spacing of the rays is needed along a circumference of radius r. With the angular increment $\Delta\theta$ given, it follows:

$$\Delta t = r \cdot \Delta\theta \qquad\qquad [3.21]$$

This relation can be rewritten when taking into account the number of samples along the diameter D of the field of measurement:

$$D = N_t \cdot \Delta t \qquad\qquad [3.22]$$

and the number of angular positions within 180 degrees, which are in case of parallel beam geometry fully sufficient:

$$\pi = N_\theta \cdot \Delta\theta \qquad\qquad [3.23]$$

Putting all together we end up with a simple equation:

$$\frac{D}{N_t} = r \cdot \frac{\pi}{N_\theta} \qquad\qquad [3.24]$$

One interpretation of this equation is that the spatial resolution decreases with increasing distance of a reconstructed pixel or voxel from the center of rotation. If we require the sampling to be adequate in angular and tangential direction up to the edge of the field of measurement, $r = D/2$, we receive a useful rule of thumb:

$$N_\theta = N_t \cdot \frac{\pi}{2} \qquad\qquad [3.25]$$

The number of angular positions should be about 50% higher than the number of pixels of the detector along one spatial direction. For instance, with a 2000 by 2000 pixel detector, approximately 3000 projection angles are necessary to allow for full spatial resolution in the whole field of measurement. Of course, each projection produces costs in terms of exposition to the object and measurement time to acquire the image. Besides the fact that longer measurements are more expensive, there are technical reasons, which call for a reduction of scan time: the risk of unwanted mechanical shifts of the sample and a drift of the focal spot size due to an elongation of metal compounds of the X-ray source caused by an increase of the tube's temperature. In other words: the faster, the better, as long as the signal-to-noise of the cross-sectional images keeps constant.

As a consequence, in micro- and submicro-CT applications, new reconstruction algorithms are under development, and the applications greatly reduce the necessary number of projection angles without any drawbacks to the image quality.

One particular approach is the use of novel iterative algorithms, which either are based on a statistical description of the X-ray imaging process. These issues will be discussed in more detail in Section 3.6.

3.5 System design (CT methods)

CT system design is strongly related to the kind of inspection task as well as to the shape and geometry of objects under inspection. However, in general and on contrary to medical systems, the objects under inspection are rotated or shifted instead of the imaging components (Fig. 3.20). Depending on reconstruction methods and required resolution (magnification), different degrees of freedom for component movement and precision have to be realized.

In case of microscopic imaging, magnifications up to a factor of 1000 must be possible, which means the capability of positioning, stability and alignment of the components in the range of submicron precision over many hours of scanning time.

Fig. 3.21 gives a short overview of the leading issues in terms of scaling, reconstruction methods and kind of data acquisition, which is related mainly to the shape of objects. As can be seen, there are various combinations within this CT landscape and depending on the chosen selection, the CT system carefully has to be designed.

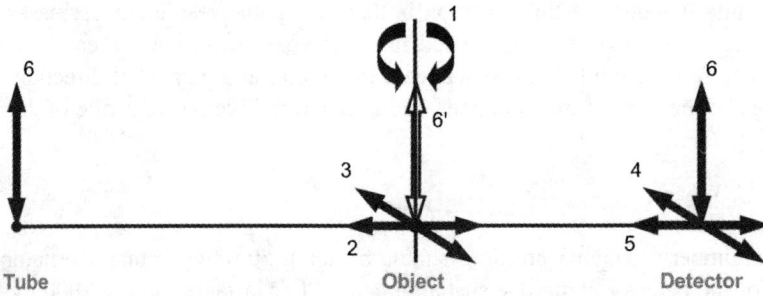

Figure 3.20 Mechanical axes: typical set-up: 1, rotation of object (projection angle); 2, horizontal translation of object (magnification); 3, lateral translation of object (system alignment); 4, lateral translation of detector (extended field of measurement); 5, horizontal translation of detector (optimization of geometry); and 6, vertical shift—synchronously during rotation.

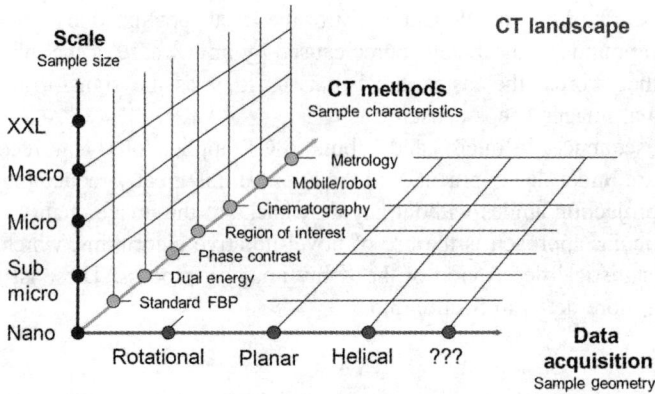

Figure 3.21 CT landscape. At least three dimensions are needed to cover all CT approaches which are available today. Depending on the size of the sample or object under investigation and the required spatial resolution, the scale can vary from several ten nanometers to several meters with cars or sea freight containers. Several essentially different data acquisition methods may be applied, often depending on the objects geometry and maybe new acquisition methods (???) will come up in future; eg, a printed circuit board shows a very high aspect ratio between the thin thickness of the plate in relation to the lateral elongation in-plane, whereas a steel tube shows a certain length, which is usually several orders of magnitude larger than the diameter of the circular cross-section.

3.6 Image reconstruction

There are two principal limitations in micro-CT reconstruction to be discussed:

Standard CT reconstruction, based on direct Fourier reconstruction or filtered back projection (FBP) suffers on direct correlation between resolution and object size which leads to limitation of resolution in case of objects, which are significantly larger than the desired spatial resolution element. With FBP methods the object has to be covered completely by the radiation field at least within the tomographic plane. For example, with a 10 cm square detector and magnification of factor 100 the sample's diameter must not exceed 1 mm, while the cross-sectional images would have a dimension of 1000 by 1000 pixels with 1 µm square each.

Second, the image quality obtained by high resolution CT systems suffers from low signal-to-noise ratios since microfocus sources are strongly limited in their output, and, on the other side, small detector pixels lead intrinsically to a worse photon statistic if the total exposure is not increased proportionally to the inverse square of the pixel size in one dimension.

To overcome this barrier, iterative reconstruction techniques today become more and more relevant. The iterative methods, which are applied today, are derived from two techniques, which have been known since the late 1970s, the algebraic reconstruction technique (Gordon et al., 1970) and statistical methods based on expectation—maximum or maximum—likelihood approaches (Lange and Carson, 1984; De Man and Fessler, 2009). In comparison, all these methods have in common that they do not apply a single-step algorithm derived from mathematically proven inversion formula of the X-ray transform (closely related to the Radon Transform). Rather, the new approaches search iteratively for the best image that fits to the measured projection data in particular and to the X-ray physics (eg, statistics, beam hardening, etc.) in general. Iterative algorithms thus differ in their strategy on how to reach the optimum solution but have in common that they perform a certain number of loops of alternated back and forward projection, accompanied by some kind of regulation, total variance minimization or statistical modeling of the photon interaction (De Man et al., 2001). The respective computation procedures involved are very time-consuming compared to the single-step analytically based filtered back projection methods of any type (generalized Feldkamp (Feldkamp et al., 1984) or Katsevich's formula for helical data acquisition (Katsevich, 2004), the family of PI-line methods (Zou and Pan, 2004) and various others).

An enormous increase of activity on the field of iterative algorithms can be observed beginning in the early 21st century, which is caused by the ever-increasing computer processing speeds (dual and quad core processors) and—what gave a very important boost to huge volume numerical data processing—by the implementation of the geometrical calculations on special-purpose graphical processing units which emerge from the gaming industry. When accelerated to computational speeds that are acceptable for practical purposes, iterative algorithms show two very attractive features: first, the algorithms deliver good image quality in those cases, where Fourier transform-based algorithms fail due to incomplete or truncated data, ie, the iterative algorithm can deal with region-of-interest measurements much more elegantly than analytical algorithms. The new generation of algebraic reconstruction methods allows for various kinds of region-of-interest reconstructions with high resolution on small regions, even within large objects, and for restricted or even irregular sampling schemes in data acquisition.

Second, because a repeated forward projection step is intrinsic to iterative algorithms, these can be easily extended to account for physical effects during a real measurement. Thus a new class of innovative artifact reduction methods has arisen, whether for noise reduction or for means against beam hardening (De Man et al., 2001).

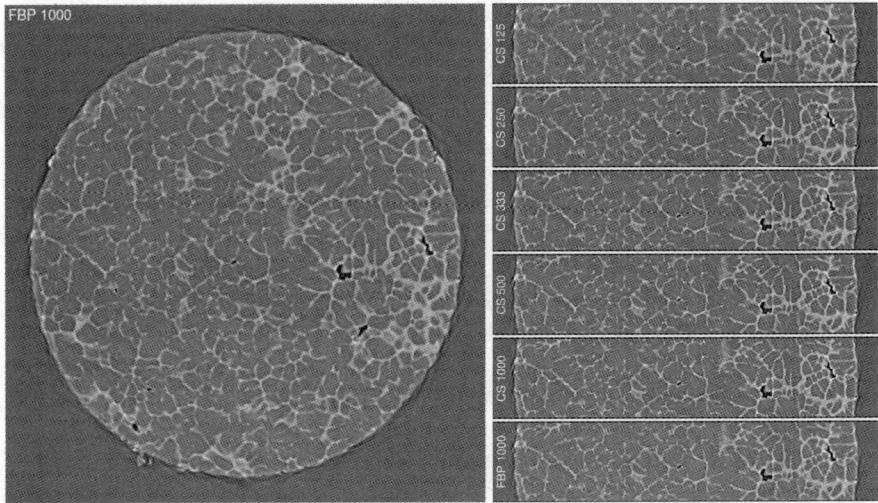

Figure 3.22 Measurement of a binary alloy of aluminum with 8 wt% of copper. The field of view has a diameter of 770 μm, the sample itself is 700 μm. A total number of 1000 projections were acquired within an angular range of 180 degree, the respective cross-sectional image achieve with a conventional filtered back projection algorithm is shown on the left side. For comparison, five compressed sensing reconstructions with varying numbers of available projections are shown on the right side, thereby a decreasing number of projections were taken into account, from bottom to top: 1000, 500, 333, 250, and 125.

As a comprehensive example, in the following we show results of a reconstruction by standard filtered back projection in comparison with an advanced compressed sensing reconstruction method. For further details, we refer to (Dittmann, 2013) and the references therein.

Fig. 3.22 shows by comparison the beneficial impact of a total variance minimization step as a particular extension of a conventional Simultaneous Algebraic Reconstruction Technique algorithm (SART) algorithm (Sidky and Pan, 2008), as described in detail in (Dittmann, 2013). The raw data were acquired within the long-term project "ma1876" at the ESRF ID19 beam line and kindly provided for these demonstrations by Pierre Lhuissier. Even after reducing the utilized number of projections to one-eight, spatial resolution and contrast detectability are still almost not affected.

3.7 Image quality

Due to the high-fidelity requirements of materials characterization for micro-CT, different image quality aspects must be considered. On the one hand, there are static characteristics of the hardware and software systems, such as detector and tube specifications but also algorithms for reconstruction and artifact suppression.

On the other hand, there are temporal effects that must be considered like the drift of the focal spot, fluctuation of the X-ray intensity, inaccuracies of mechanical rotation, respectively translation or even improper detector behavior, eg, blinking pixels during the scan. A characterization of each single aspect is helpful in the design phase of the micro-CT but less in the final application where the result is a superposition of all these effects.

Today's state-of-the-art micro-CT machines try to cope with all of these effects. Some do more and some do less, but at challenging applications with resolutions under 0.7 µm the quality and reliability of results are decreased dramatically by each single distortion, resulting in multiple measurement efforts and long stabilization periods.

For these applications an optimized combination of components with individual performance instead of a general-purpose micro-CT design leads to the desired excellent image quality. During the last years, most of the detector manufacturers have established characterization methods for the behavior of their components based on American Society for Testing and Materials (ASTM) and European Standard (EN) standards providing comparability. The driving force was the introduction of digital detector arrays to radiographic inspection that is already a mass application. Quantitative measures like modulation transfer function (MTF) and detective quantum efficiency are standards available for industrial detectors. Further characteristics and measures can be read in, eg, (Sukowski et al., 2009). In the field of X-ray tubes the standardization has not come so far yet, but features like focal spot size can be traced by resolution patterns as shown in Fig. 3.2. The determination is based on the contrast ratio resulting in the imaged structures being defined correspondingly to the MTF at 20% contrast. On the JIMA RT-RC-02 test pattern with a well-defined size, the focal spot corresponds to the double size of the pattern if the contrast reaches 20%.

For both detector and source used for cone beam CT the 2-D resolution must be considered; Fig. 3.23 shows how an elliptical focal spot size can affect the resolution in one direction (vertical) while the other (horizontal) is imaged well.

Beside these hardware component features the application of micro-CT for materials characterization suffers from many artifacts related directly to the CT modality

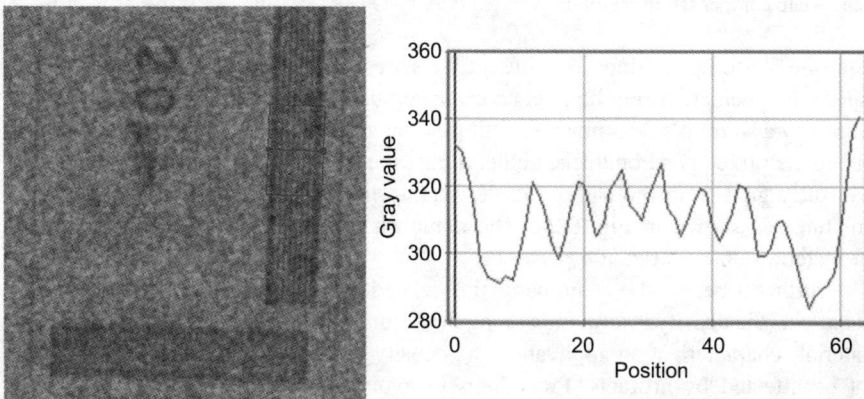

Figure 3.23 Radiograph of the JIMA test pattern using an elliptical focal spot size. Note that the horizontal lines are not resolved at all.

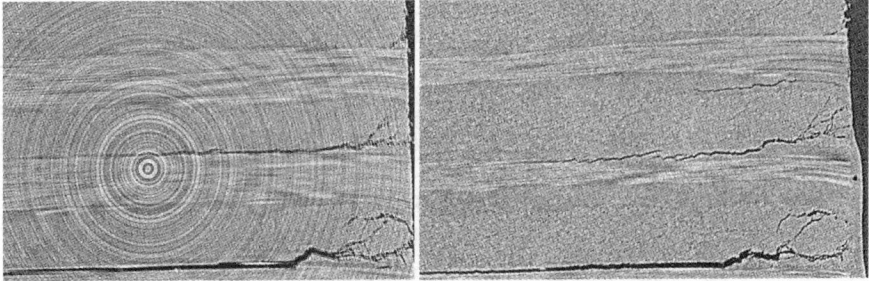

Figure 3.24 Cross-section of a CFRP sample reconstructed from the raw data (left); reconstructed using ring artifact suppression software (right).

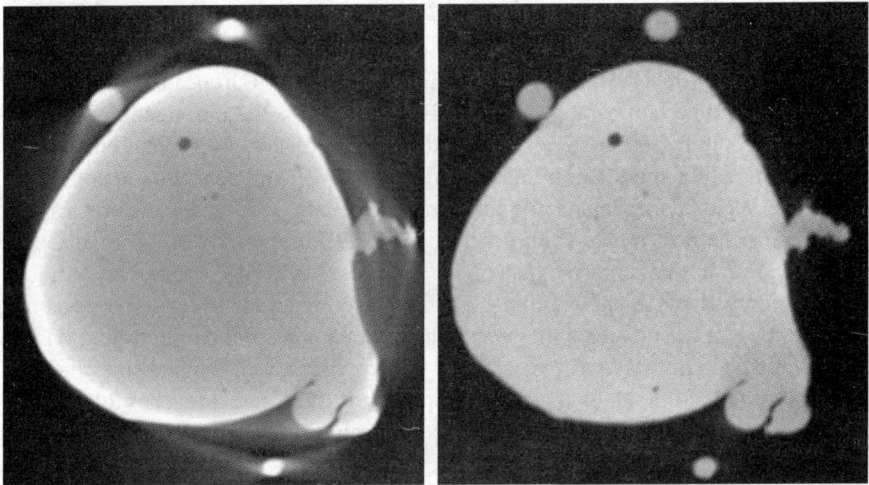

Figure 3.25 Cross-section of a nickel alloy reconstructed from the raw data (left). For comparison, a reconstruction after application of an iterative artifact reduction software (IAR) is shown (right). The density signal at the samples edges represented by the gray value is strongly affected in the left image. Apparently, the beam hardening effect can be reduced significantly by appropriate compensation methods.

requiring for compensation. The ring-artifacts are the most evident interfering signals, especially when analyzing homogeneous materials with lowest contrasts. The error is caused by a slight pixel-to-pixel deviation of the response to the same X-ray intensity. Due to their fixed position to each other during the whole scan the error is often higher than the signal from the sample itself, leading to the generation of axial ring-type structures as shown in Fig. 3.24. The demonstrated suppression method is crucial for materials characterization purposes.

A further aspect is the beam hardening related to the polychromatic X-ray spectra that affects the density information along the volume. Contrary to a standard NDT for a materials characterization application the density is a main characteristic that should not be affected by artifacts. Procedures to avoid or suppress the beam hardening artifacts are required for materials characterization as shown by Fig. 3.25.

Temporal effects must be considered at least as all above-mentioned static effects, when a long measurement time is required. The most dominant temporal effect beside vibrations or sample dislocation is the spatial instability of the focal spot. This effect is not only depending on the source itself but also on the mechanical setup that might be also affected by the emitted heat of the tube's focusing unit.

The drift behavior can be tracked experimentally using a well-defined object like a small ball and image it highly magnified during a long period. An example for such a measurement is shown in Fig. 3.26. The plotted graphs correspond to the drift of the spot in x- and y-direction measured at a voxel size of 0.6 μm resulting in a drift of approximately 4 to 5 voxels.

Compensation of this drift behavior leads to a significant enhancement of CT quality enabling the recognition of finest material structures as shown in Fig. 3.27.

In addition to the described methods enabling high-fidelity imaging, so-called image quality indicators (IQI) are used to quantify average system features in the resulting volume-like dimensional accuracy. Typically, these IQI's are precisely fabricated and quantified objects that offer the ability to reference the resulting dimension in the volume to the real dimension of the IQI. Mostly, they consist of two or three ruby spheres mounted on a holder in biangular or triangular orientation. Due to their size, ie, several millimeters, the application for micro-CT with resolutions under one micron is mostly limited.

3.8 Radiation exposure

Many people are concerned about the radiation exposure they could receive by an X-ray examination they have to undergo. While the tomography of a human being as well as all living beings is not covered by the field of NDT and evaluation, it is worthwhile to shortly discuss the potential harm that is caused to the samples when investigated by X-ray microtomography.

Theoretically, one can prove that the amount of radiation necessary to achieve a certain signal-to-noise ratio inside a voxel is related to the size of the voxel (Fuchs and Kalender, 2003). A reduction of the voxel size by a factor x, which corresponds to an x-times better spatial resolution, leads to an increase of dose at least by x^3. The exposition delivered by micro-CT methods with spatial resolutions below 10 μm can reach several 100 mGy compared to measurements with 200 up to 500 μm voxel size.

On the other hand, there is virtually no evidence that a moderate X-ray dose can damage inorganic materials like metals or stones. Summing up, the issue of radiation exposure is more important than ever since the trend to image small organic or biological structures is strong. The problem will be encountered in two ways. First, the exposure for a specific sample is measured precisely and documented. Second, the impact of X-ray exposure on such critical materials is to be tested experimentally to develop a deeper understanding of the influence of ionizing radiation on very small but complex materials.

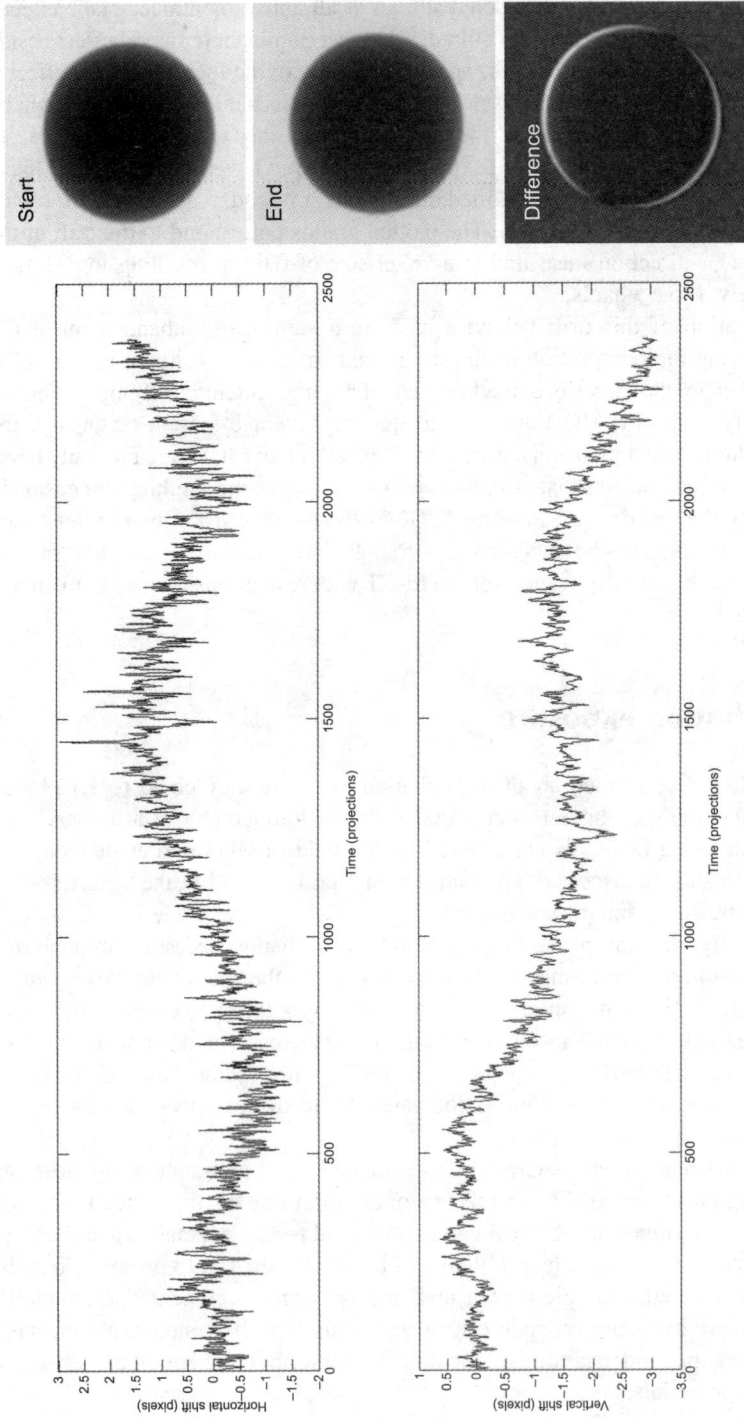

Figure 3.26 Focal drift measurement in *x*- and *y*-direction over a scan time of 4 h using a 0.85 mm ball positioned on the CT manipulator (left). The radiographs of the initial and final state (right) can be used to extract the total drift by evaluating a subtraction image. The white penumbra effect shows the resulting drift.

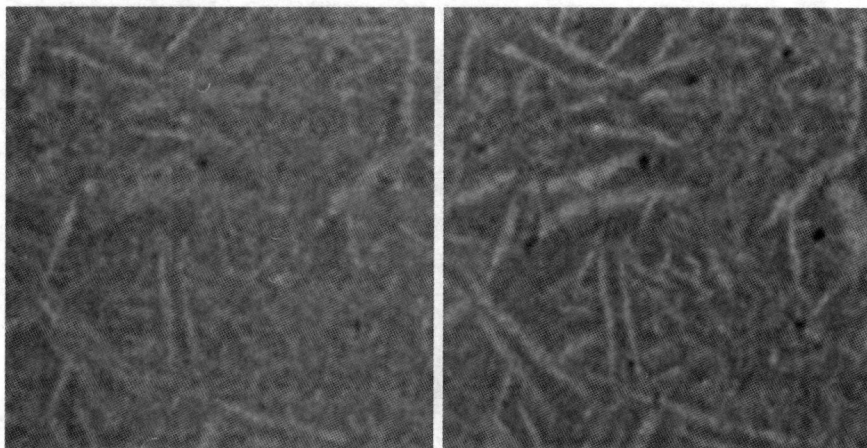

Figure 3.27 Cross-section of α-Al alloy scanned with a resolution of 1.4 μm, reconstructed directly from the raw data (left), and reconstructed using a stabilized manipulator and a drift compensation (right).

3.9   Examples of important and/or frequent applications for materials characterization

Materials characterization requires maximum CT image quality from one scan. In contrast to today's NDT applications, where the focus is on inline integration of sufficient but robust and fast CT inspection for as many parts as possible, even very long scan times are acceptable for highly resolving imaging of microstructures. Micro-CT as a modality becomes a complementary method in the laboratory, besides classical metallography that is mostly 2-D, requiring for cutting, embedding and polishing of single slices. Moreover, micro-CT allows for a 3-D analysis even without any preparation efforts for some types of samples. A very good example is the structural analysis of alloys used for historical organ pipes shown in Fig. 3.28. X-ray microtomography enables a nondestructive analysis on the organ pipes and allows for the reconstruction of historical alloy manufacturing technique.

The introduction of micro-CT to metallography revolutionized the materials characterization domain. Nevertheless, there are applications that ignore the 3-D information due to a lack in adaption of new standards. Porosity is the most prominent feature that is still often misinterpreted in the 2-D regime. Especially for materials with low porosities the 2-D porosity can differ up to one order of magnitude from the volume porosity, as shown in Fig. 3.29, depending on the cross-section taken into account for the analysis.

To overcome this uncertainty, the metallography relies as far as possible on large fields of view using sample sizes of several centimeters. This is contrary to the micro-CT application, where the diameter of the sample also relates to the achievable resolution; thus the micro-CT is not widespread in the metallography domain yet.

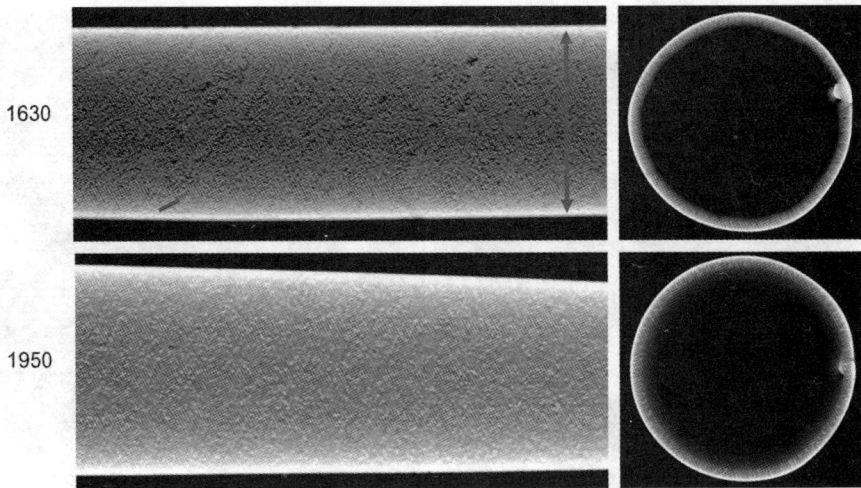

Figure 3.28 X-ray microtomography of lead. A comparison of two organ pipes with 10 mm diameter at a spatial resolution of 7 μm: longitudinal cut (left) and cross-section (right). The different metallic fine structure is clearly visible, as well as the soldering seam of the two samples which were manufactured in the year 1630 and 1950, respectively.

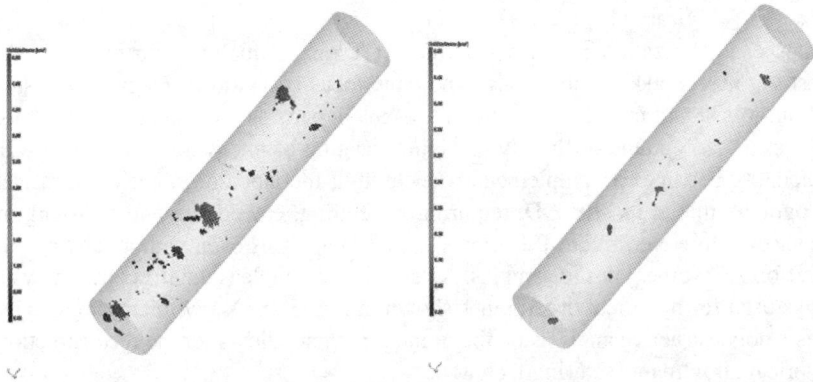

Figure 3.29 Reconstructed 3-D volumes of two different specimens for tensile testing with different porosity. A 2-D metallographic analysis along a cross-section at any single longitudinal position is not representative for porosity determination. Thus a reliable material analysis requires an examination by means of micro-CT.

The predominant adaption of micro-CT to metallography already took place where the nondestructive analysis, ie, in situ, is required or where the sample type and size limits the metallographic preparation capabilities or efforts. A main domain is fatigue analysis on tensile samples that are scanned before and after, or even during the tensile test.

In particular, for new materials with structurally oriented elements like the GFRP shown in Fig. 3.30 the 3-D structure is of relevance. The internal strain behavior can be tracked by micro-cracks using micro-CT. But also for classical metal alloys the micro-CT offers a very good approach for research and analysis in fatigue of materials.

Another main micro-CT domain away from the industrial laboratory use is the 3-D investigation of fragile samples, eg, biological with 3-D functional structures like insects or plants. Standard slicing preparation methods lead to enormous efforts with a high risk of structural distortion due to the implied force during preparation. In

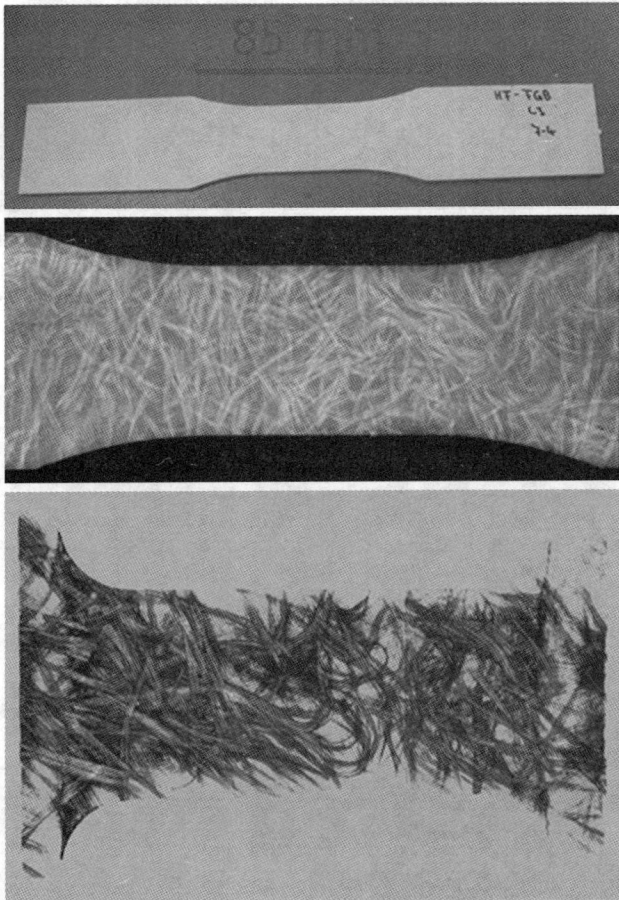

Figure 3.30 Measurements of a test sample made out of glass-fiber reinforced polymer (short fibers). The sample is used for mechanical, ie, destructive testing of the tension finally leading to failure. The measurement of approximately 3700 projection images were performed within 5 h with 120 kV tube voltage and a size of the resolution element of 26 μm (geometrical magnification factor 7.3).

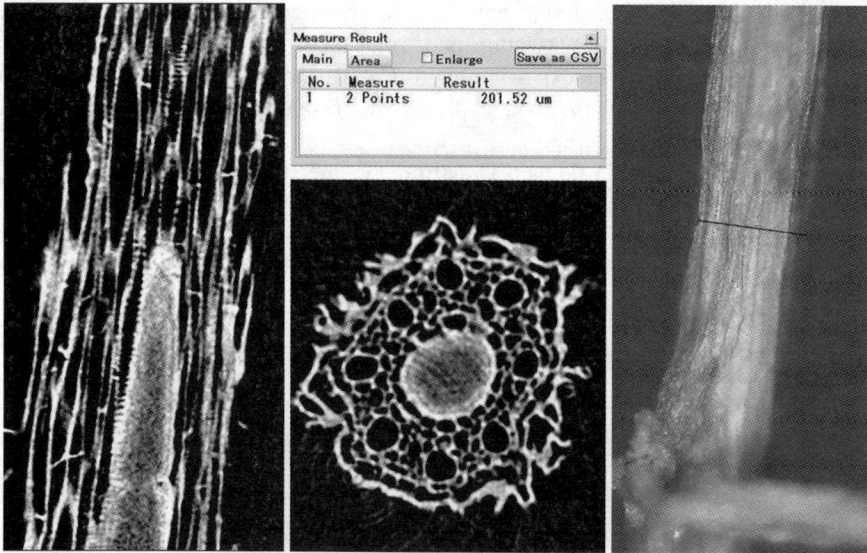

Figure 3.31 CT cross-sections of a root structure (diameter approximately 200 μm), generated at 400 nm voxel size in sagittal (left) and axial (middle) orientation. Scaling has been proven by light-optical microscopy (right). The fine vacuoles have a diameter between 10 and 30 μm. Note the rough surface providing large interaction area and proper fixation of the plant in the soil.

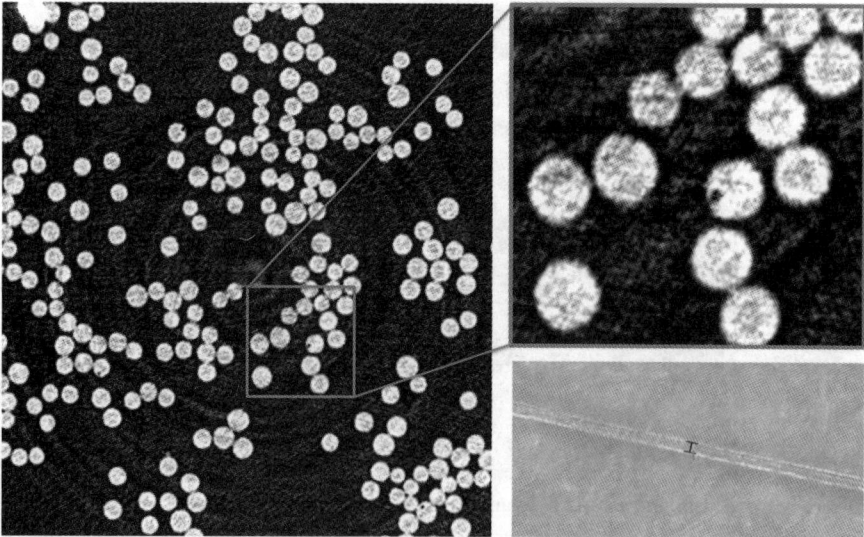

Figure 3.32 Cross-section of a glass-fiber sample scanned at a voxel size of 400 nm (left). Some of the filaments with an approximate size between 7 and 15 μm contain a significant amount of porosity (right).

Fig. 3.31 the micro-vacuoles of a root structure with a diameter of 0.2 mm are shown in a 400 nm resolved cross-section.

Also, for material analysis this high resolution capability can be useful as shown in Fig. 3.32. The single resolved glass-fibers of a diameter between 7 and 15 μm are analyzed on porosities providing important material information.

3.10 Conclusions and future trends

The main goal in microcomputed tomography is to further increase the spatial resolution while simultaneously reducing scan times and improving the contrast-to-noise ratio.

This will be achieved by advanced X-ray sources with even smaller focal spot sizes and by advanced flat panel detector systems with smaller pixel sizes. Moreover, there is a trend to total new detector systems which work in a photon counting mode, and thereby reduce image noise, nonlinearities of the measurement, and artifacts due to the polychromacity of the radiation.

3.11 Further literature

For further information on basics and methods of computed tomography imaging, we recommend the following books: Buzug (2008), Herman (2009), and Natterer (1986).

References

Buzug, T.M., 2008. Computed Tomography: From Photon Statistics to Modern Cone-Beam CT. p.s.l, ISBN 978-3540394075.

De Man, B., Fessler, J.A., 2009. Statistical iterative reconstruction for x-ray computed tomography. In: Censor, Y., Jiang, M., Wang, G. (Eds.), Biomedical Mathematics: Promising Directions in Imaging, Therapy Planning, and Inverse Problems, p. s.l.

De Man, B., et al., October 2001. An iterative maximum-likelihood polychromatic algorithm for CT. IEEE Transactions on Medical Imaging 20 (10), 999−1008.

Dittmann, J., 2013. Tomographic Reconstruction From Few Projections Based on the Theory of Compressed Sensing (Master Thesis). 1. Julius-Maximilians-University, Würzburg, Germany.

Feldkamp, L.A., Davis, L.C., Kress, J.W., 1984. Practical cone-beam algorithm. Journal Optical Society of America 6, 612.

Fuchs, T., Kalender, W., April-June 2003. On the correlation of pixel noise, spatial resolution and dose in computed tomography: theoretical prediction and verification by simulation and measurement. Physica Medica XIX (2), 153−164.

Goodman, J.W., 1996. Introduction to Fourier Optics, second ed.

Gordon, R., Bender, R., Herman, G.T., December 1970. Algebraic reconstruction techniques (ART) for three-dimensional electron microscopy and x-ray photography. Journal of Theoretical Biology 29 (3), 471−481.

Hanke, R., Boebel, F., 1992. Determination of material flaw size by intensity evaluation of polychromatic X-ray transmission. NDT&E International 25 (2), 87–93.

Hanke, R., Fuchs, T., 2008. Task-driven Design of X-ray Systems for Industrial Inspection. IEEE Nuclear Science Symposium Conference Record, 523–527.

Hemberg, O., Otendal, M., Hertz, H.M., 2003. Liquid-metal-jet anode electron-impact X-ray source. Applied Physics Letters 83.

Herman, G.T., 2009. Fundamentals of Computerized Tomography: Image Reconstruction From Projections, second ed. Springer, Dordrecht, ISBN 978-1-85233-617-2.

Holzner, C., et al., 2010. Zernike phase contrast in scanning microscopy with X-rays. Nature Physics 6, 883–887.

Hounsfield, G.N., 1973. Computerized transverse axial scanning (tomography): Part 1. Description of system. British Journal on Radiology 1016–1022.

Hubbell, J.H., Seltzer, S.M., 1989. Tables of X-Ray Mass Attenuation Coefficients and Mass Energy-Absorption Coefficients [Online]. National Institute of Standards and Technology [Zitat vom: 10. August 2015.]. http://www.nist.gov/pml/data/xraycoef/.

Hubbell, J.H., 1982. Photon mass attenuation coefficients and energy-absorption coefficients from 1 keV to 20 MeV. International Journal of Applied Radiation and Isotopes 33, 1260–1290.

Katsevich, A., 2004. Improved exact filtered back-projection algorithm for spiral CT. Advances in Applied Mathematics 681–697.

Lange, K., Carson, R., April 1984. EM reconstruction algorithms for emission and transmission tomography. Journal Computer Assisted Tomography 8 (2), 306–316.

Mayo, S.C., 2002. Quantitative X-ray projection microscopy: phase-contrast and multi-spectral imaging. Journal of Microscopy 207, 79–96.

Nachtrab, F., et al., 2011. Quantitative material analysis by dual-energy computed tomography for industrial NDT applications. Nuclear instruments and methods in physics research, Section A. Accelerators, Spectrometers, Detectors and Associated Equipment 633, 159–162.

Natterer, F., 1986. The Mathematics of Computerized Tomography. B.G. Teubner, Stuttgart, ISBN 0-471-90959-9.

Otendal, M., 2006. A Compact High-Brightness Liquid-Metal-Jet X-Ray Source, Doctoral Thesis. Department of Applied Physics, Royal Institute of Technology, Stockholm, Sweden, 2006.

Salamon, M., et al., 2008a. Comparison of different methods for determining the size of a focal spot of microfocus X-ray tubes. Nuclear Instruments and Methods in Physics Research A 591 (6), 54–58.

Salamon, M., et al., 2008b. Realization of a computed tomography setup to achieve resolutions below 1 μm. Nuclear Instruments and Methods in Physics Research A 591 (6), 50–53.

Salamon, M., et al., 2009. Upcoming challenges in high resolution CT below 1 micron. Nuclear Instruments and Methods in Physics Research A: Accelerators, Spectrometers, Detectors and Associated Equipment 607 (8), 176–178.

Sidky, E.Y., Pan, X., 2008. Image reconstruction in circular cone-beam computed tomography by constrained, total-variation minimization. Physics in Medicine and Biology 53 (17).

Stahlhut, P., et al., 2013. Laboratory x-ray microscopy using a reflection target system and geometric magnification. Journal of Physics Conference Series 463 (10), 2013.

Stahlhut, P., et al., April 2014. A laboratory X-ray microscopy setup using a field emission electron source and micro-structured reflection targets. Nuclear instruments and methods in physics research, Section B. Beam Interactions with Materials and Atoms 324, 4–10.

Stahlhut, P., 2012. Aufbau und Charakterisierung eines Röntgenmikroskops mit Reflektionstarget. Diploma thesis. Chair of X-ray microscopy, Julius-Maximilians-Universität, Würzburg, Germany.

Sukowski, F., et al., 2009. Virtual detector characterization with Monte-Carlo-simulations. Nuclear Instruments and Methods in Physics Research Section A: Accelerators Spectrometers Detectors and Associated Equipment 607 (1), 253−255.

Vogel, M., 2015. Röntgen 3.0. Physik Journal 4, 18.

von Ardenne, M., 1939. Zur Leistungsfähigkeit des Elektronen-Schattenmikroskopes und über ein Röntgenstrahlen-Schattenmikroskop. Naturwissenschaften 27, 485−486.

Zabler, S., et al., 2012. High-resolution X-ray imaging for lab-based materials research. Wels, Austria. In: Conference on Industrial Computed Tomography (ICT).

Zabler, S., Fella, C., Dietrich, A., August 2012. High-resolution and high-speed CT in industry and research. In: SPIE Conference: Developments in X-Ray Tomography, VIII, pp. 13−15. Bd. 8506.

Zernike, F., 1935. Das Phasenkontrastverfahren bei der mikroskopischen Beobachtung. Physikalische Zeitschrift 36, 848−851.

Zou, Y., Pan, X., 2004. Exact image reconstruction on PI-lines from minimum data in helical cone-beam CT. Physics in Medicine and Biology 49, 941−959.

X-ray diffraction (XRD) techniques for materials characterization

4

J. Epp
Foundation Institute of Materials Science, Bremen, Germany

4.1 Introduction

The discovery of X-rays by Wilhelm Conrad Roentgen in 1895 allowed important innovations in all scientific disciplines, making the development of new medical and technical applications possible (Roentgen, 1895). In particular, the research on X-ray diffraction (XRD) by crystals initiated by Laue, Friedrich, and Knipping in 1912 opened new possibilities in the study of crystalline materials (Friedrich et al., 1913). Since then, these methods have been further developed to become very powerful tools in the fields of materials science and engineering. The experimental methods based on X-ray that are used in materials science and engineering can be divided into three main categories (Spieß et al., 2009). X-ray fluorescence spectroscopy is widely used for qualitative and quantitative chemical analysis, in particular, in electron microscopes. The X-ray radiography is an imaging technique based on the registration of the intensity passing through an object by using films or detectors which allow making its internal structure visible due to the local variation of the absorption. One of the major developments of the last decades in this field is the X-ray computer tomography. Finally, the XRD methods are based on the ability of crystals to diffract X-rays in a characteristic manner allowing a precise study of the structure of crystalline phases. Recorded diffraction patterns contain additive contributions of several micro- and macrostructural features of a sample. With the peak position, lattice parameters, space group, chemical composition, macrostresses, or qualitative phase analysis can be investigated. Based on the peak intensity, information about crystal structure (atomic positions, temperature factor, or occupancy) as well as texture and quantitative phase analyses can be obtained. Finally, the peak shape gives information about sample broadening contributions (microstrains and crystallite size) (Dinnebier and Billinge, 2008).

In the field of materials science and engineering, several applications were developed to become state of the art techniques, in particular qualitative and quantitative phase analyses, investigations of crystallographic textures, and residual stress measurements.

The present chapter first presents a condensed overview of the generation of X-rays as well as the theory of diffraction of X-rays by crystals. Afterward, a short survey of the hardware for XRD measurements is given. The methods of phase analysis, residual stress measurements, and texture investigations of polycrystalline materials are then described with practical examples, and finally, special methods and future trends are presented.

Materials Characterization Using Nondestructive Evaluation (NDE) Methods
http://dx.doi.org/10.1016/B978-0-08-100040-3.00004-3

4.2 Principles of X-ray diffraction techniques

4.2.1 Generation of X-ray radiation

X-rays are high-energy electromagnetic waves with a wavelength between 10^{-3} and 10^1 nm (Spieß et al., 2009). The generation of X-rays is generally achieved by the use of sealed tubes, rotating anodes or synchrotron radiation sources. Sealed tubes and rotating anodes, which are used in laboratory equipment, both produce X-rays by the same principle. Electrons generated by heating a tungsten filament in a vacuum are accelerated through a high potential field and then directed to a target which then emits X-rays. The incident electrons induce two effects leading to the generation of X-rays: the first is the deceleration of the electrons leading to the emission of X-ray photons with a broad continuous distribution of wavelength, also called *Bremsstrahlung* (Schwartz and Cohen, 1987). The second is the ionization of the impinged atoms by ejecting electrons from the inner shells. In order to get a more stable state, electrons from outer shells "jump" into these gaps. The difference between the electron energies of the inner shell and of the incoming electron is emitted in the form of photons, with a characteristic energy depending on the initial and final shell position of the electrons and on the material as shown exemplarily in Fig. 4.1 (Schwartz and Cohen, 1987). The characteristic radiation requires minimum excitation potential of the electrons to be emitted, which depends on the target material.

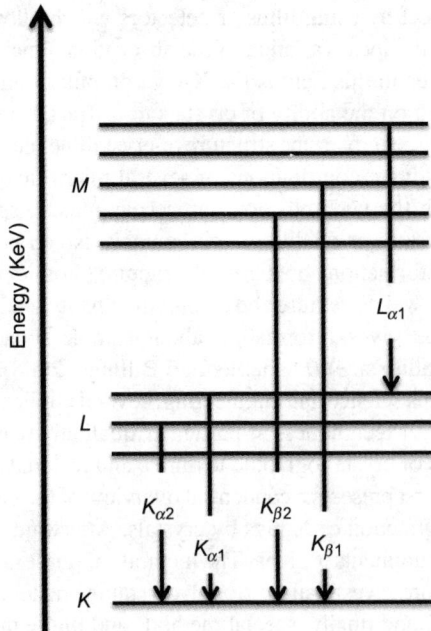

Figure 4.1 Schematic of the atomic energy levels and emission of characteristic X-ray radiation.

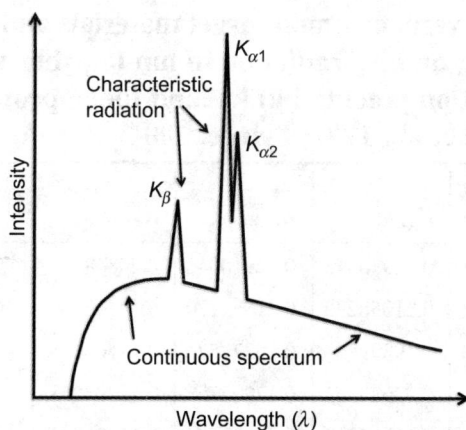

Figure 4.2 Intensity over wavelength distribution of the X-ray radiation produced by a sealed-tube showing the continuous and the characteristic spectrum.

The radiation coming out of a sealed tube or a rotating anode is therefore a super-imposition of a continuous spectrum and of characteristic radiations as presented schematically in Fig. 4.2. In general, XRD methods only use the characteristic radiation with the highest intensity, the K_α radiation, and remove most of the remaining radiation by using appropriate filters or monochromators. The filtering is based on the nonlinear absorption of the filter material regarding the wavelength, leading to absorption edges. According to the filter material, the absorption edge is situated at a different wavelength allowing a strong absorption of the continuous spectrum as well as of the K_β radiation while letting most of the K_α intensity passing through. There are appropriate filter materials for all targets. A list of the common target materials with the wavelength of their characteristic radiations K_α and K_β, together with the minimum excitation potential and the required filters is given in Table 4.1.

The production of X-ray radiation in synchrotrons is based on a different principle and is generated by bunches of electrons or positrons which circulate at relativistic energies close to the speed of light. High energy electrons are produced by a high power electron gun and injected in a booster ring for acceleration. The electrons are then introduced in a storage ring where the synchrotron radiation is produced. Modern synchrotron storage rings are generally not circular but have a polygon shape where the electron beam is guided from a straight section to another by dipole magnets (Fitzpatrick and Lodini, 2003). The generation of the radiation takes place in the straight regions where so-called insertion devices are placed. In third generation synchrotrons, wigglers or undulators are generally used, which consist of a periodic arrangement of magnets. The electrons are then forced to follow a sinusoidal path in the orbital plane of the ring leading to the emission of photons. The energy of the emitted photons covers a full spectrum from the infrared to γ-rays (Fitzpatrick and Lodini, 2003). The radiation produced by synchrotron emission has several advantages compared to the conventional method to produce X-rays. First of all, the energy of the radiation can be selected in a wide range that allows the use of the most appropriate radiation for each specific

Table 4.1 **List of several common target materials and corresponding wavelength of K_α and K_β radiation in nm together with the minimum excitation potential in kV and the appropriate filter material (Hölzer et al., 1997; Prince, 2004)**

Target	$K_{\alpha 1}$	$K_{\alpha 2}$	$K_{\alpha\ mean}$	K_β	Excitation potential	Filter
Cr	0.22897263	0.22936513	0.22910346	0.20848881	5.98	V
Mn	0.21018543	0.21058223	0.21031770	0.19102164	6.54	Cr
Fe	0.19360413	0.19399733	0.19373520	0.17566055	7.11	Mn
Co	0.17889961	0.17928351	0.17902758	0.16208263	7.71	Fe
Ni	0.16579301	0.16617561	0.16592054	0.15001523	8.33	Co
Cu	0.15405929	0.15444274	0.15418711	0.13922346	8.98	Ni
Mo	0.07093000	0.07135900	0.07107300	0.06322880	20.0	Zr

problem. The high energy range (up to 500 keV or more) permits to measure through samples of several mm or cm in thickness. The most important advantage of the synchrotron radiation is the very high brilliance of the generated beam. Therefore, synchrotron radiation is particularly indicated for short measurements of high quality as needed for in situ investigations. However, it requires the application for beam time at one of the synchrotron radiation facilities available over the world, and therefore only a few preselected measurements can be performed in each experiment.

4.2.2   Diffraction of X-ray by crystalline materials

When X-ray photons reach matter, several types of interactions can take place leading to different absorption and scattering effects, which will not be treated here. An elastic (coherent) scattering, also called Rayleigh scattering, occurs between the photons and the electrons surrounding the atomic nuclei. In this case, the energy of the scattered wave is unchanged and it retains its phase relationship to the incident wave (Dinnebier and Billinge, 2008). As a consequence, the X-ray photons impinging on all atoms of an irradiated volume are scattered in all directions (Noyan and Cohen, 1987). However, due to the periodic nature of a crystalline structure, constructive or destructive scattered radiation will result, leading to characteristic diffraction phenomena which can be studied to investigate the crystal structure of materials.

The principle of the methods is based on the diffraction of X-rays by periodic atomic planes and the angle or energy-resolved detection of the diffracted signal. The geometrical interpretation of the XRD phenomenon (constructive interferences) has been given by W.L. Bragg (Bragg, 1913). Fig. 4.3 gives the details about the geometrical condition for diffraction and the determination of Bragg's law. Bragg's law is given in Eq. [4.1].

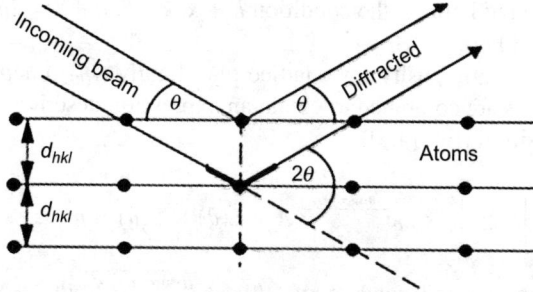

Figure 4.3 Geometrical condition for diffraction from lattice planes (Spieß et al., 2009).

$$n\lambda = 2d_{hkl} \sin(\theta) \qquad [4.1]$$

In Eq. [4.1], n is the order of diffraction, λ the wavelength of the incident beam in nm, d_{hkl} the lattice spacing in nm and θ the angle of the diffracted beam in degree. In a polycrystalline, untextured material with fine grains, diffraction occurs for each lattice plane and direction that satisfies the Bragg's law in the case of constructive interferences. This results in the occurrence of diffraction cones appearing in the form of so-called Debye rings or diffraction rings if detected by a plane detector (Fig. 4.4).

The total intensity diffracted by a considered unit cell is described by the summation of the intensity scattered from the individual atoms (Noyan and Cohen, 1987). The diffracted intensities $I_{(hkl)}$ are directly proportional to the square of the crystallographic structure factor $F_{(hkl)}$, which is a complex quantity (Eq. [4.2]) (Will, 2006).

$$F_{(hkl)} = \sum_{j=1}^{N} f_j \times \exp\left(2\pi i \left(hx_j + ky_j + lz_j\right)\right) \qquad [4.2]$$

with f_j the form factor or atomic scattering factor of atom j, hkl the Miller indices of the diffracting planes and xyz the relative atomic positions in the unit cell. The summation j runs over all atoms in one unit cell (Will, 2006).

According to the crystal symmetry, different extinctions of interferences will occur, leading to different diffraction patterns. For example, in the case of body-centered lattices, diffraction occurs if the condition $h + k + l = 2n$ is satisfied,

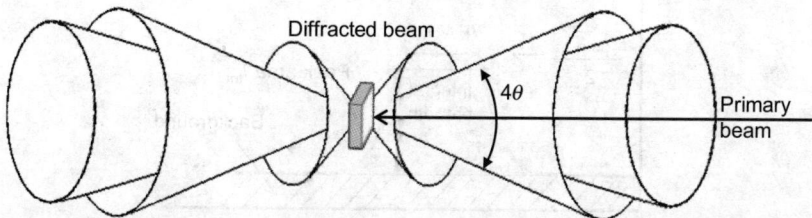

Figure 4.4 Diffraction cones in transmission and reflection occurring for a polycrystalline material.

while in face-centered lattices, the condition $h + k, k + l, h + l = 2n$ has to be satisfied (Spieß et al., 2009).

The total diffracted intensity for a lattice plan family ($I_{(hkl)}$) depends on several factors. These terms are combined to obtain an expression describing the total intensity at any 2θ position (Eq. [4.3]).

$$I_{(hkl)} = K \times \left| F_{(hkl)} \right|^2 \times f_a e^{\frac{-B \cdot \sin^2(\theta)}{\lambda^2}} \times A \times L(\theta) \times P(\theta) \times m \qquad [4.3]$$

With K, a constant independent of 2θ, $f_a e^{-B \cdot \sin^2(\theta)/\lambda^2}$ the temperature factor describing the average displacement of atoms from their mean position due to temperature, A the absorption factor, $L(\theta)$ the Lorentz factor which is equal to $1/\sin(2\theta)$, $P(\theta)$ the polarization factor which is equal to $(1 + \cos^2(2\theta))/2$ and m the multiplicity describing the number of equivalent planes that can diffract at a given Bragg angle (Spieß et al., 2009; Noyan and Cohen, 1987).

As real materials always contain imperfections, the intensity distribution of the signal diffracted by considered family of *hkl* planes can be altered. The shape of the diffracted signal is influenced by different factors, and the resulting signal is then a convolution of the following contributions (Guinebretière, 2006):

- The instrumental broadening which depends on the X-ray source, primary and secondary optics, detector, etc.;
- Composition heterogeneity within the analyzed crystallites (leading to a distribution of lattice constants and so to a possible broadening);
- The size of the coherently diffracting domains (also called crystallite size);
- Crystal defects like dislocations, stacking faults, twins, etc.;
- Inhomogeneous strains and microstrains.

In general, diffraction data are represented as intensity distribution as a function of the 2θ angle. The information content that can be extracted is represented in Fig. 4.5.

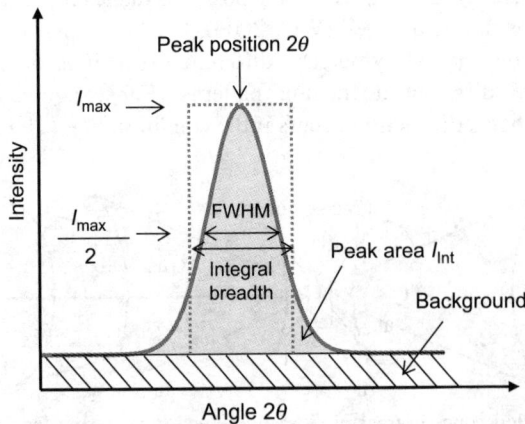

Figure 4.5 Diffraction peak and information content that can be extracted.

After background subtraction, the maximum peak intensity I_{max} can be defined as well as the integrated intensity I_{int} (area under the peak). The peak position can be determined by several methods (center of gravity, fit of different mathematical function, etc.). The peak width can be generally characterized either by the full width at half maximum (FWHM) corresponding to the peak breadth at half of the maximum intensity, or by the integral breadth (IB), which corresponds to the width of a rectangle of the same maximal and integrated intensity as the considered peak. Depending on the purpose of the measurements, the different peak parameters are used.

As a consequence of absorption and scattering effects, X-rays are weakened along their path in the material. The intensity loss is exponential and follows the general law of Beer–Lambert. The transmitted intensity I can be calculated by Eq. [4.4] with I_0 being the initial beam intensity, x is the thickness of the absorbing layer, and μ is the mass absorption coefficient depending on the wavelength of the X-rays and on the material (Klug and Alexander, 1974). The depth penetration of X-rays produced in laboratory equipment is typically in the range of a few micrometers to a maximum of several tens of micrometers for common wavelengths in metals (Spieß et al., 2009).

$$I = I_0 \times \exp(-\mu \times x) \tag{4.4}$$

4.2.3   Special methods

4.2.3.1   Energy-dispersive X-ray diffraction

The methods presented previously in this chapter are all for the commonly used angle-dispersive measurements, meaning that a (quasi)-monochromatic beam is used while the angular position of the diffraction peaks is measured. An alternative method called energy-dispersive consists of the use of a white beam (polychromatic beam over the complete available wavelength range) and the detection of diffraction peaks at a fixed 2θ angle by using energy-resolved detectors. In this case, the Bragg's law has to be modified as given in Eq. [4.5] (Spieß et al., 2009).

$$E_{hkl} = n \times \frac{h \times c}{2 \times \sin(\theta)} \times \frac{1}{d_{hkl}} \tag{4.5}$$

This method has been mainly developed in synchrotron applications as high beam intensity over a large range of energy is available. One advantage of the method is that no detector movement is needed to record a large number of peaks. As a consequence, rapid measurements can be performed, for example, for in situ analysis. A second advantage is based on the varying penetration depth of the signal due to the covered energy range in the case of measurements in reflexion mode. Thanks to this property, the varying diffraction peaks at different energy levels can be assigned to different depth positions, and therefore analysis of depth profile can be performed without material removal and without sample or detector movement. One main application is the depth-resolved residual stress analysis (Spieß et al., 2009).

4.2.3.2　Small angle scattering

The method small angle X-ray scattering allows the investigation of periodic structures within a considered material. In general, measurements are performed in transmission with a monochromatic fine-focused beam at thin samples. The considered range of diffraction angles is thereby smaller than $10°$. The principle is still based on the measurement of a diffracted beam by using a 2-D detector. However lattice plans are not considered for diffracting the X-ray beam, but rather, small periodic structures within a material like precipitates, fibers, etc. The phenomenon of diffraction takes place due to variations of electron density at the boundaries of investigated structures. Information about structures with dimensions between 5 and 150 nm can typically be collected. Specific applications can even go to larger structure sizes.

4.2.3.3　In situ X-ray diffraction

One major field of special application of the XRD methods is for the investigation of processes under nonambient and temporally changing conditions. These in situ XRD methods were developed over several decades as they present many advantages compared to other techniques like resistivity measurements or dilatometry. In particular, in situ XRD allows to investigate phase-specific properties like lattice parameters, strains, orientations, phase contents, etc., in a direct manner. Sample environments can be as varied as needed for the investigations, such as high pressures and more generally mechanical loading, temperature variations in high temperature ranges, or below room temperature, etc. However, one major barrier for these investigations is the time resolution. With the rapid progress of microelectronics and computers, the methods of in situ XRD investigations could be developed to become more and more accurate with high time resolutions. In particular, the construction of third-generation synchrotrons opened new possibilities in studying very fast processes with time resolutions up to 30 Hz or higher (Curfs, 2002).

In the area of steel research, several phenomena occurring during heat treatment were investigated using standard angle-dispersive methods with synchrotron radiation, like lattice parameter fluctuations in austenite and transformation to bainite (Babu et al., 2007) or phase transformations during welding (Elmer et al., 1996). In situ measurements during mechanical loading were used for the investigation of austenite stability in TRIP-steel (Kruijver et al., 2002).

As already explained in Section 4.2.3.1, methods based on energy-dispersive measurements are also available, allowing recording of energy-dispersive signals from white X-ray beams without any need of detector displacement. Time and depth-resolved texture, residual stress, and phase transformation investigations of rapid processes can be performed, such as the martensitic transformation of filler material for welding (Kromm et al., 2010) or the generation of residual stresses during laser hardening with a measuring frequency of 10 Hz (Kostov et al., 2012).

However, it is not always possible to perform every experiment at a synchrotron as generally only several days of beam time per year are accorded to each project. Therefore, methods of in situ XRD with laboratory equipment are also of great interest. Due to much lower signal intensity and energy, the methods of laboratory in situ

XRD are much more difficult to apply than at synchrotron. Major problems include time resolution and undesired surface modification during heat or surface treatment as the depth penetration of laboratory X-ray is low (Recke and Hirsch, 2007). Several processes could already be investigated, such as thermal residual stress relaxation of steels coated by chemical vapour deposition (CVD) by using a position sensitive detector (Tomala, 1998) or the formation of nitride layers and the generated stresses during gas nitriding treatment (Günther, 2004). Finally, phase transformations and carbide dissolution kinetics during isothermal holding and continuous heating (Epp et al., 2007) as well as the martensitic transformation during rapid quenching (Rocha et al., 2005) could be recently investigated thanks to a 2-D detector and rotating anode, allowing faster measurements.

4.2.4 Hardware for X-ray diffraction measurements with laboratory equipment

Modern lab diffractometers are computer-controlled and equipped with different hardware depending on their scope of action. The instruments are all composed of an X-ray source, primary and secondary optics, a goniometer, a sample holder, and a detector. Generally, stationary equipment is used in labs. However, in particular in engineering applications, the investigation of very large or heavy components can be required, and therefore mobile X-ray diffractometer, which can be directly placed on the part to investigate, is available. Such equipment will, however, not be treated in the present section.

4.2.4.1 X-ray source

The different possibilities for the production of X-ray radiation and the commonly used wavelength in laboratory equipment were already presented in Section 4.2. The two main aspects of the X-ray production are the intensity of the radiation and the size/shape of the beam coming out of the anode according to the purpose of the measurements. For the generation of X-ray photons, a high-voltage acceleration of electrons coming out of a tungsten filament is necessary. Generally acceleration voltage between 20 and 60 kV are used. As 99% of the energy used for the production of X-ray is dissipated in heat and only 1% is available for the generation of the radiation, heat development is a major issue limiting the brilliance of the X-ray beam (Spieß et al., 2009). Continuous water cooling of the anode is therefore mandatory, and only low current below 60 mA can generally be used in standard X-ray tubes, limiting the beam intensity. Higher current and consequently higher beam intensity can be achieved by using rotating anodes, as the point of the electron impingement is continuously changed at high speed (up to 25,000 rpm) and therefore the local heat development can be limited.

Common X-ray tubes and rotating anodes allow the use of two types of focus shape depending on the outlet window which is used (sealed tubes generally have four outlet windows, one at each side of the tube: two with point and two with line focus). For texture, residual stress measurements and phase analyses with high spatial resolution point focus should be chosen, while line focus can be used for phase analysis or residual stress measurements when a large area can be measured as much higher signal intensities are achievable, due to larger irradiated area on the sample.

In the last decades, microfocus tubes using air cooling were developed, allowing very small focus size and high beam intensity. These are particularly indicated for measurements with very high spatial resolution ($\emptyset$ from 10 to 200 µm).

4.2.4.2 Goniometer

All laboratory equipment is equipped with a goniometer, which is the central part of the diffractometer allowing to move the X-ray source, the sample and the detector relative to each other in a very precise manner. In general, Bragg—Brentano geometry is used, which means that the distance between the sample and the detector is constant for all θ angles. For the investigation of powder or massive samples with laboratory equipment, the reflexion mode is generally used as the strong absorption of the X-ray beam by the material does not allow transmission measurements. Two major basic types of goniometer are widespread: θ/θ goniometers, where the sample is fixed (Fig. 4.6(a)), while the X-ray source as well as the detector moves, and $\theta/2\theta$ goniometers for which the X-ray source is at a fixed position while the sample and the detector are moving (Fig. 4.6(b)).

For residual stress and texture measurements, additional rotation axes are generally required in order to position the sample as needed for these investigations. The in-plane rotation of a sample (azimuth angle φ) and the tilt-angle (also called pole angle χ) are commonly available (so-called four-circles goniometers). The χ-rotation can be either achieved by eulerian cradle, as shown in Fig. 4.7, by special tilting devices or by using a robot arm.

According to the sample geometry and the purpose of the measurement, specific sample holders (eg, to achieve automatic sample positioning, automatic sample changing, continuous sample rotation or translation, etc.), or even a controlled sample environment for in situ investigations can be used. In the case of powder samples, amorphous polymer or glass plates can be used to carry the powder. For solid samples, in particular

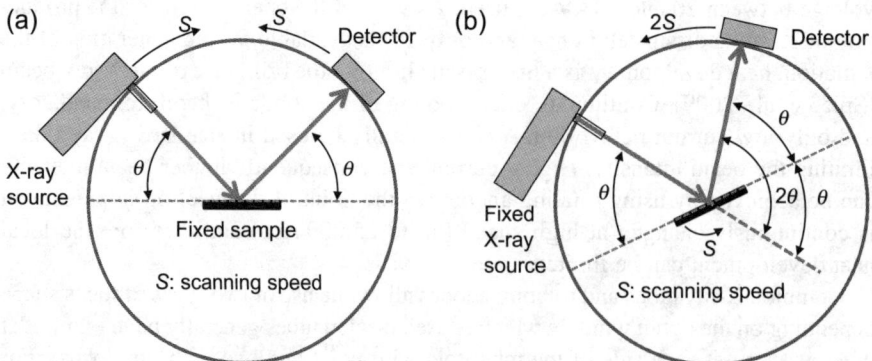

Figure 4.6 Principle of (a) θ/θ goniometers and (b) of $\theta/2\theta$ goniometers.

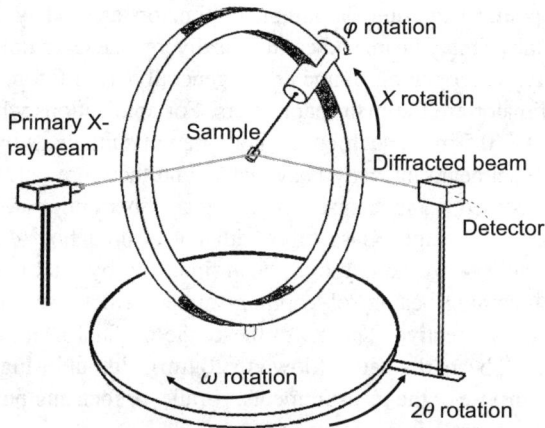

Figure 4.7 Principle of a four-circles goniometer.

for engineering applications, heavy samples and complex geometries are common. Therefore, adapted sample positioning devices have to be used.

4.2.4.3 Primary optics

In order to define the size and form of the primary beam as well as to obtain the highest possible signal intensity or to select the wavelength precisely, special optical devices can be used.

With the use of single crystal monochromators, a specific wavelength can be defined and a monochromatic beam, generally $K_{\alpha 1}$, can be selected, while $K_{\alpha 2}$ and K_β can be removed. Several crystals are available, allowing for the selection of a defined wavelength (see Ref. Spieß et al., 2009). Monochromators have also been developed to allow a parallelization of the beam. These optics, commonly named Göbel mirrors, are multilayer coated optic systems and consist in a succession of strongly and weakly scattering materials arranged in a defined manner. They are particularly indicated for the investigation of irregularly shaped samples and for applications where high beam intensity or high angular resolution is needed.

In order to define the size and the shape of the divergent X-ray beam coming out of the anode, different optical devices can be employed. In the case of line focus, slit systems comprising antiscattering slits and divergence slits are generally used, allowing reducing the beam size and its divergence. According to the number, the position and the opening of the slits, the beam size reaching the sample can be defined.

Another common type of optics are Soller slits, which are generally composed of 20–40 thin plates of metals with high absorption, piled in parallel with a small gap and with a defined length. Divergent beam parts are absorbed and only parallel or slightly divergent beam parts are coming through the slits. With this type of optic, the size of the beam is not directly modified, but only its divergence. They can be applied to reduce the divergence in vertical or in horizontal direction. As a consequence, the total beam intensity is strongly reduced. This type of optic is indicated for measurements where high angular resolution of the peaks is required.

In the case of point focus sources, pinhole collimators are widely used, in order to define the size of the primary beam and control its divergence by removing a part of the incoming intensity. The opening diameter ($\varnothing$ generally from 0.5 to 3 mm) and the length of the collimator are the main parameters. For applications where small beam size is required ($\varnothing$ < 0.5 mm), such collimators are not suitable as only a very small portion of the original beam intensity is available, and therefore only very low signal intensity can be recorded. In such cases, special glass fiber optics are generally used. In monocapillaries, one single glass tube with the desired hole diameter is used, which allows an increase in the outgoing beam intensity by total reflexion of a part of the divergent incoming beam. Polycapillary optics, which use a large number of glass fibers, are more effective. Thanks to these, the original beam diameter can be reduced to very small beam diameters (down to 20 μm) without losing much intensity. According to the purpose of the measurements, parallel or focusing polycapillaries can be used, leading to a parallel or a focused beam.

4.2.4.4 Secondary optics

Secondary optics are devices placed between the sample and the detector and are used to define the diffracted beam. Several components described for the primary optics can also be used as secondary optics. Systems of slits are generally used to reduce the divergence of the beam and lead to narrow peaks but with intensity lost. As well, Soller slits can be used to reduce the beam divergence and achieve higher angular resolution. Secondary monochromators can be used in order to remove undesirable wavelengths such as K_β radiation or fluorescent radiation (for example, in the case of Cu-K_α radiation used for the investigation of Fe-based material). Filters to reduce the K_β radiation (see Section 4.2) can generally be placed in front of the detector.

4.2.4.5 Detectors

Several types of detectors are available for the detection of the diffracted beam intensity. The working principle of the detectors is based on the conversion of impinging X-ray photons into another signal which can be analyzed. Several technologies are available for the detectors. Gas detectors are based on the ionization of a gas by the incoming X-ray photons that generate voltage pulses. Solid detectors use the phenomenon of fluorescence of special materials and either convert into voltage pulses or into visible light that can be recorded by a charged coupled device (CCD) camera. Finally, semiconductor technologies are also available for the detection of X-rays. These have a very good energy resolution and are therefore widely used in an energy-dispersive measurement mode (Guinebretière, 2006).

These detectors can be classified in three categories, according to the size of their active detecting area. Zero-dimensional detectors are point detectors, which means a very small angular range is recorded at one detector position. Point detectors are generally either proportional counters (gas ionization) or scintillation counter (fluorescent crystal). One-dimensional detectors (position sensitive detectors (PSD)) have a large active detecting area in 2θ (from a few ° to 120° 2θ) but a limited height

(Guinebretière, 2006). Such detectors are often based on gas ionization in combination with a counter wire allowing the spatial localization of the incoming X-ray photons by analyzing the time delay of the impulses at both ends of the wire (Spieß et al., 2009). Two-dimensional detectors are in form of plane or curved plates that can measure a large part or even complete diffraction rings (Spieß et al., 2009). The working principle of such detectors can be either as PSD by using multiwire techniques or based on CCD camera technologies (Spieß et al., 2009). With the use of 1-D and 2-D detectors, time can be gained because a large angular range can be recorded simultaneously. In the case of 2-D detectors, the large portion of diffraction rings that can be measured can allow analyzing texture or residual stresses from one single measurement and can be very useful in the investigation of materials with very large grains or in microdiffraction, leading to spotty rings (He et al., 2000).

4.3 Applications

4.3.1 Qualitative and quantitative phase analysis

Materials for engineering applications are generally optimized regarding their mechanical properties by controlling the present phases and their distribution. Therefore, complex multiphase materials result. In particular, for the development of new alloys or of new production technologies, but also in the context of failure analysis and identification of the present phases, determination of the respective phase contents are of great importance.

4.3.1.1 Measurement of diffraction patterns by X-ray diffraction

As explained in the previous sections, different lattice planes are in diffracting conditions, and varying intensities of the diffracted signal occur according to the crystal structure and space group of the present phases. As a consequence, each phase produces a characteristic diffraction pattern that allows its identification. Moreover, when several phases are present in a system, the characteristic patterns of all phases are superimposed and the intensity of the diffraction peaks of the phases are respectively proportional to their amounts. Therefore, XRD methods are widely used for the identification of present phases (qualitative analysis) and for the determination of their respective amounts (quantitative analysis).

In order to perform a qualitative/quantitative phase analysis, a diffraction pattern covering a large 2θ range has to be measured in order to record as much diffraction peaks as possible. For this, the different hardware presented in the previous section can be used. Generally, so-called coupled scans are performed, which means that the sample is always at half the position of the detector in the case of a $\theta/2\theta$ goniometer, or that the X-ray tube and the detector are always at the same angle θ in the case of a θ/θ goniometer. In special cases, uncoupled scans can be performed where the θ and 2θ angles are independently moved. One main application of this strategy is the grazing incidence method, where the primary beam is reaching the sample by a constant

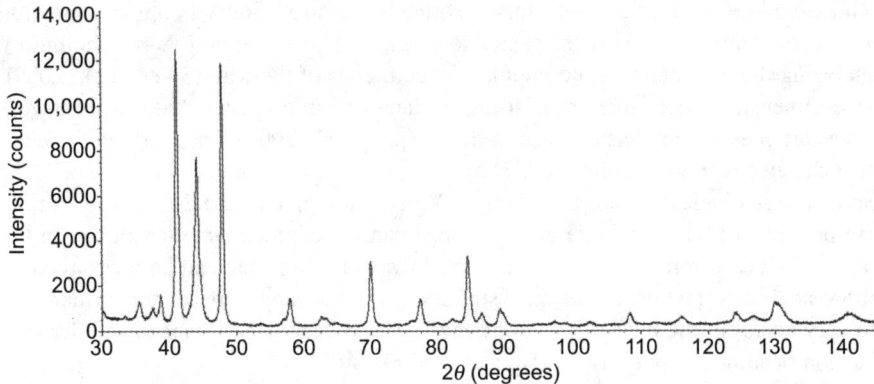

Figure 4.8 Diffraction pattern of a nitrided tool steel X40CrMoV5-1.

and very flat angle (generally between $1°$ and $5°$) while only the detector is scanning. This method is applied for the investigation of thin films in order to reduce the penetration depth of the X-ray beam inside the sample by increasing the beam path.

In order to achieve reliable phase analyses, high quality diffraction patterns are required. Depending on the hardware and on the sample to analyze, the measurement of a diffraction pattern can take between several minutes and 100 h or more. When point detectors are used, continuous scans (continuous scanning of the detector with a defined speed) or step scans (measurements at discrete positions with a defined step size and measurement time for each step) can be performed. In the case of step scans, the step size has to be small enough in order to achieve well-defined peak shapes.

An example of a complex diffraction pattern is shown in Fig. 4.8. The sample is a tool steel of type X40CrMoV5-1 treated by a nitriding treatment leading to the formation of a white layer. The measurement was performed with a $\theta/2\theta$ diffractometer using Cu-K_α radiation with a line focus, divergent slits as primary optics and as secondary optic a Soller slit, a LiF monochromator with a scintillation counter. The measurements were performed in a 2θ range from $30°$ to $145°$ with a step size of $0.02°$ and 20 s/step, leading to a total measurement time of 32 h.

The diffraction pattern contains plenty of peaks over the complete 2θ range, showing varying and partially overlapped intensity. In order to identify the present phases and quantify their respective amounts, the different methods of qualitative and quantitative phase analysis can be employed.

As already remarked in Section 4.2.2, the penetration depth of X-rays is generally limited to a small surface layer of some micrometers. When measurements in deeper layers are required, a material removal is needed. This can be the case, for example, when the very surface of a sample has been modified and is not representative of the bulk material or when depth profile of residual stresses or phase contents is needed. For this, it has to be ensured that the material is not influenced by the layer removal operation. Therefore, local electrochemical etching of the surface is often used, as no mechanical load and no pronounced heat development occur.

4.3.1.2 Qualitative phase analysis

Once a diffraction pattern has been measured, a qualitative phase analysis can be performed. In order to identify the present phases, a comparison of the present diffraction peaks with known data from a database has to be performed. The main available database is provided by the International Center for Diffraction Data (ICDD) and is based on different scientific sources (http://www.icdd.com/). According to the special needs of the users, different databases, called powder diffraction files (PDF), can be purchased: PDF-2 for nonorganic materials, PDF-4 for organics, PDF-4 for minerals or PDF-4+, which contains all entries as well as additional information about atomic coordinates of the phases, etc. These databases contain at the present time up to 500,000 entries. For each documented phase, lattice structure, space group, lattice parameters, and the corresponding position, intensity and indexation (Miller's indices) of diffraction peaks are available.

In the past, the data were available as a system of classified cards (joint committee on powder diffraction standards (JCPDS)) which could be consulted and the peak information had to be compared manually in order to identify the present phases. The databases are now available electronically and can be integrated in the software packages of most of the providers of diffraction equipment. Thanks to this, the comparison of the present peaks is computer-assisted so that the identification of present phases is much more convenient.

After the measurement of the diffraction pattern, a data treatment can be required in order to achieve reliable results. In general, the background has to be subtracted from the measured pattern. Further pattern treatment, such as mathematical removal of $K_{\alpha2}$ radiation or smoothing operations, might be indicated. Depending on the software, the peak search can be performed by different methods: manual marking or automatic identification by intensity threshold, peak shape specification, and first or second derivatives (Spieß et al., 2009).

In order to start phase identification, several input data are required in order to achieve reliable results. First of all, the nature of the analyzed sample has to be known and inputted into the software as inorganic, organic or mineral. Then, the chemical elements which should be taken into account for the possible phases have to be selected. For this the knowledge about the chemical composition of the analyzed sample is required and therefore at least a qualitative element analysis should be available (better to have a quantitative full analysis). All elements which are not present can then be discarded, while available elements can be marked as "mandatory" or as "possibly present" for the considered phases.

The measured peak data are automatically compared with the entries of the database that fulfill the criterion, which results in a list of possible phases. For each phase, quality marks giving information about the quality and the reliability of the data are specified and should be considered in order to evaluate the results. For every proposed phase, theoretical peak positions and the associated theoretical intensity are shown superimposed to the measured pattern. In order to evaluate whether a proposed phase is present or not, several approaches can be used. A list containing the number of matching and of nonmatching peaks is given from the automatic comparison of the present peaks with the theoretical peaks of a phase from the database. A figure of merit is calculated on this base in order to evaluate globally how good each phase matches, with the smallest value being the best figure. After preselection of possible phases based on this information, a visual (manual) control has to be performed. Deviations from ideal peak positions due to solid solution or strains as well as intensity variations have to be taken into account. In the case of fine powder samples, texture effects as well as macroscopic strains can be avoided, and therefore the theoretical patterns can match the measurements quite well. On the other hand, solid samples and engineering components can exhibit strong textures, large grains (and therefore irregular peak shape due to poor diffraction statistic) or peaks shifts due to high macroscopic strains. The final selection of the present phases can therefore only be performed by the user on the base of the proposed results and of his knowledge about the investigated sample and the measurement conditions. In an ideal case, all measured diffraction peaks should be assigned to a phase.

As an example, the diffraction pattern of a nitrided tool steel previously presented in Fig. 4.8 has been analyzed with the software EVA (Bruker-AXS, Karlsruhe, Germany) including the PDF-2 database. With the knowledge of the chemical composition of the surface layer and of the core, the elements N, O, Cr, and Fe were taken into account for possible phases. Fig. 4.9 present the diffraction patterns with the identified phases. All major peaks were assigned to Fe_3N and Fe_4N nitrides, while small peaks corresponding to CrN were also identified. Due to the thickness of the nitride layer and the penetration depth of the radiation, the α-Fe matrix below the white layer could also be identified. Finally, as a consequence of the formation of a thin oxide layer at the surface, small peaks corresponding to Fe_3O_4 were also identified. For each identified phase, the positions and the intensity of the theoretical peaks are given. Additional

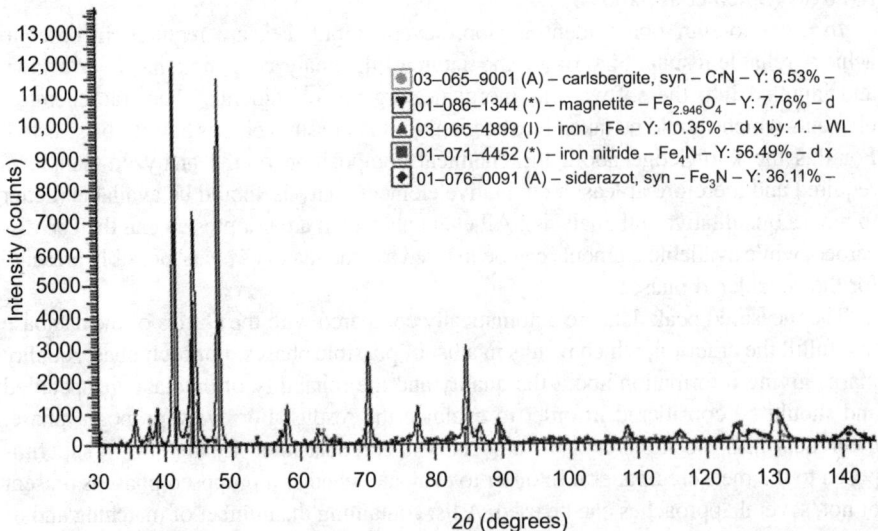

Figure 4.9 Qualitative phase analysis of a nitrided tool steel X40CrMoV5-1.

information like the reference numbers of the identified phases, chemical formula, lattice parameters, etc. are available for the different phases.

4.3.1.3 Quantitative phase analysis

Once the present phases are known, quantitative phase content analysis can be performed. In the case of multiphase compounds, the determination of phase contents can generally not be performed directly on the base of the measured integrated intensities as if the phases have different mass absorption coefficients, the intensity evolution is not linear with the increasing amount (Klug and Alexander, 1974). According to the investigated sample and the purpose of the analysis, different methods can be used:

- Method with external standard
- Method with internal standard
- Method of intensity ratio
- Rietveld method

Method with external standard

The method with external standard consists of comparing the intensity of the reflexion measured at a multiphase material with the intensity of a pure sample of the considered phase measured under the same experimental condition. The volume fraction of the considered phase in the multiphase material (V_i^s) can be calculated by the intensity ratio of the integrated intensity of a given (hkl)-peak of the considered phase within the investigated sample $I_i^s(hkl)$ and of the pure phase $I_i^p(hkl)$, by taking into account the mass absorption coefficients of the pure phase (μ^p) and of the sample (μ^s) as given in Eq. [4.6].

$$\frac{I_i^s(hkl)}{I_i^p(hkl)} = V_i^s \times \frac{\mu^p}{\mu^s} \qquad\qquad [4.6]$$

One advantage of this method is that it allows determining the amount of amorphous or of not-considered phases. However, it requires the experimental determination of the mass absorption coefficient of the analyzed sample (Spieß et al., 2009).

Method with internal standard

The method with internal standard is well indicated for the analysis of compounds with more than two phases and is based on the addition of a standard phase with defined volume or mass in the investigated compound. This is, of course, only possible for powder samples. As a consequence, the knowledge of the absorption coefficient is not required anymore. The mass concentration of the phase of interest W_i is then proportional to the intensity ratio of considered $I_i(hkl)$ and added phase $I_j(hkl)$ by taking into account a factor K_{ij} and the mass concentration of the added phase W_j (Eq. [4.7]) (Dinnebier and Billinge, 2008).

$$W_i = K_{ij} \times W_j \frac{I_i(hkl)}{I_j(hkl)} \qquad\qquad [4.7]$$

The factor K_{ij} can be determined experimentally by measurement of known mixtures of the standard and of the considered phase. In order to be able to use this method, it has to be ensured that the peaks of the added standard are not overlapped to other peaks of the compound.

One variation of this method is the reference intensity ratio (RIR) which is based on the intensity ratio with the corundum (113) peak (Dinnebier and Billinge, 2008). One major advantage of this method is that once the required factors have been determined for the phase of interest, no standard phase is longer needed in the compound (Dinnebier and Billinge, 2008).

Method of intensity ratio

The method of intensity ratio consists in calculating directly the intensity ratios of the reflexions of each present phase in a sample by correcting them with factors taking into account multiplicity and other parameters for each reflexion. This method is well indicated for solid samples for which the addition of an internal standard is not possible. Moreover, the knowledge of the absorption coefficient of the phases is not required. In order to calculate the amount of each phase, the total amount of the considered phases has to be known or to be scaled to 1. The calculation of the volume fraction of phase $i(V_i)$ in the measured material can be calculated on the base of the integrated intensity of a peak of this phase $I_i(hkl)$ and of other present phases $I_j(hkl)$ by taking into account phase- and peak-specific factors ($R_x(hkl)$) as given in Eq. [4.8] (Spieß et al., 2009). The factors $R(hkl)$ can be calculated or determined experimentally. With this method, different peak combinations can be used, allowing calculating mean phase content and so possible errors due to texture effects or other factors can be reduced.

$$V_i = \left(\sum_j \frac{I_j(hkl)}{I_i(hkl)} \times \frac{R_i(hkl)}{R_j(hkl)} \right)^{-1} \qquad\qquad [4.8]$$

One common application of this method is the quantitative analysis of retained austenite and martensite in hardened steel. For this, tables containing R-factors can be found in the literature for the different diffraction peaks of martensite and retained austenite depending on the used wavelength (ASTM, October 2000).

Rietveld method

Contrary to the previous methods, the Rietveld method is a whole pattern method. This means that the evaluation is based on the simultaneous analysis of several peaks. Here, the whole measured pattern is refined with a calculated pattern taking into account several structural, microstructural, and experimental parameters (Will, 2006). The refinement is performed by minimization of the function S given in Eq. [4.9].

$$S = \sum_i u_i |y_i obs - y_i calc|^2 \qquad [4.9]$$

In this equation, $y_i obs$ the measured and $y_i calc$ the calculated intensities at each 2θ position i and u_i, a weighting factor taken from the experimental error margins, which are assumed to be proportional to the square root of the count rate $y_i obs$ following Poisson counting statistics (Will, 2006).

The calculation of $y_i calc$ at each position i is a function of instrumental contribution, reflexions of all present phases, backgrounds, etc. All factors depending on each phase of the analyzed sample which are taken into account for the refinement are as follows (Spieß et al., 2009): position of all atoms in the elementary lattice; temperature factor; occupation factor; space group of the lattice; lattice parameters; texture; crystallite size; microstrains; and phase contents. Moreover, several instrumental parameters are also taken into account in order to separate the contribution of instrument and sample. These are: 2θ shifts (error of the instrument); instrumental profile; profile asymmetry; background; wavelength (emission profile); sample positioning error; and absorption (Spieß et al., 2009).

Two distinct strategies are available for profile refinement. The first is based on describing the peak shape by mathematical function such as Gaussian, Lorentzian, Voigt, Pseudo-Voigt, Modified Cox-Hasting Voigt function, Pearson VII, etc. (Will, 2006). These functions can be used for profile refinements if instrumental details are not known. However, these methods are purely based on mathematical fitting and do not allow to extract directly microstructural information from analyzed diffraction patterns, like crystallite size and microstrain. The second strategy available for Rietveld profile refinement is the fundamental parameters approach. With this method, profile calculation is done by convolution of emission profile (W), all instrumental contributions (G) and sample contribution (P), as given in (Eq. [4.10]) (Spieß et al., 2009).

$$Y(2\theta) = (W \otimes G) \otimes P \qquad [4.10]$$

All geometrical features of the instrument are described by functions that are convoluted to each other. By correct calculation of the instrumental function, contribution of sample properties to the diffraction pattern can be analyzed (Kern et al., 2004). The instrumental function can also be measured by using a standard without sample broadening. In general, the standard SRM660a (LaB_6) is used, as the large crystallites of about 2 µm do not lead to a significant broadening of the peaks.

At the end of a refinement, it is necessary to check whether the results are reliable and whether they meet certain standard criteria. The overall best criterion for the refinement is difference plots between observed and calculated data. When large discrepancies are present, the different parameters taking into account for the refinement have to be checked. The calculated criterion Residuals weighted profile (Rwp) gives a reliable information about the fit quality. Thereby, the smallest Rwp value represents the best refinement. It is calculated as shown in Eq. [4.11]. The parameter Rexp (Eq. [4.12]) represents the minimum expected Rwp depending on the number of experimental points (N) and the number of refined parameters (Q). The ratio of both parameters gives the goodness of fit (GOF), which can also be used as a criterion of the refinement quality (Eq. [4.13]).

$$Rwp = \sqrt{\frac{\sum\limits_{i} u_i(y_iobs - y_icalc)^2}{\sum\limits_{i} u_i(y_iobs)^2}} \qquad [4.11]$$

$$Rexp = \sqrt{\frac{N - Q}{\sum\limits_{i} u_i(y_iobs)^2}} \qquad [4.12]$$

$$GOF = (Rwp/Rexp)^2 \qquad [4.13]$$

With this method, complex multiphase materials can be analyzed in order to determine the respective phase contents. Also, it can be used for the evaluation of crystallographic texture, crystallite size, strains and microstrains. However, in order to get reliable results, precise information about the crystal structure of the present phases, including the atom coordinates within the lattice, is required.

An example of a quantitative phase analysis by the Rietveld method is shown in Fig. 4.10 for the previously presented nitrided tool steel X40CrMoV5-1. By taking into account the phases identified by qualitative phase analysis, a refinement was performed with the Rietveld-Software TOPAS V4.2 from Bruker-AXS (Bruker-AXS, 2008). A fundamental parameter approach was used after definition of the instrument by measurement of LaB_6 powder.

The refinement is of very good quality so that the measured pattern and the calculated pattern are almost superimposed and cannot be differentiated. The resulting

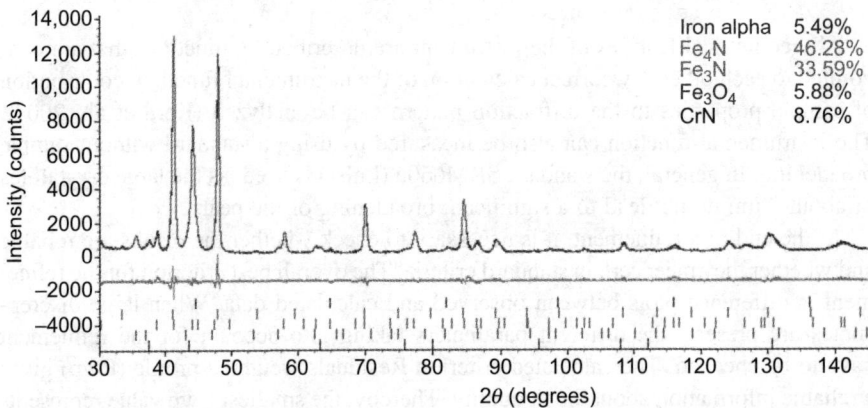

Figure 4.10 Quantitative phase analysis of a nitrided tool steel X40CrMoV5-1 by the Rietveld method.

difference curve is shown over the 2θ angle below the patterns. Only few zones are present where the line is not flat. The Rwp value for this fit is 8.03% while the GOF amounts to 2.01. The result of the quantitative analysis is shown within the diagram for the considered phases: 46.5% Fe_4N, 33.5% Fe_3N, 9.0% CrN, 6.0% Fe_3O_4, and 5.5% α-Fe. The positions of all diffraction peaks for each phase are shown at the bottom of the pattern.

With this powerful method, very complex multiphase materials can be investigated and quantitative analysis can be performed. However, the method has to be used carefully as many parameters can be refined without physical meaning that would lead to good refinements but erroneous results. Therefore, parameter constraints and control are mandatory for such analyses.

4.3.2   Residual stress analysis

Residual stresses are in general key factors for the later service properties and performance of engineering components. They are, therefore, always more in the focus of research activities as well as of quality management of industrial production. Beside material removal techniques that will not be treated here, XRD is one of the most used methods to characterize residual stresses.

4.3.2.1   Definition of residual stresses

Residual stresses are mechanical stresses which remain at room temperature in a work piece free of any mechanical loading and under homogeneous and temporal stable temperature field. Forces resulting from the residual stresses are balanced over the work piece. Residual stresses are acting forces per unit area (σ^{RS} in N/mm^2 or MPa).

Residual stresses can be classified in three different categories according to their scope of action (Macherauch et al., 1973). Residual stresses of the first kind (σ^I) are the volume average of the position-dependent residual stresses taken over all crystallites and phases within the considered bulk. These represent the macroscopic material balanced over the entire work piece and are also called macroresidual stresses (Eq. [4.14]) (Macherauch et al., 1973). Residual stresses of the second kind (σ^{II}) are deviations from the first-kind residual stresses, for example within grains of different phases (Eq. [4.15]). They are also called homogeneous microstresses (Macherauch et al., 1973). Residual stresses of the third kind (σ^{III}) are local variations at submicroscopic scale, for example due to dislocations or other lattice defects (Eq. [4.16]). They are also called inhomogeneous microstresses (Macherauch et al., 1973). An illustration of the scope of action of the three kinds of residual stresses is given in Fig. 4.11 for a two-phase material composed of α and β grains.

$$\sigma^I = \frac{1}{V_{macro}} \int\limits_{V_{macro}} \sigma(x)dV_{macro} \qquad\qquad [4.14]$$

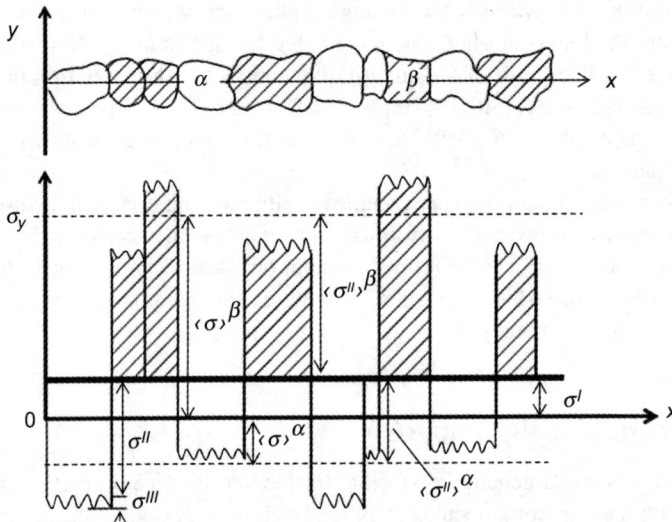

Figure 4.11 Residual stress distribution over several grains of a two-phase material showing the three kinds of residual stresses.

$$\sigma^{II} = \frac{1}{V_{\text{grain}}} \int_{V_{\text{grain}}} \sigma(x) - \sigma^I dV_{\text{grain}} \qquad [4.15]$$

$$\sigma^{III}(x) = \sigma(x) - \sigma^I - \sigma^{II} \qquad [4.16]$$

In general, the stress state inside a specimen is defined by a symmetric tensor of third order (σ_{ij}) expressed as a function of a coordinate system arbitrarily defined with i the direction of acting force and j the area on which the force is applied (Eq. [4.17]). The stresses acting on planes in orthogonal direction to those are called normal stresses while the stress components acting parallel to the considered planes are called shear stresses.

$$\sigma_{ij} = \begin{pmatrix} \sigma_{11} & \sigma_{12} & \sigma_{13} \\ \sigma_{21} & \sigma_{22} & \sigma_{23} \\ \sigma_{31} & \sigma_{32} & \sigma_{33} \end{pmatrix} \qquad [4.17]$$

4.3.2.2 Residual stress measurement

Measurement principle

The methods of XRD for measuring of residual stresses have been well established for more than 30 years (Macherauch et al., 1973). As it is easily accessible, laboratory XRD measurements of residual stress can be applied to a wide range of crystalline materials. Moreover, this method is phase sensitive, which means that the residual stresses present in different phases of a material can be determined separately, which might be of interest. One of the major aspects of XRD residual stress analysis is that in most cases, no standard or precise reference values are required for the determination of residual stresses.

The method is based on the measurement of the lattice spacing of considered lattice plans (*hkl*) allowing the determination of strains in a given direction. Fig. 4.12 presents the basic principle of XRD stress measurement showing a stress-free lattice with a lattice spacing d_0^{hkl} leading to a diffraction peak at given 2θ angle. If the material is under stress, the lattice spacing d_i^{hkl} can increase or decrease, which leads to a shift of the recorded peak position. The residual stresses of first and of second kind can be evaluated based on this principle.

In order to perform residual stress measurements by XRD, the sample orientation regarding the coordinate system of the equipment used is of major importance. Therefore, the sample coordinate system (*S*) and an arbitrary laboratory coordinate system (*L*) defined by the orientation within the measuring equipment should be considered (Fig. 4.13). The associated polar angle ψ and azimuth angle φ describe the orientation of the sample in the coordinate system.

The strain $\varepsilon^{hkl}(\varphi, \psi)$ of defined *hkl* planes at any φ and ψ angle is defined by Eq. [4.18] with d_0^{hkl} being the strain free lattice spacing and $d^{hkl}(\varphi, \psi)$ the lattice spacing measured at any φ and ψ angle.

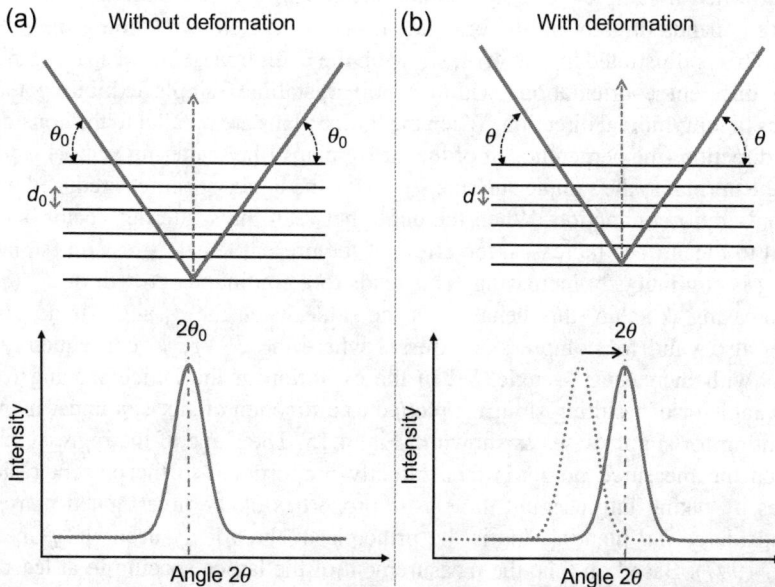

Figure 4.12 Schematic description of diffraction at given *hkl* planes and corresponding signal recorded by an X-ray diffractometer: (a) for a stress-free lattice; (b) for a lattice under compressive residual stresses.

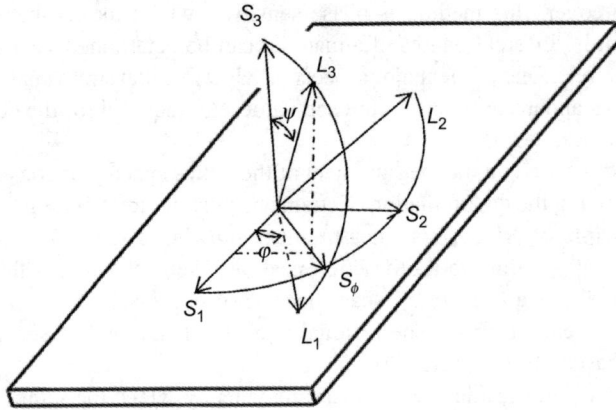

Figure 4.13 Definition of the laboratory coordinate system (L), the sample coordinate system (S) and the associated polar angle ψ and azimuth angle φ (Noyan and Cohen, 1987).

$$\varepsilon^{hkl}(\varphi,\psi) = \frac{d^{hkl}(\varphi,\psi) - d_0^{hkl}}{d_0^{hkl}} \qquad [4.18]$$

In order to perform a residual stress measurement by XRD, the evolution of lattice spacing due to the effect of stresses is measured for varying sample orientation. Indeed, when a sample exhibits residual stresses in a given direction, the lattice spacing depends directly on the orientation of the crystallite regarding the present stress. This is illustrated in Fig. 4.14, showing the evolution of lattice spacing $d^{hkl}(\varphi, \psi)$ for different ψ-orientations within a polycrystalline sample exhibiting tensile stresses in longitudinal direction. When the lattice plans are parallel to the considered stress direction (the perpendicular of the lattice plans [the scattering vector] is parallel to the normal to the sample surface; $\psi_1 = 0°$), the lattice spacing is reduced by the Poisson's ratio contraction. When the angle between the scattering vector and the normal to the surface increases, the effect of the present tensile stress on the lattice spacing is continuously increasing. This leads to a continuous growth of $d^{hkl}(\varphi, \psi)$ for increasing ψ value (this behavior is the same when the ψ_i angle is negative). This is also valid for compression stresses where the $d^{hkl}(\varphi, \psi)$ consequently decreases with increasing ψ-angle. When the evolution of the lattice spacing (or of the 2θ angle or of resulting strain) is plotted as a function of $\sin^2\psi$, a linear increase is resulting for simple cases as shown in Fig. 4.15. The slope of the regression line between the measured points is then directly proportional to the present residual stresses by taking into account the elastic properties of the investigated phase for the considered *hkl* planes. This is the principle of the widely used $\sin^2\psi$ method (Hauk, 1997). Based on this, the measurement of the lattice spacing in at least two different ψ_i angles allows the calculation of the stress. As only elastic strains influences the lattice spacing, only elastic strains leading to stresses can be measured by this method (Prevey, 1986). Also, it has to be mentioned that the residual stress

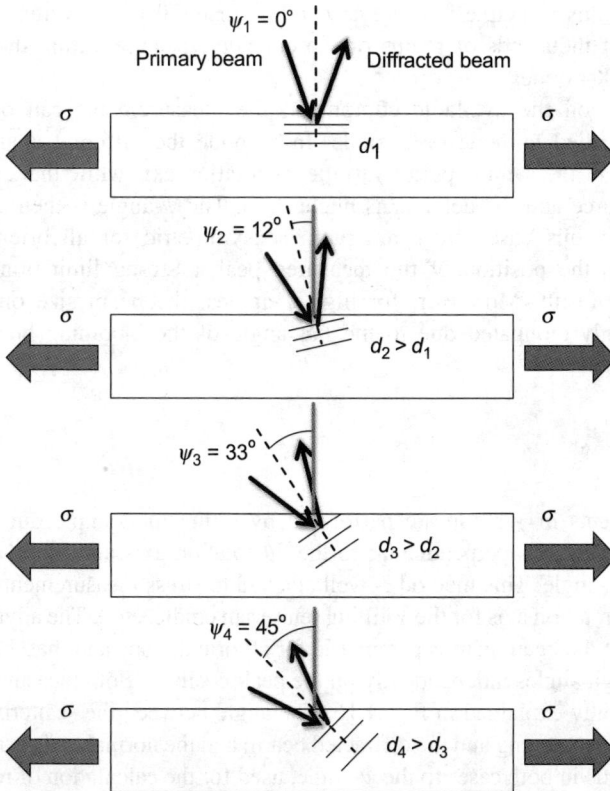

Figure 4.14 Evolution of lattice spacing as a function of the orientation regarding longitudinal tensile stresses.

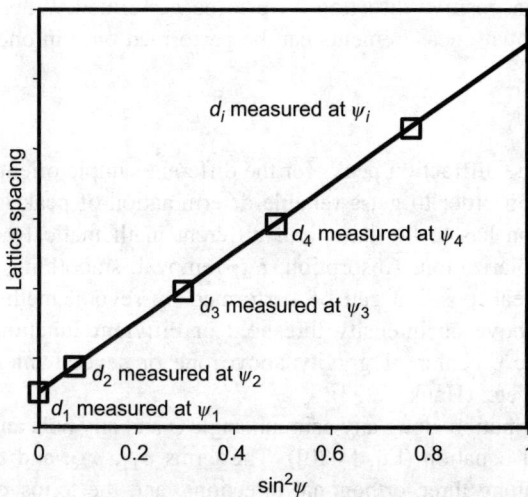

Figure 4.15 Evolution of lattice spacing as a function of $\sin^2\psi$ for tensile residual stresses.

measured by this principle in a polycrystalline material is an average value of the stress state of thousands of grains of the considered phase within the penetration depth of the X-ray beam.

Depending on the available equipment, the measurements can be performed either in so-called ω-mode or χ-mode. In ω-mode the different ψ-angles are set by a rotation of the sample parallel to the 2θ rotation axis while the angle between the X-ray source and the detector is unchanged. The ψ-angle is then calculated by Eq. [4.19]. In this case, the beam path is asymmetric for all orientations, and depending on the position of the measured peak a strong limitation of possible ψ-angles can result. Moreover, for high ψ-angles, the beam size on the sample can be strongly elongated due to the flat angle of the incoming beam what can lead to problems.

$$\psi = \frac{2\theta}{2} - \omega \qquad\qquad\qquad\qquad [4.19]$$

Measurements in χ-mode are performed by tilting the sample out of the plane around a rotation axis perpendicular to the 2θ rotation axis and the ψ-angle is then equal to the χ-angle. This method is well adapted to stress measurement and requires an additional rotation axis for the χ-tilting (eurlerian cradle, etc.). The advantage of this method is that the beam path is symmetric for all orientations and that the sample can be tilted to high angles independently on the peak position. Both measurement modes are schematically explained in Fig. 4.16. The angle between the scattering vector (S) (bisector of the incoming and the diffracted beam) and the normal to the sample surface (N) corresponds in both cases to the ψ-angle, used for the calculation of residual stress from the measurements.

In order to determine the complete residual stress tensor, measurements along three different azimuth angles φ (generally $0°$, $45°$, and $90°$) with polar angles (ψ) in negative and in positive direction have to be performed. When principal stress directions are known, measurements can be performed only in one or two azimuth angles.

Data analysis

After recording the diffraction peaks for the different sample orientation, data treatment is required in order to get a reliable determination of peak location. A background subtraction has to be done, and different mathematical correction can be used (Lorentz, Polarization, Absorption, $K_{\alpha2}$-removal, smoothing, etc.). After this, the analysis of peak position can be performed by several methods like peak fit using parabola above an intensity threshold or different functions (Pearson VII, Pseudo-Voigt, etc.), center of gravity above one or several intensity thresholds, cross-correlation, etc. (Hauk, 1997).

The strain distribution along any azimuth angle φ and any pole angle ψ is given by following general equation (Eq. [4.20]). The terms σ_{11}, σ_{22}, and σ_{33} represent the normal stresses along three orthogonal directions, and the terms σ_{12}, σ_{13}, and σ_{23} are the shear stresses as given in Eq. [4.17] (Hauk, 1997).

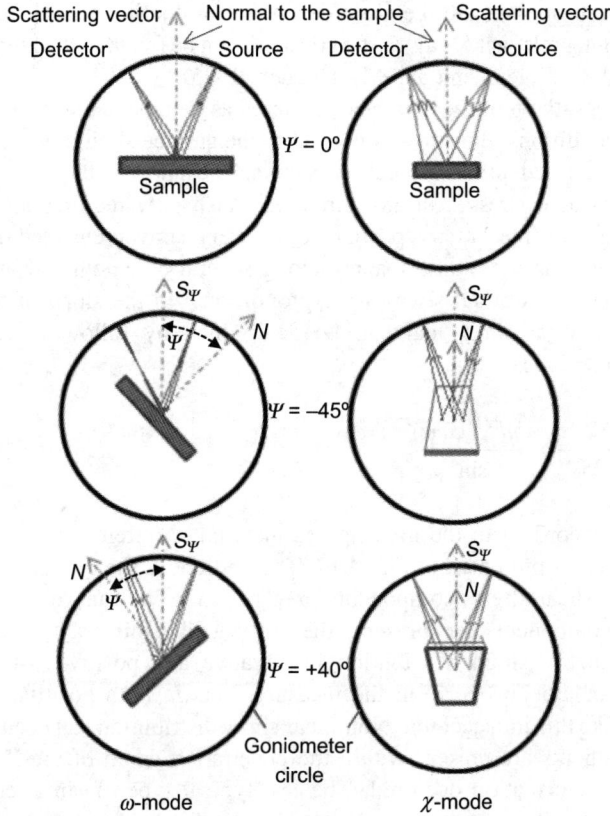

Figure 4.16 Principle of residual stress measurements in ω-mode (left side) and χ-mode (right side) (Macherauch and Zoch, 2014).

$$\varepsilon^{hkl}(\varphi, \psi) = s_1^{hkl}(\sigma_{11} + \sigma_{22} + \sigma_{33}) + \frac{1}{2}s_2^{hkl}\left[(\sigma_{11} \cos^2\varphi + \sigma_{22} \sin^2\varphi - \sigma_{33})\sin^2\psi\right.$$

$$\left. + \sigma_{33}\right] + \frac{1}{2}s_2^{hkl}\left[\sigma_{12} \sin 2\varphi \sin^2\psi + \sigma_{13} \cos\varphi \sin 2\psi\right.$$

$$\left. + \sigma_{23} \sin\varphi \sin 2\psi\right]$$

$$[4.20]$$

The terms S_1^{hkl} and $1/2S_2^{hkl}$ are the X-ray Elastic Constants and can be calculated from the hkl specific elastic constants as follows: $S_1^{hkl} = -v^{hkl}/E^{hkl}$ and $1/2S_2^{hkl} = (1 + v^{hkl})/E^{hkl}$, with v^{hkl} being the Poisson's ratio and E^{hkl} the elastic modulus in MPa of the considered phase for defined hkl planes (Spieß et al., 2009). It has to be kept in mind that even if the investigated polycrystalline material is isotropic regarding its macroscopic elastic properties, the elastic properties of every single grain (or crystallite) are not necessarily isotropic due to crystal anisotropy and the elastic properties are therefore depending on the investigated lattice planes. Tables containing relevant X-ray elastic constants can be found in the literature (Noyan and Cohen, 1987; Eigenmann and Macherauch, 1995).

In most cases, the term σ_{33} can be considered as being equal to zero, since due to equilibrium conditions, the stresses normal to the surface should be zero within the shallow penetration depth of laboratory X-rays. The $\sin^2\psi$ method can then be used to measure residual stresses in any direction. As it only requires an approximate value of the strain-free lattice spacing, it is a very convenient method. In simple cases, by measuring the lattice spacing along several pole angles ψ at a single azimuth angle φ, the residual stresses are proportional to the slope of the measured lattice spacing over $\sin^2\psi$ and can be determined by following equation (Eq. [4.21]) (Hauk, 1997).

$$\sigma_\varphi = \frac{1}{1/2 S_2^{hkl}} \times \frac{\delta d^{hkl}(\varphi, \psi)}{\delta \sin^2\psi} \qquad [4.21]$$

However, according to the investigated material, different type of $\sin^2\psi$ curves can be obtained as presented in Fig. 4.17. The first one is the typical linear distribution without shear-stress components σ_{31}/σ_{32} in a normal direction. If such shear-stress components are present, the $\sin^2\psi$ will exhibit a typical ψ-splitting due to asymmetric strain distribution in negative and positive ψ-tilting. In this case, it is mandatory to perform the measurements in both positive and negative ψ-angles to take this into account. Nonlinear $\sin^2\psi$ distribution can occur when residual stress gradients are present within the penetration depth of the X-ray beam or when plastic deformation occurred. The last typical type of $\sin^2\psi$ curve exhibits an oscillating distribution indicating the presence of inhomogeneous residual stress state in the different direction, generally as a consequence of crystallographic texture. For these last two cases, if the nonlinearity is pronounced, the standard $\sin^2\psi$ method cannot be used anymore and other special methods have to be used for the stress calculation (see Ref. Hauk, 1997).

When a multiphase material is investigated, all phases with an amount larger than 10% should be measured. In order to calculate macroscopic residual stresses in a

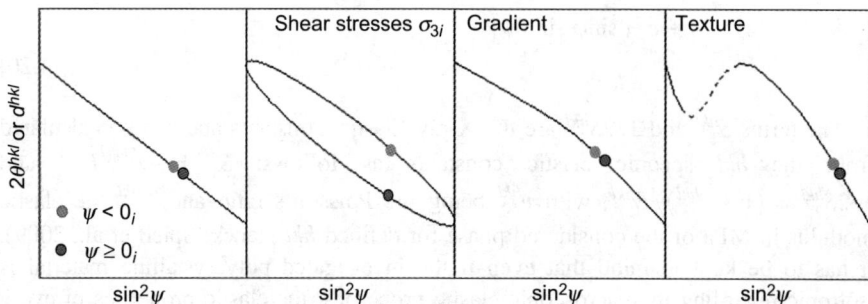

Figure 4.17 Different types of $\sin^2\psi$ distributions depending on the investigated material.

multiphase material, a weighted average of residual stresses in all phases can be calculated by Eq. [4.22], with σ_{ij}^I being the macrostresses along a given direction, f_i is the volume fraction of phase i, and σ_{ij}^i is the measured residual stress along a given direction in phase i (Fitzpatrick and Lodini, 2003). For this, the knowledge of the respective phase contents is required.

$$\sigma_{ij}^I = \sum f_i \times \sigma_{ij}^i \qquad [4.22]$$

It has to be remarked that in multiphase materials, the stress component σ_{33} cannot always be considered to be zero (Behnken, 2003). Indeed, the macroscopic stress σ_{33} has to be zero at the surface, but different phases might exhibit σ_{33} residual stresses of the second kind with opposite signs if the microstructural periodicity is lower than the penetration depth of X-rays (Behnken, 2003). If a σ_{33} component is present, the slope of the $\sin^2\psi$ plot measured in a given direction is not only proportional to the normal stresses σ_{11} or σ_{22} anymore, but to $(\sigma_{11} - \sigma_{33})$ or $(\sigma_{22} - \sigma_{33})$. For the calculation of σ_{33}, very precise determination of the stress-free lattice spacing d_0 is then required (Hauk, 1997).

In order to perform residual stress measurements by XRD, several issues should be considered:

- The used diffractometer has to be aligned carefully and calibrated with known standards.
- Generally, positive and negative ψ from $-45°$ to $+45°$ in at least nine steps should be measured.
- Diffraction peaks without overlapping with other phases should be used.
- The measurement time should be long enough to measure well-defined peaks, and the measured 2θ range has to be large enough to measure the background at both sides of the peak.
- Diffractions peaks at high 2θ angles ($>100°$) should be chosen for the measurements as errors due to misalignment of the diffractometer are strongly reduced.
- The sample preparation should not affect the residual stress state.
- This method is valid for monophase materials. If a multiphase material is investigated and the amount of secondary phases is significant ($>10\%$), measurements should be performed in all phases. Moreover, nonzero second kind residual stress components normal to the surface (σ_{33}) can be present in the different phases.
- When strong crystallographic texture or strong gradients are present, large errors can result.
- When the grain size is very large compared to the beam size, errors can result. In order to improve the results, sample oscillation should be used.
- If electrochemical layer removal is used, data correction should be used.
- Detailed information can be found in the norms (EN 15305:2008, 2008) and (ASTM E2860-12, 2012).

Examples of application

An example of a residual stress measurement performed at a turbine blade made of Ti6Al4V is presented. The measurement was performed with Nickel-filtered Cu-Kα-Radiation in longitudinal direction at the root where a shot-peening treatment was applied. The measurements were performed in ω-mode due to accessibility

(a)   (b)

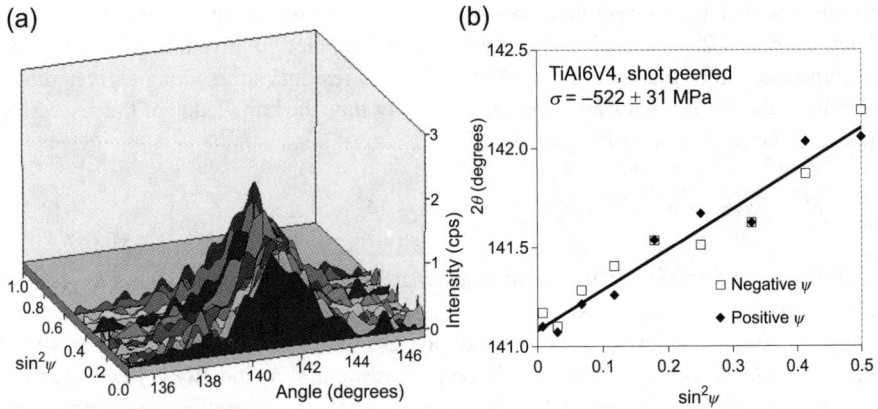

Figure 4.18 Residual stress measurement at a turbine-blade of TiAl6V4: (a) diffraction peaks for all ψ-angles; (b) $\sin^2\psi$ plot.

reason. The {213} diffraction peak of α-Ti was measured with 19 different ψ-angles from $-45°$ to $+45°$. The measured diffraction peaks are shown in Fig. 4.18(a) for the different ψ-angles and the resulting $\sin^2\psi$ plot is given in Fig. 4.18(b). Slight scattering of the single peak positions can be observed, but no splitting or curved distribution is present. Therefore, the linear regression of the data set gives a reliable normal residual stress value of -522 ± 31 MPa.

A second example of XRD residual stress measurements concerns the investigation and the control of undesired distortion during the manufacturing of machine components like bearings. Cylindrical and tapered rings made from the steel-grade 100Cr6 were investigated after soft turning and subsequent heat treatment. As known from earlier investigations, the combined effect of clamping and machining can introduce inhomogeneous residual stresses around the periphery of the rings (Epp and Hirsch, 2010). XRD residual stress measurements were executed by Bragg–Brentano diffractometers (type F, Siemens AG, Germany) equipped with scintillation counters. The measurements were performed with vanadium-filtered Cr-Kα-radiation with 35 kV and 35 mA and a primary beam diameter of 2.4 mm. The lattice planes {211} were measured along 11 ψ angles from $-45°$ to $+45°$. The detector steps were 0.05° with a measurement time of 3 s/step. In order to have an almost complete surface characterization, residual stress measurements were done at 36 points in the middle height around the external periphery (every 10°) by automatic ring rotation, as shown in Fig. 4.19.

Due to different machining parameters (clamping tool and feed-rate), different distributions of residual stresses were measured around the outer circumference of the rings, as presented in Fig. 4.20(a) where a periodic distribution can be distinguished. Variation A and B were clamped with a 6-point clamping tool with respectively 0.1 and 0.4 mm feed-rate while the variations C and D were machined with the same respective feed-rates but with a clamping using a three jaw chuck. The inhomogeneous

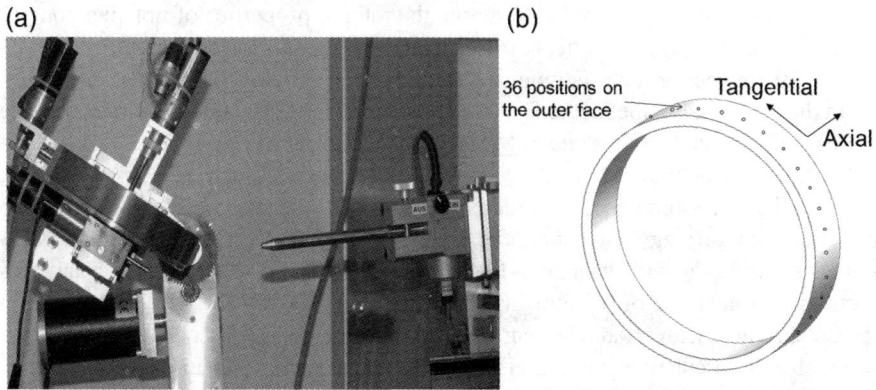

Figure 4.19 (a) Picture of the experimental set-up for automatic measurement of residual stresses around the circumference of a ring in axial direction; (b) sketch of a ring and positions measured by X-ray diffraction.

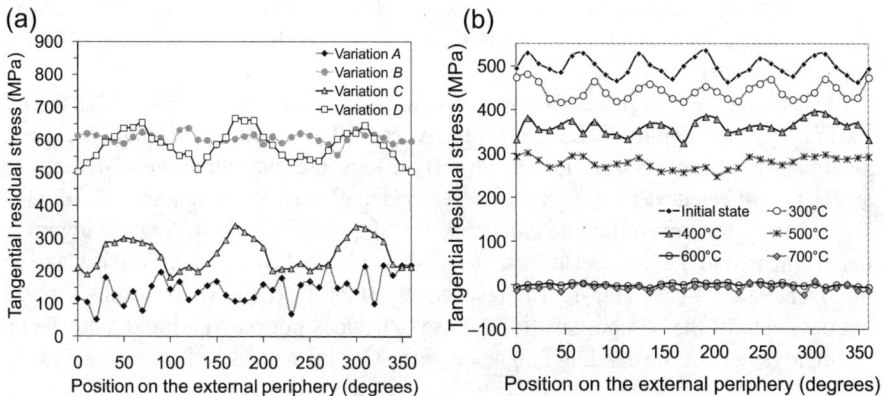

Figure 4.20 (a) Residual stress distribution for different machining parameters (Epp and Hirsch, 2010); (b) residual stress distribution measured at different rings after heating to different temperatures and cooling back to room temperature (Epp et al., 2011).

distribution is a distortion potential which can be released during subsequent heat treatment, leading to distortion. The thermally induced residual stress relaxation of a ring with an initial sixth order periodicity of the residual stress distribution is shown in Fig. 4.20(b), showing a progressive loss of periodicity and a general decrease of the residual stress level (Epp et al., 2011).

Finally, a last example of applications of XRD residual stress investigations concerns the effect of alternative peening methods on the fatigue properties of high-strength carburized steel gears. Indeed, it is well known that compressive residual stresses act like loading stresses and can therefore reduce the local loading

condition what is widely used to improve the fatigue properties of notched components. In order to take the effects of residual stresses on fatigue properties into account, the concept of local fatigue strength has been developed (Winderlich, 1990).

In the present case, the methods of water jet peening (WJP) and laser shock peening
(LSP) are compared to shot peening (SP) as standard and as duo-process and investigated in terms of induced residual stresses and fatigue properties (Epp and Zoch,
2014). XRD measurements of residual stresses and retained austenite have been
executed with a Bragg−Brentano diffractometer (type MZ VI E, GE Inspection Technologies) equipped with vanadium-filtered Cr-Kα-radiation. The primary beam was
focused to a diameter of 0.3 mm by a focusing polycapillary in order to measure in
the notched area with a radius of curvature of 1.5 mm. As a consequence of the small
beam size, an oscillating translation in notch direction was used in order to improve
the grain statistic. A position sensitive detector recorded the diffracted beam of {211}
lattice planes of martensite and {220} of retained austenite. Measurements were performed in χ-mode with 15 tilt-angles from $-45°$ to $+41°$ perpendicularly to the root.
In order to establish depth profiles, measurements were conducted at different preselected depths after electropolishing. The electropolishing was carried out using an elec-
trolyte solution containing 80% H_3PO_4 and 20% H_2SO_4. Correction of the effect of layer
removal on the residual stress values in the depth was performed according to Moore
and Evans (Totten et al., 2002).

SP leads to very high compressive residual stresses (up to -1650 MPa) with a
maximal affected depth of about 170 μm. After WJP, maximum residual stresses of
about -1000 MPa with very low depth (50 μm) are present, while the LSP leads to
very large affected depths (<1 mm) and residual stresses of about -1100 MPa
(Fig. 4.21(a)). In terms of fatigue properties, the SP process shows the highest improvement compared to the heat treated state (+47%) while the LSP and the WJP leads to an
improvement of +15% and +23%, respectively (Fig. 4.21(b)). These results showed
that the depth of the compressive residual stresses does not play an important role on
the improvement of the bending fatigue properties for the considered notched geometry

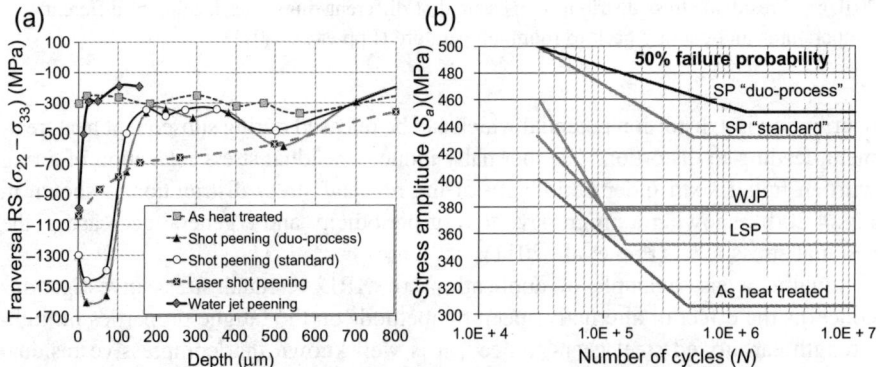

Figure 4.21 (a) Residual stress depth profiles measured for different surface peening treatments;
(b) S−N curves obtained for different surface peening treatments (Epp and Zoch, 2014).

but that only the maximum compressive residual stress at the surface was the major factor (Epp and Zoch, 2014).

4.3.3 Texture measurements

In polycrystalline materials for engineering applications, several production steps are required to reach the final state with the desired properties. All the process steps, starting from material casting, to hot or cold forming processes, machining processes as well as heat treatment or joining operations can lead to the formation of a more or less strong crystallographic texture. This means that the present phases are not randomly oriented within the microstructure, but that they present a preferred orientation regarding their crystallographic structure. Depending on the structure of the considered phases and the process step responsible for the creation of the texture, different preferred orientations can be present. The knowledge about the presence of a crystallographic texture in a polycrystalline material can be of major importance. Indeed, it generally influences several macroscopic material properties due to the anisotropic properties of the crystal structure itself. If the orientation of the crystallites is not random, the macroscopic properties will therefore not be isotropic but will present variations in the different directions. Such properties are mechanical properties like elastic constants, magnetic properties, chemical properties, etc. Moreover, information about texture can be required for the reliable investigation of phase contents and residual stresses.

4.3.3.1 Texture measurements by X-ray diffraction

Due to the phase selective measurements and the possibility of precise sample orientation with the available equipment, XRD has been developed to become one of the major methods for the characterization of the crystallographic texture of materials. The method is based on the evaluation of the intensity variation of one or several diffraction peaks of a given phase along a large amount of sample orientation. For this, the sample is typically placed in a four-circle diffractometer with eulerian cradle (Fig. 4.7), and a defined 2θ range covering one diffraction peak is recorded in a similar manner as it would be the case in residual stress measurements. However, not the position of the diffraction peak is of interest but the peak intensity, as the intensity recorded in a given direction is directly proportional to the volume fraction of crystallites in diffracting condition in this direction. Measurements in transmission mode can be performed, but for solid samples, most applications use the reflexion mode. In texture analysis in reflexion mode, the previously defined angles are redefined as: $\omega = \theta$; $\chi = \alpha$; and $\varphi = \beta$.

The base of XRD texture analysis is the measurement of pole figures. For this, the same coordinate system as presented in Fig. 4.13 is applied. A combination of φ- and ψ-angles is used to measure the considered diffraction peak and its intensity variation. As a list of all measured intensities would be difficult to interpret, a graphical representation allows the evaluation of the results. With the help of a stereographic projection (Fig. 4.22), the intensity variation for the investigated φ- and ψ-angles can be

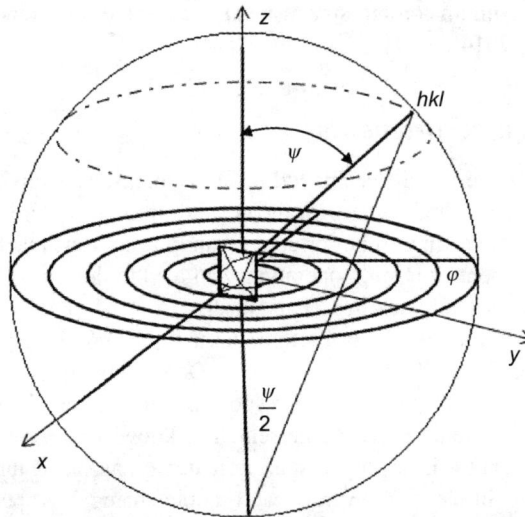

Figure 4.22 Principle of a stereographic projection for the determination of pole figures.

represented for all combination of these angles. For each considered lattice plans (*hkl*), an own pole figure can be measured.

For the measurement of pole figures by XRD, different strategies can be used. Generally, combinations of α- and β-angles in small discrete steps of several degrees are used with a complete rotation of the sample ($0 < \beta < 360°$) and maximum tilt angles (α) of $70°{-}80°$. This limitation is due to the fact that at a higher angle, the grazing incidence of the X-ray beam leads to a very elongated beam size on the sample. High absorption and defocusing effects can then occur leading to decreasing reliability of the values. In general, absorption and defocusing effects should be corrected by using a reference sample without texture (powder) in order to compensate intensity variations occurring at large tilt angles. Other measurement strategies using, for example, φ- and ψ-continuous scanning can be used (Spieß et al., 2009). Due to the large number of sample orientations that have to be evaluated, the measurement of a single pole figure can take up to several days. Therefore, position sensitive detectors or 2-D detectors are particularly indicated for texture analysis. According to the crystal symmetry and the sample symmetry (due to the process step responsible for the created texture), varying ranges of the pole figure have to be measured in order to obtain representative texture measurement (Spieß et al., 2009).

One major field of application of texture measurements is the analysis of cold-rolled metal sheets, which has already been used for several decades. For example, online (in-process) texture analysis methods were already developed and implemented in the 1980s in the production line at Hoesch Stahl AG in Dortmund, Germany (Kopineck and Otten, 1987). The technique was based on energy-dispersive measurements in transmission mode using the continuous spectrum of a tungsten-anode to measure portions of several pole figures. For this, two detectors were used, allowing them to measure

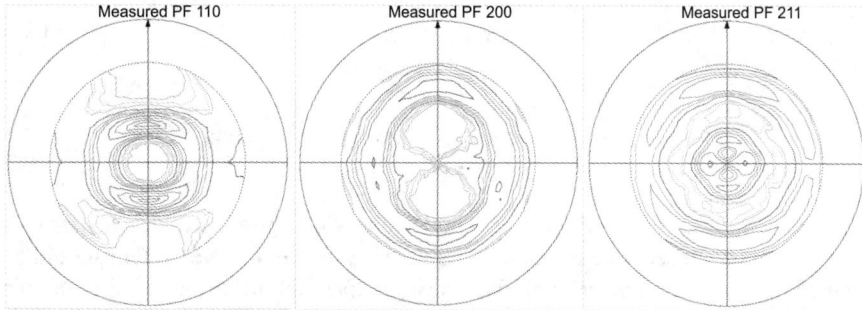

Figure 4.23 Pole figures measured at a cold-rolled α-Fe thin sheet.

simultaneously at two β-angles. In preliminary investigations, it had been demonstrated that the intensity of Fe-α-{211} and Fe-α-{220} peaks are directly correlated to the plastic strain ratio (Lankford coefficient, or R-value) of the sheets, and this ratio is an indicator of formability of recrystallized low-carbon steel as in deep-drawing processes. These texture measurements therefore allowed an online quality control of the steel-sheet production.

An illustration of typical pole figure measurements is presented in Fig. 4.23 showing the {110}, {200}, and {211} pole figures measured in the core at a cold-rolled thin sheet of armco-Iron (α-Fe) with of 100 μm thickness. Angle-dispersive measurements were carried out on an ETA 3003 (GE Inspection Technology, Ahrensburg, Germany) with Cr-kα radiation anode with a current of 40 mA and a voltage of 40 kV. The beam was collimated to a diameter (Ø) of 2 mm and the diffracted signal was measured by using a position sensitive detector with an angular range of 10° 2θ and a resolution of 0.1° within 21 s for each α and β combination. The diffraction peaks were recorded from 0° to 70° with an angular spacing of 5° in α and from 0° to 350° with an angular spacing of 10° in β. The total time for measuring these three pole figures was 35 h.

Iso-intensity lines are represented, indicating the zones of intensity concentration. For the three pole figures, different zones with high intensity can be observed. The symmetry of the pole figures shows that the measurement of a β-range of 0°−90° would have been sufficient to describe the texture.

4.3.3.2 Evaluation of texture measurements

Based on measured pole figures, a quantitative analysis of the crystallographic can be performed by using the orientation distribution function (ODF) (Bunge, 1982). The ODF, $f(g)$ gives the volume fraction of crystallites $dV(g)$ which are contained within the range dg of the crystallographic orientation g regarding a defined coordinate system measured within the considered Volume V (Eq. [4.23]). The ODF is scaled to 1 over all sample's orientations (Eq. [4.24]) (Bunge, 1982). It has to be remarked that the determination of the ODF requires the measurement of several pole figures. The higher the crystal symmetry, the less pole figures are needed.

$$f(g)dg = \frac{dV(g)}{V}$$ [4.23]

$$\int_g f(g)dg = 1$$ [4.24]

In order to describe ODFs, Euler's angles are widely used, which consists of a system of three consecutive rotations. First, the crystal coordinate system should be oriented in such a manner that the axes are parallel to those of the sample coordinate system. Then, the crystal coordinate system is rotated about the Z'-axis through the angle φ_1. Second, the X'-axis (in its new orientation) is rotated through the angle Φ. Finally, a third rotation around the new axis Z' by the angle φ_2 is following. Any crystallite orientation g can be described by these three angles (Bunge, 1982).

The ODF has to be calculated on the basis of the measured pole figures. For this, different methods can be used: the direct method, the independent component analysis, and the series expansion method which is the most widely used method (Spieß et al., 2009).

In the series expansion method, the ODF is calculated via a series of generalized spherical harmonic functions $T_\lambda^{mn}(g)$ as given in Eq. [4.25]. The $T_\lambda^{mn}(g)$ are known functions and can be calculated for all orientations g, while the unknown series expansion coefficients C_λ^{mn} can then be determined (Bunge, 1982).

$$f(g) = \sum_{\lambda=0}^{\lambda_{max}} \sum_{m=-\lambda}^{+\lambda} \sum_{n=-\lambda}^{+\lambda} C_\lambda^{mn} T_\lambda^{mn}(g)$$ [4.25]

As already reported previously, the plastic strain ratio (R-value) of metal sheets was identified to be directly related to diffraction peak intensity. A deeper correlation is possible by the use of fourth order series expansions coefficient of the ODF (Banabic et al., 2000). Indeed, during cold-rolling of cubic materials an orthorhombic texture is generated. The lowest-order nonrandom approximation is then the fourth order ($\lambda = 4$) that describes the influence of a texture on the elastic properties of the material by the knowledge of the three fourth order series expansion coefficients C_4^{1x}. This can be used for the evaluation of the R-values in sheet material (Banabic et al., 2000).

From the ODF, a convenient manner to describe and compare the sharpness of a texture of different samples is to calculate the texture index J (Eq. [4.26]), which can be calculated from the coefficients of the series expansion. The index can be between 1 for a texture-free sample and infinite for a single crystal (Bunge, 1982).

$$J = \int_g [f(g)]^2 dg$$ [4.26]

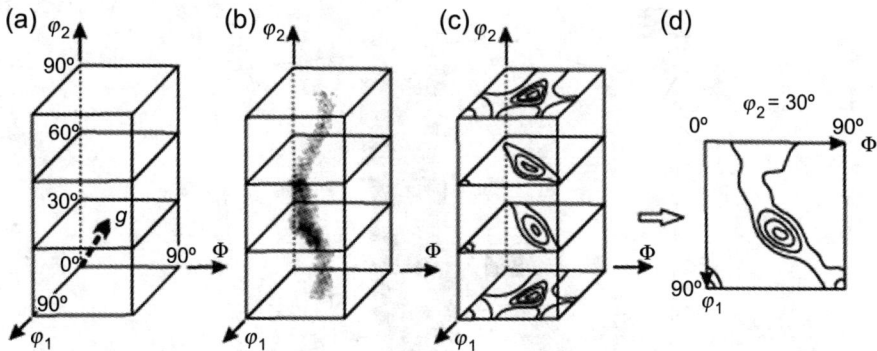

Figure 4.24 Representation of ODFs in Euler space: (a) Euler space; (b) scatter-plot of the 3-D ODF representation; (c) 3-D ODF representation as a succession of 2-D projections; and (d) Example of a 2D-projection at constant φ_2 (Spieß et al., 2009).

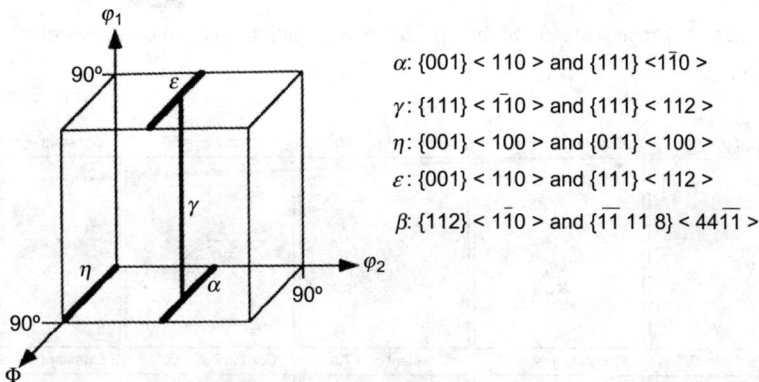

α: $\{001\} < 110 >$ and $\{111\} <1\bar{1}0 >$

γ: $\{111\} < \bar{1}\bar{1}0 >$ and $\{111\} < 112 >$

η: $\{001\} < 100 >$ and $\{011\} < 100 >$

ε: $\{001\} < 110 >$ and $\{111\} < 112 >$

β: $\{112\} < 1\bar{1}0 >$ and $\{\bar{1}\bar{1}\ 11\ 8\} < 44\bar{1}\bar{1} >$

Figure 4.25 Ideal fiber-texture components of rolled body-centered-cubic metals (Spieß et al., 2009).

The ODF can be visualized as 3-D representation of iso-intensity surface in the Euler space or as a succession of 2-D projections of iso-intensity lines at different positions in the Euler space. The different possibilities of ODF representation are illustrated in Fig. 4.24.

According to the crystal structure of the investigated phase, different typical textures can be observed by using the representation in the Euler space. As an example, the ideal fiber components for rolled body-centered-cubic (BCC) metals are presented in Fig. 4.25.

The ODF calculated by means of series expansion from the three pole figures measured at a cold-rolled α-Fe thin sheet presented in Fig. 4.23 is shown as 3-D representation in Fig. 4.26 and as 2-D projections of φ_1 in Fig. 4.27. Compared

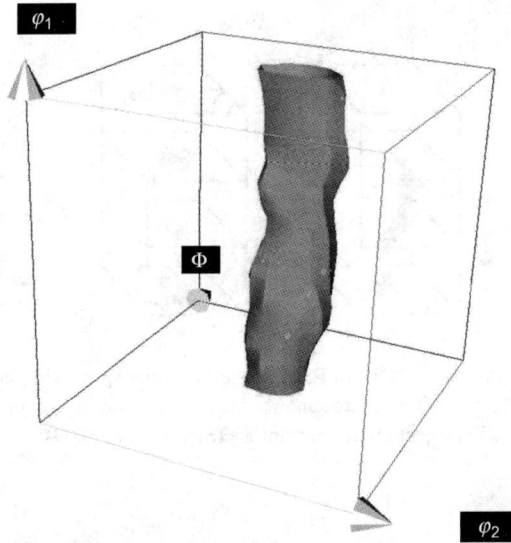

Figure 4.26 3-D representation of the orientation distribution function of a cold-rolled α-Fe thin sheet.

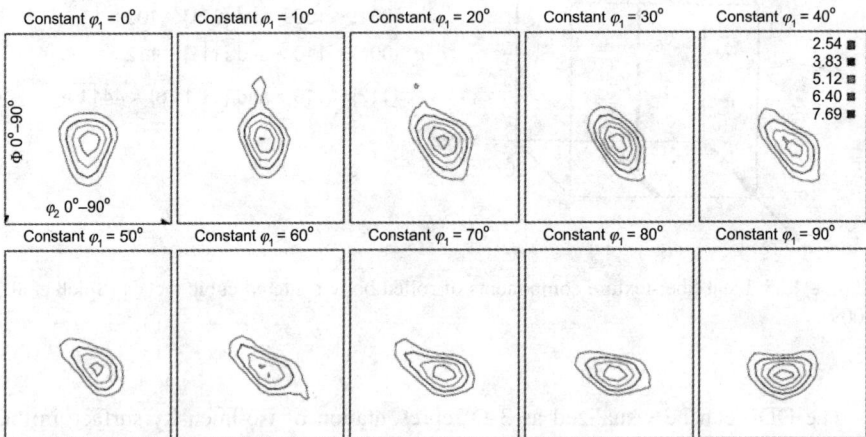

Figure 4.27 Orientation distribution function of a cold-rolled α-Fe thin sheet represented as 2-D projections at φ_1 between $0°$ and $90°$.

to the known texture components in BCC metals given in Fig. 4.25, it can be observed that an almost ideal γ-fiber is present, as expected for cold-rolled BCC materials.

Finally, another common method for the representation of crystallographic texture is the so-called inverse pole figure. The inverse pole figure is a sort of opposite to the pole figure and it shows how a defined direction in the sample coordinate system is distributed in the crystal coordinate system (Bunge, 1982).

4.3.4 In situ X-ray diffraction experiments during quenching of a steel ball bearing

One recent example of an application of in situ XRD experiments is the investigation of martensitic transformation during rapid quenching of ball-bearing steel using the angle-dispersive method with synchrotron radiation (Epp, 2014). The experiments were performed at european synchrotron radiation facility (ESRF) on Beamline ID11 in transmission mode with a monochromatic beam (energy of 71 keV) and varying heat treatment parameters. By variations of the austenitizing temperature, different carbon contents in solution were set in hypereutectoid steel, leading to different behavior during quenching. The temperature was controlled by the use of a type-K thermocouple welded on the surface of the samples.

After austenitizing at different temperatures, increasing amount of carbon is in solid solution in the austenite leading to a decreasing martensite start temperature and a modified transformation kinetic. With a measuring frequency of 1 Hz, complete diffraction

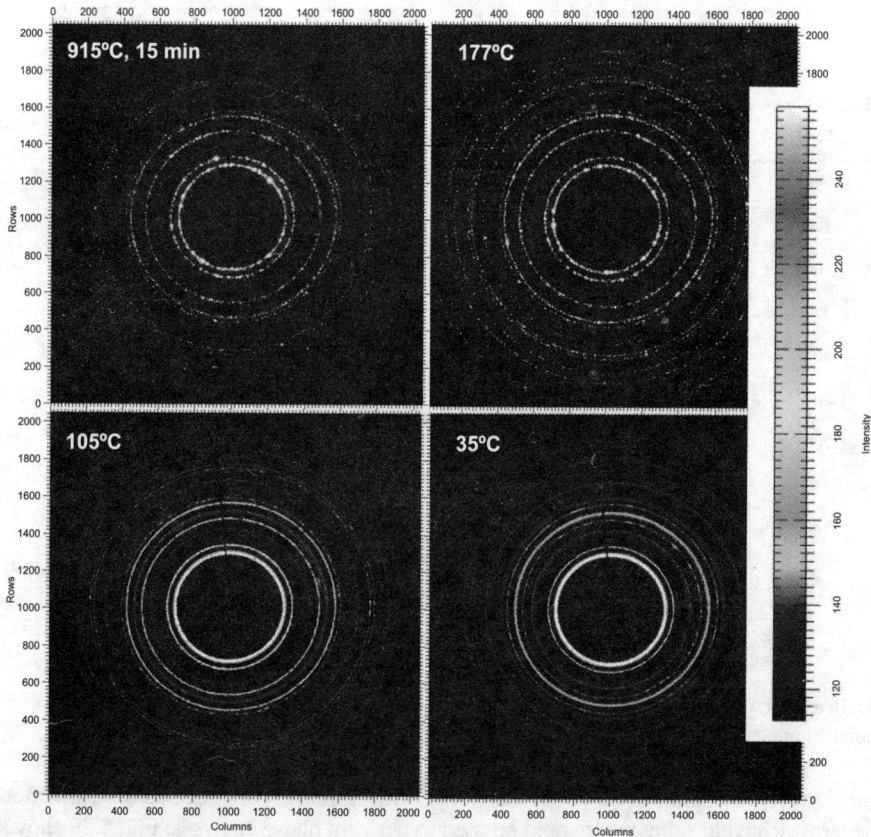

Figure 4.28 2-D diffraction frames taken at different temperatures during quenching (Epp, 2014).

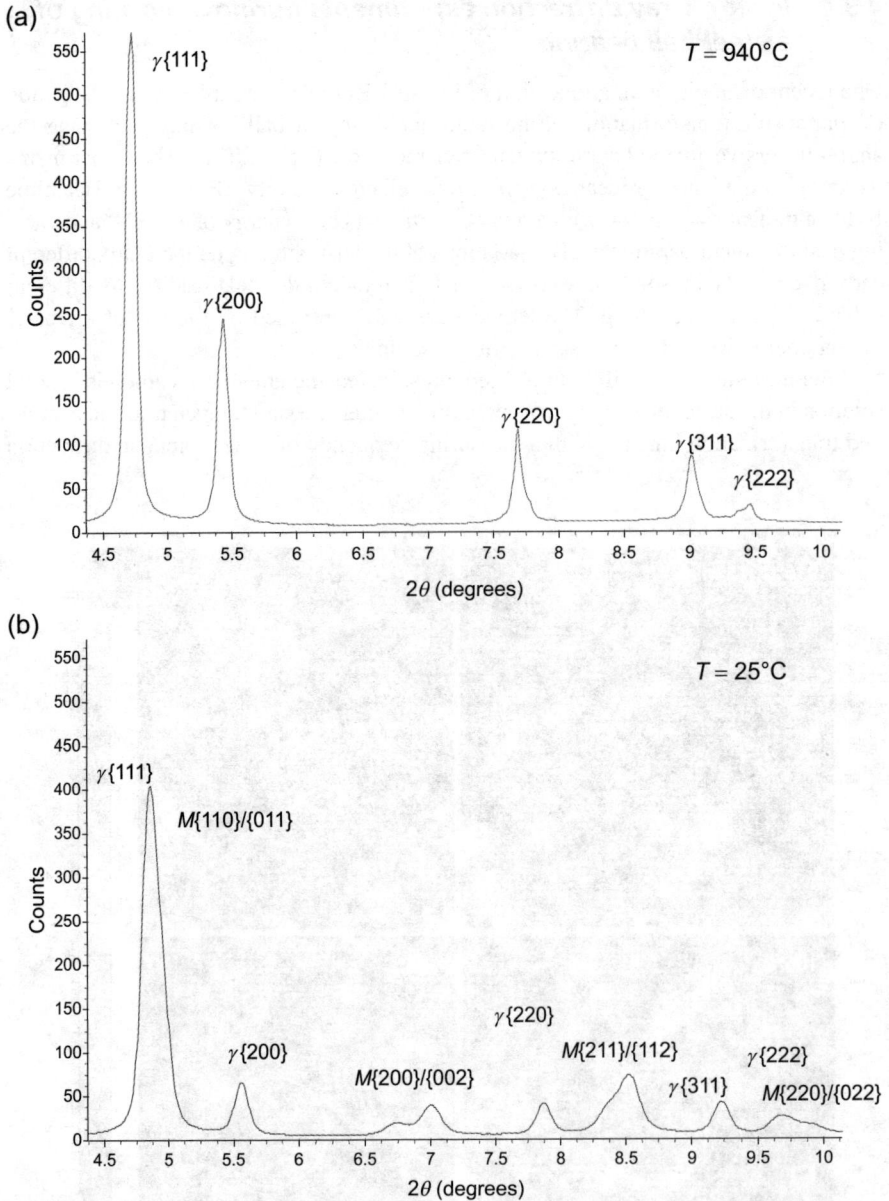

Figure 4.29 Integrated diffraction patterns (a) taken at austenitizing temperature and (b) after quenching.

rings were measured (Fig. 4.28) and after integration to obtain classic diffraction patterns, Rietveld refinement could be used to perform phase analyses. Fig. 4.29 shows two integrated diffraction patterns at austenitizing temperature (a) and after quenching to room temperature (b), showing the good signal quality allowing precise quantitative

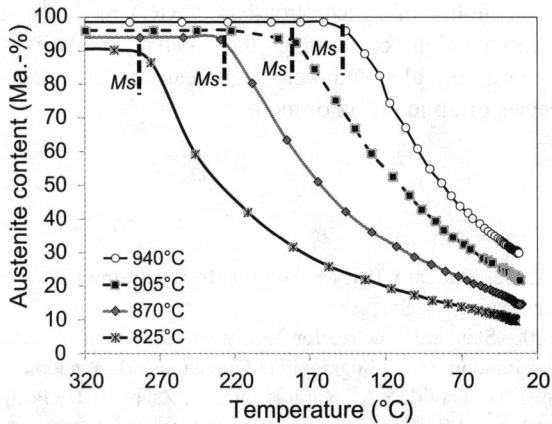

Figure 4.30 Evolution of austenite content during quenching of four samples after austenitizing at different temperatures (Epp, 2014).

analyses. By using Rietveld refinements, the temperature-dependent phase contents could be determined and the transformation kinetics could be described (Fig. 4.30) (Epp, 2014). Further, time- and temperature-resolved evolution of stresses and phase specific density could be characterized (Epp, 2014).

4.4 Conclusions and future trends

Since the discovery of X-rays and the first experiments on XRD by crystals, XRD methods have become very powerful state of the art techniques for advanced material characterization. In the last decades, developments in computer technologies, microelectronics, and data analysis have allowed to make these techniques broadly available and convenient to apply. As a consequence, the knowledge about crystal structure, microstructure, and properties of components could have been strongly improved, allowing the design of high performance materials and components.

On one side, hardware developments (like X-ray source, optics, or detectors) leading to increasing precisions and decreasing measurement time enables increasingly effective and reliable investigations. One innovation in this field is the liquid-metal X-ray source, allowing to reach high beam intensity as the heat development is strongly reduced.

On the other side, new developments of the presented methods are still ongoing, especially for complex materials, geometries, and applications. For example, the advancements of the energy-dispersive methods or high resolution XRD are opening new possibilities of investigations.

Further, the analyses under nonambient conditions or of transient processes by in situ or in-process measurements with XRD methods have an enormous potential for further improvement of materials and manufacturing processes.

Finally, the accessibility of synchrotron-based XRD methods pushes the limits of material characterization concerning the spatial resolution (down to the nanometer-sized beam), but also with very high beam intensity, allowing measuring with high frequencies of up to 30 Hz or more.

References

ASTM E2860-12, 2012. Standard Test Method for Residual Stress Measurement by X-Ray Diffraction for Bearing Steels.

ASTM, October 2000. Standard Practice for X-Ray Determination of Retained Austenite in Steel with Near Random Crystallographique Orientation. Designation: E975−13.

Babu, S.S., Specht, E.D., David, S.A., Karapetrova, E., Zschack, P., Peet, M., et al., 2007. Time-resolved X-ray diffraction investigation of austenite and transformation to bainite. In: Proceedings of the First International Symposium on Steel Science.

Banabic, D., Bünge, H.J., Pöhlandt, K., Tekkaya, A.E., 2000. Formability of metallic materials. In: Banabic, D. (Ed.), Plastic Anisotropy, Formability Testing and Forming Limits. Springer Verlag, Heidelberg.

Behnken, H., 2003. Mikrospannungen in vielkristallinen und heterogenen Werkstoffen. Shaker Verlag, Aachen.

Bragg, W.L., 1913. The diffraction of short electromagnetic waves by a crystal. Proc Camb Philos Soc 17, 43−57.

Bruker-AXS, 2008. TOPAS V4: General Profile and Structure Analysis Software for Powder Diffraction Data. User's manual. Bruker-AXS, Karlsruhe (Germany).

Bunge, H.J., 1982. Texture Analysis in Materials Science. Butterworths, London.

Curfs, C., 2002. Time-Resolved Studies of the Self-Propagating High-Temperature Synthesis of Compounds from the System Aluminium-Nickel-Titanium-Carbon [Ph.D. Thesis]. Université Joseph Fourier, Grenoble (France).

Dinnebier, R.E., Billinge, S.J.L., 2008. Powder Diffraction, Theory and Practice. The Royal Society of Chemistry.

Eigenmann, B., Macherauch, E., 1995. Röntgenographische Untersuchung von Spannungszuständen in Werkstoffen- Teil II. Mater Werkst 3, 148−160.

Elmer, J.W., Wong, J., Fröba, M., Waide, P.A., Larson, E.M., 1996. Analysis of heat-affected zone phase transformations using in situ spatially resolved X-ray diffraction with synchrotron radiation. Metall Mater Trans A 27A, 775−782.

EN 15305:2008, 2008. Non-Destructive Testing - Test Method for Residual Stress Analysis by X-Ray Diffraction.

Epp, J., Hirsch, T., 2010. Residual stress state characterization of machined components by X-ray diffraction and multiparameter micromagnetic methods. Exp Mech 50-1, 195−204.

Epp, J., Surm, H., Kessler, O., Hirsch, T., 2007. In situ X-ray phase analysis and computer simulation of carbide dissolution of ball bearing steel at different austenitizing temperatures. Acta Mater 55-17, 5959−5967.

Epp, J., Surm, H., Hirsch, T., Hoffmann, F., 2011. Residual stress relaxation during heating of bearing rings produced in two different manufacturing chains. J Mater Process Technol 211, 637−643.

Epp, J., Zoch, H.-W., 2014. Comparison of water jet peening and laser shock peening with shot peening for the improvement of fatigue properties of case hardened steel gears. In:

Wagner, L. (Ed.), Proceedings of the Twelfth International Conference on Shot Peening, September 15–18, 2014, Goslar, Germany, pp. S214–S219.

Epp, J., 2014. Investigation of triaxial stress state in retained austenite during quenching of a low alloy steel by in situ X-ray diffraction. Adv Mater Res 996, 525–531.

Fitzpatrick, M.E., Lodini, A., 2003. Analysis of Residual Stress by Diffraction Using Neutron and Synchrotron Radiation. Taylor and Francis, London.

Friedrich, W., Knipping, P., von Laue, M., 1913. Interferenzerscheinungen bei Röntgenstrahlen. Ann Phys 41, 971–990.

Günther, D., 2004. Einfluss unterschiedlicher Chromgehalte auf die Ausbildung von Eigenspannungen und Phasenzusammensetzung beim Gasnitrieren von Stählen [Ph.D. thesis]. University of Bremen.

Guinebretière, R., 2006. Diffraction Des Rayons X Sur Échantillons Polycristallins, second ed. Lavoisier, Paris.

Hölzer, G., Fritsch, M., Deutsch, J., Härtwig, M., Förster, E., 1997. $K_{\alpha1,2}$ and $K_{\beta1,3}$ X-ray emission lines of the 3d transition metals. Phys Rev A 56, 4554–4568.

Hauk, V., 1997. Structural and Residual Stress Analysis by Nondestructive Methods. Elsevier Science, Amsterdam.

He, B.B., Preckwinkel, U., Smith, K.L., 2000. Fundamentals of two-dimensional X-ray diffraction. Adv X-ray Anal 43, 273–280.

Kern, A., Coelho, A.A., Cheary, R.W., 2004. Convolution based profile fitting. In: Mittemeijer, E., Scardi, P. (Eds.), Diffraction Analysis of the Microstructure of Materials, Springer Series in Materials Science, vol. 68, pp. 17–50.

Klug, H.P., Alexander, L.E., 1974. X-Ray Diffraction Procedures for Polycrystalline and Amorphous Materials, second ed. John Wiley & Sons, New York-Sydney-Toronto.

Kopineck, H.-J., Otten, H., 1987. Texture analyzer for on-line rm-value estimation. Textures Microstruct 7, 97–113.

Kostov, V., Gibmeier, J., Wilde, F., Staron, P., Rössler, R., Wanner, A., 2012. Fast in-situ phase and stress analysis during laser surface treatment: a synchrotron X-ray diffraction approach. Rev Sci Instrum 83, 1–11.

Kromm, A., Kannengiesser, T., Gibmeier, J., 2010. In situ observation of phase transformations during welding of low transformation temperature filler material. Mater Sci Forum 638–642, 3769–3774.

Kruijver, S., Zhao, L., Sietsma, J., Offerman, E., van Dijk, N., Margulies, L., et al., 2002. In situ observations on the austenite stability in TRIP-steel during tensile testing. Steel Res 73 (6/7), 236–241.

Macherauch, E., Wohlfahrt, H., Wolfstieg, U., 1973. Zur zweckmaßigen Definition von Eigenspannungen. Härterei-Technische Mitt 28-3, 201–211.

Macherauch, E., Zoch, H.-W., 2014. Röntgenografische Eigenspannungsbestimmung in Praktikum in Werkstoffkunde, 12th ed. Springer.

Noyan, I.C., Cohen, J.B., 1987. Residual stress, measurement by diffraction and interpretation. Springer Verlag.

Prevey, P.S., 1986. X-ray diffraction residual stress techniques. In: Metals Handbook, 10 Metal Parks. ASM, pp. 380–392.

Prince, E., 2004. Volume C - Mathematical, Physical and Chemical Tables, International Tables for Crystallography, third ed. Kluwer Acad. Publ., Dordrecht.

Recke, S., Hirsch, T., 2007. Röntgenographische In-situ-Messungen der Eigenspannungen zwischen 750°C und 900°C. Optimierung der Messeinrichtung. MP Mater Test 10, 509–514.

Rocha, A., da, S., Hirsch, T., 2005. Fast in situ X-ray diffraction phase and stress analysis during complete heat treatment cycles of steel. Mater Sci Eng A 395, 195–207.

Roentgen, W.C., 1895. Über eine neue Art von Strahlen. In: Sitzungsbericht der Würzburg Physik und Medicin Gesellschaft.

Schwartz, L.H., Cohen, J.B., 1987. Diffraction from Materials. Springer Verlag, Berlin.

Spieß, L., Teichert, G., Schwarzer, R., Behnken, H., Genzel, C., 2009. Moderne Röntgenbeugung, second ed. Teubner Verlag, Wiesbaden.

Tomala, V., 1998. In-situ Eigenspannungsmessungen bei hohen Temperaturen in CVD-beschichteten Stahlsubstrat-Verbundwerkstoffen [Ph.D. thesis]. University of Bremen.

Totten, G., Howes, M., Inoue, T. (Eds.), 2002. Handbook of Residual Stress and Deformation of Steel. ASM International.

Will, G., 2006. Powder Diffraction: The Rietveld Method and The Two Stage Method. Springer Verlag.

Winderlich, B., 1990. Das Konzept der lokalen Dauerfestigkeit und seine Anwendung auf martensitische Randschichten, insbesondere Laserhärtungsschichten. Mat.-wiss U Werkst 21, 378−389.

Further reading

Bunge, H.J., 1982. Texture Analysis in Materials Science. Butterworths, London.

Dinnebier, R.E., Billinge, S.J.L., 2008. Powder Diffraction, Theory and Practice. The Royal Society of Chemistry.

Hauk, V., 1997. Structural and Residual Stress Analysis by Nondestructive Methods. Elsevier Science, Amsterdam.

Noyan, I.C., Cohen, J.B., 1987. Residual Stress, Measurement by Diffraction And Interpretation. Springer Verlag.

Spieß, L., Teichert, G., Schwarzer, R., Behnken, H., Genzel, C., 2009. Moderne Röntgenbeugung, second ed. Teubner Verlag, Wiesbaden.

Microwave, millimeter wave and terahertz (MMT) techniques for materials characterization

5

C. Sklarczyk
Fraunhofer Institute for Nondestructive Testing (IZFP), Saarbrücken, Germany

5.1 Introduction

Electromagnetic waves (EMW) in the frequency range of microwaves, in the millimeter range and in the terahertz range (MMT) can be used to nondestructively characterize many materials. There is no unique definition of the term microwaves. It can encompass decimeter and centimeter waves corresponding to a wavelength from 1 m to 10 cm and 10 cm to 1 cm, respectively. The frequencies of these ranges are 0.3–3 GHz and 3–30 GHz, respectively. The millimeter waves extend from 1 cm to 1 mm (30–300 GHz), while the terahertz waves (THz) have wavelengths less than 1 mm (300 GHz). The upper limit of the range of THz waves—also termed far infrared—is not clearly defined.

In contrast to conventional ultrasound testing, no coupling agent is needed to couple the EMW into the test object to be investigated. Principally the distance between the test object and the test device is not limited.

The interior of metallic materials cannot be tested by MMT waves since the high electrical conductivity generates a total reflection of the incident waves thus reducing the penetration depth to a few μm or less.

5.2 Principles of MMT techniques

5.2.1 Propagation of electromagnetic waves and interaction with the material

For all EMW and thus also for MMT the well-known relation (Ida, 1992; Born and Wolf, 1986)

$$\lambda = c/f \qquad [5.1]$$

holds, where λ, wavelength in the material; c, propagation velocity in the material and f, frequency.

Materials Characterization Using Nondestructive Evaluation (NDE) Methods
http://dx.doi.org/10.1016/B978-0-08-100040-3.00005-5

The relations between these quantities c and λ in the material as well as in vacuum or approximately in air (then denoted c_0 and λ_0) is given by

$$c = c_0 / \sqrt{\mu_r \varepsilon_r} \quad \text{and} \quad \lambda = \lambda_0 / \sqrt{\mu_r \varepsilon_r} \qquad [5.2]$$

ε_r, relative dielectric permittivity (dielectric constant); $c_0 \approx 2.998 \times 10^8$ m/s, speed of light in free space; and μ_r, relative magnetic permeability ($\mu_r \approx 1$ for nearly all materials relevant in the MMT domain).

The interaction of MMT waves with a material object to be tested is described by the well-known physical laws of optics and is determined by the permittivity ε_r of the object under test (OUT), which is a complex quantity given by

$$\varepsilon_r = \varepsilon_r' - i\varepsilon_r'' \qquad [5.3]$$

with $i = \sqrt{-1}$. ε_r' is related to the electromagnetic energy stored in the material of the OUT while ε_r'' is related to the dissipation (losses) of energy within the OUT. In metals and semiconductors the electrical conductivity is to be taken into account:

$$\varepsilon_r'' = \varepsilon_{rd}'' + \sigma / (2\pi f) \qquad [5.4]$$

σ, electrical conductivity; f, frequency; ε_{rd}'', dielectric part of ε_r''.

The permittivity is related to the refractive index n by

$$\varepsilon_r \mu_r = n^2 \qquad [5.5]$$

Generally the numerical values of the refractive index in the optical and in the MMT range are not identical.

In dielectric materials the penetration depth of an EMW is a function of ε_r' and ε_r'' (Venkatesh and Raghavan, 2005):

$$\delta = \frac{\lambda_0 \sqrt{\varepsilon_r'}}{2\pi \varepsilon_r''} \qquad [5.6]$$

In an electrically conductive material it is approximately given by (Ida, 1992):

$$\delta = 1 / \sqrt{\pi f \mu_0 \mu_r \sigma} \qquad [5.7]$$

μ_0, magnetic permeability of free space.

From Eq. [5.7], it can be recognized that the penetration depth in the MMT range is very small in all common metals (eg, in copper only about 2 μm at 1 GHz). Therefore, metallic materials can be inspected with MMT only at their surface.

A plane EMW impinging on the interface between two dielectric layers with the thicknesses d_1 and d_2 and the refractive indices n_1 and n_2 is partly reflected and it partly

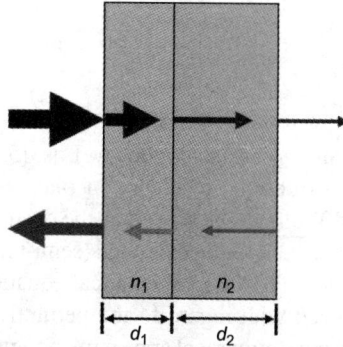

Figure 5.1 Reflection and transmission of electromagnetic waves (direction of propagation depicted by arrows) at dielectric layers. d_1 and d_2, thicknesses; n_1 and n_2, refractive indices. Source: Jonuscheit, J., 2014b. Zerstörungsfreie Analyse. QZ 59 (5), 94–96.

transmits through the interface between the media (Fig. 5.1). The reflection coefficient for the simple case of two plane infinitely extended layers and normal incidence can be derived from the Fresnel formulae (Born and Wolf, 1986; Marion and Heald, 1980)

$$r_{12} = \frac{E_r}{E_i} = \frac{\sqrt{\varepsilon_{r1}} - \sqrt{\varepsilon_{r2}}}{\sqrt{\varepsilon_{r1}} + \sqrt{\varepsilon_{r2}}} \qquad [5.8]$$

E_r, E_i, electric field strength of reflected and incident wave, respectively. The indices 1 and 2 stand for the two media, respectively. The transmission coefficient is given by

$$t_{12} = \frac{E_t}{E_i} = \frac{2\sqrt{\varepsilon_{r1}}}{\sqrt{\varepsilon_{r1}} + \sqrt{\varepsilon_{r2}}} \qquad [5.9]$$

E_t, electric field of transmitted wave. The intensities of the reflected and transmitted waves are given by

$$R = |r|^2 \quad \text{and} \quad T = \frac{\sqrt{\varepsilon_{r2}}}{\sqrt{\varepsilon_{r1}}} \cdot |t|^2 \qquad [5.10]$$

where because of energy conservation, $R + T = 1$.

For a dielectric layer between two other layers, Eqs. [5.8] and [5.9] become (Fig. 5.1)

$$r = \frac{r_{12} + r_{23}e^{2j\beta}}{1 + r_{12}r_{23}e^{2j\beta}} \qquad [5.11]$$

$$t = \frac{t_{12}t_{23}e^{j\beta}}{1 + r_{12}r_{23}e^{2j\beta}} \qquad\qquad [5.12]$$

with $\beta = \frac{2\pi}{\lambda_0}\sqrt{\varepsilon_{r2}}h$ and h, thickness of the layer.

The expressions for r_{23} and t_{23} are analogous to Eqs. [5.8] and [5.9].

The more complex case of oblique incidence of plane waves is described by the Fresnel equations (Born and Wolf, 1986) as well as the occurrence of additional layers. The equations given above are valid for an extended (semi-infinite) test object. In air the relationship $\varepsilon_{r1} \approx 1$ holds. Usually the mathematical equations to describe the multi-layer systems cannot be solved with regard to the permittivity, conductivity or layer thickness. To get these quantities, numerical approximation methods have to be applied.

For wideband measuring systems like ground penetrating radar (GPR, see Section 5.2) and in case of thick layers the echo from the back wall of the layer can be separated from the entry echo. Then the permittivity ε_{r2} of an unknown material can be determined from the measuring voltage of the MMT device if the permittivity ε_{r1} of the first layer material is known. In case of air the following equation holds (Shangguan et al., 2014):

$$\varepsilon_{r2} = \left(\frac{A_m + A_2}{A_m - A_2}\right)^2 \qquad\qquad [5.13]$$

where A_2, amplitude of the surface reflection; A_m, amplitude of the incident signal measured over a thin and flat metal plate that has been placed on the surface of the unknown material. MMT waves are totally reflected by metallic materials ($r \approx 1$). Here it is assumed that the measuring amplitude is linearly proportional to the electric field strength and the object is positioned in the far field of the antenna to avoid near-field effects.

Therefore, in order to calibrate the measuring system the material to be character-ized can be replaced by the metallic object or a thin metallic sheet can be placed on the surface of the object under test.

Additional and deeper layers can be reconstructed, too, based on the first recon-structed layer (layer stripping) providing that the signals from the respective interfaces can be clearly separated.

In reflection mode the thickness of the layers can be determined by (Al-Qadi et al., 2006; von Aschen et al., 2011)

$$d_l = \frac{c_0 \cdot t_l}{2\sqrt{\varepsilon_l}} \qquad\qquad [5.14]$$

where ε_l is the relative permittivity of the layer and t_l the time of flight in the layer.

However, in more complex layered systems multiple reflections of the MMT waves occur at the interfaces resulting in interferences. Then the evaluation of echoes like in Eq. [5.13] is no more possible. Based on Eqs. [5.11] and [5.12], there exist mathemat-ical methods to calculate the reflection and transmission of plane EMW in multilayered test objects (Orfanidis, 2014). Here again, the parameters are the layer thickness, complex permittivity, electrical conductivity and complex magnetic permeability. As mentioned above the last quantity can be equaled to one ($\mu_r = 1$) in nearly all electrically insulating or low-conductive materials.

5.2.2 Antennas

5.2.2.1 Properties of antennas

The MMT waves are radiated and received by antennas. For distances which are much bigger than the wavelength and the dimensions of the antenna, ie, in far or Fraunhofer field, the electric and magnetic field are perpendicular to each other and to the direction of propagation. For very big distances the curved waves in the vicinity of the MMT transmitter can be approximated mathematically by a planar wave. For smaller distances (near field) the relationships between these variables are more complex and depend on the type of antenna. The far field roughly begins at

$$y_f = R^2 / \lambda \tag{5.15}$$

where R, size of antenna; λ, wavelength and extends to infinity, with the axial intensity uniformly decreasing with y^{-2} (inverse square law function; McMaster and McIntire, 1986).

One important property of antennas is their gain, which is usually defined as the ratio of the power produced by the antenna from a far field source on the antenna's beam axis to the power produced by a hypothetical lossless isotropic antenna.

A frequently used type, mainly in the far field, is the horn antenna which exists in many versions and frequency bands. Examples are shown in Fig. 5.2.

Figure 5.2 Antennas: double-ridged horn (0.7–18 GHz) (right); circular horn for W-band (70–110 GHz) (left).

5.2.2.2 Open-ended waveguide

Open-ended waveguides like the coaxial probe (Fig. 5.3) or the rectangular waveguide probe (Fig. 5.4) are often used for characterizing solid and liquid materials in the MMT range with regard to their permittivity. Usually they are in contact to the OUT but they can also be used with a small lift-off distance like an air gap. It must be taken into account that this distance acts like an additional layer which influences the reflection and transmission behavior of the OUT (Ganchev et al., 1995). In connection with a vector network analyzer (see Section 5.2.4.4) the coaxial sensor can be applied in a very broad frequency band.

5.2.2.3 Microstrip antenna

Stripline antennas are often used in telecommunications as they are relatively inexpensive and have a flat geometry. The most common type of microstrip antenna is a narrowband rectangular patch. Fig. 5.5 gives a scheme of this antenna. Microstrip sensors have repeatedly been used to measure the moisture, eg, the density-independent moisture content of powdered materials (Zhang and Okamura, 2006).

5.2.3 Properties of dielectric materials

5.2.3.1 Moisture

The permittivity of most solid materials only slightly varies as function of frequency in the MMT range. The most important substance with a different behavior is liquid water, whose molecule exhibits a considerable dipole moment which strongly couples to the electromagnetic field. For pure water at room temperature, the real part of permittivity is given by $\varepsilon_r \approx 80$ at low microwave frequencies. This value is much bigger than for most other materials. Furthermore water exhibits a strong maximum

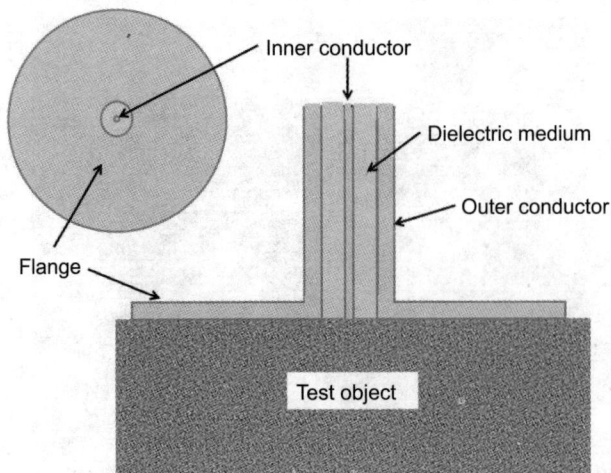

Figure 5.3 Scheme of a coaxial probe.

Figure 5.4 Open rectangular waveguide with attached microwave sensor.

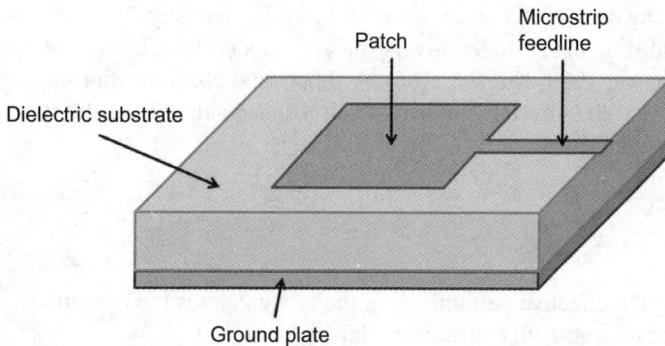

Figure 5.5 Scheme of a microstrip antenna.

of the imaginary part at about 10 GHz (Fig. 5.6) resulting in a pronounced attenuation of microwaves in water. These peculiarities of the permittivity of water result in a changed reflection and attenuation behavior of microwaves in moist materials and are the basics for moisture measurement with the help of microwaves. Usually, some calibration procedure is needed to get the moisture value from the microwave measuring values.

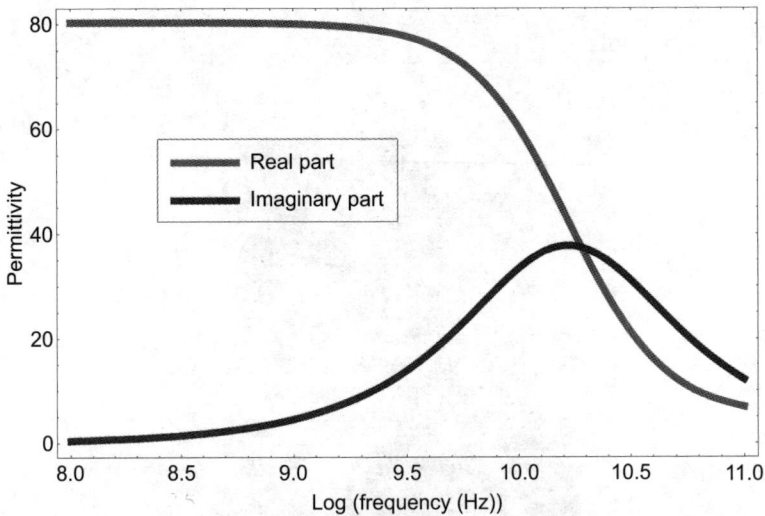

Figure 5.6 Real and imaginary parts of permittivity of pure liquid water in the frequency range from 100 MHz to 100 GHz.

Material properties like density or porosity or some mechanical properties are normally correlated with the permittivity of these materials. Usually the relationship between the material property to be measured and the permittivity has to be determined experimentally with the help of a calibration procedure. Here the permittivity is determined as function of the known material property. Normally the permittivity increases with increasing density and decreases with increasing porosity. To model the manifold effects of moisture in substrates, diverse mixture models have been developed (Nyfors and Vainikainen, 1989; Kupfer, 2005). A simple and useful relationship between the permittivity and the moisture content which is independent from the structure of the material is given by

$$\varepsilon_{rm}^{a} = \sum_{i} f_i \cdot \varepsilon_{ri}^{a}$$ [5.16]

where ε_{rm} is the effective permittivity of the mixture, ε_{ri} is the permittivity, and f_i the volume fraction of the ith constituent with $\sum_{i} f_i = 1$.

To get the volumetric water content ψ from permittivity, empirical calibration functions can be posed. For example, the Topp equation gives this relation for a wide range of soils (Topp et al., 1982; Jol, 2009):

$$\psi = -0.053 + 0.0293\varepsilon_r - 0.00055\varepsilon_r^2 + 0.0000043\varepsilon_r^3$$ [5.17]

MMT waves, which are radiated by an antenna, are subjected to geometric spreading and other physical phenomena known from optics concerning reflection, refraction, diffraction, and polarization. The refraction at the interface between the media one and two is described by Snell's law:

$$\frac{\sin \theta_1}{\sin \theta_2} = \frac{c_1}{c_2} \qquad [5.18]$$

where θ_i is the angle measured from the normal of the boundary and c_i is the velocity of propagation in the ith medium.

5.2.3.2 Anisotropy

Anisotropic materials such as fiber-reinforced plastics can be characterized with polarized MMT waves as the permittivity depends on the orientation of fibers with regard to the polarization of the MMT. In most cases the MMT waves emitted from the MMT sensor are already linearly polarized so that this wave property can be used to characterize the material. Fig. 5.7 shows the variation of the transmission amplitude versus rotation angle for a unidirectional glass fiber-reinforced composite. If there is no specimen between the transmitting and receiving antenna the amplitude amounts to 1 in this figure. In the 0° position the electric field vector is oriented perpendicularly to the fiber direction.

As the EMW penetrates into the material, it is split into one component the field vector of which is parallel to the fiber direction and another, perpendicular component. Both components exhibit different propagation velocities in the material due to their different refractive indices. At leaving the material they recombine and interfere, resulting in a periodical behavior of the transmission amplitude (Fig. 5.7).

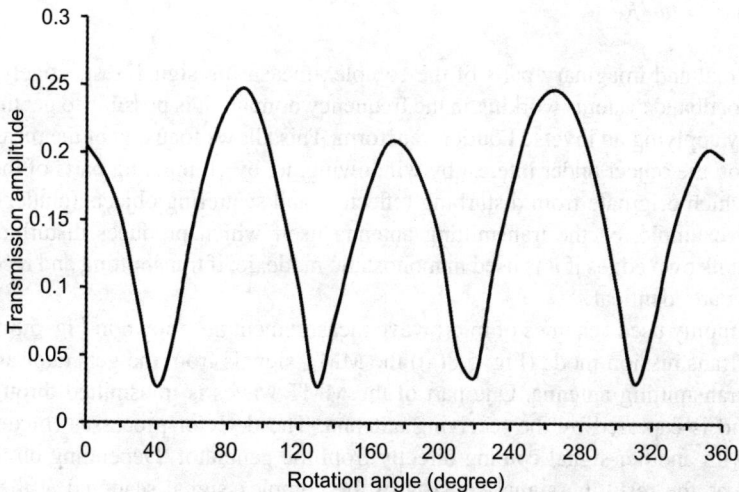

Figure 5.7 Dependence of the transmission amplitude from the rotation angle for a plate made of glass fiber-reinforced composite (thickness, 19 mm), network analyzer system in frequency range between 75 and 100 GHz, normal incidence.

5.2.4 Methods of wave generation, radiation and detection

5.2.4.1 Principles

Historically the components for microwave and millimeter wave generation, transmission and reception were mostly big and expensive. In the field of nondestructive testing (NDT) usually a low power level is sufficient, which lets the OUT unchanged and undamaged. Here modern solid-state sources such as the gallium arsenide (GaAs) Gunn diodes or the silicon IMPATT diodes provide coherent microwave radiation (McMaster and McIntire, 1986).

Usually microwave detectors for NDT are made of semiconductors including Schottky and GaAs diodes and are based on signal rectification. They can be used as demodulators for lower-frequency amplitude modulation applied to the microwaves (the carrier).

The homodyne and heterodyne downconversion techniques are more sophisticated detection methods and are mentioned below. In the THz range mostly other concepts are applied for signal generation and detection (see Section 5.2.4.4).

5.2.4.2 Measuring systems

From a mathematical point of view, it is favorable to describe the wave equation of the MMT as a complex quantity with real and imaginary parts. This helps to determine the most common measuring quantities magnitude M and phase φ, respectively, loss tangent (dissipation factor) $\tan \varphi$:

$$M = \sqrt{Re^2 + Im^2} \tag{5.19}$$

$$\tan \varphi = Im/Re \tag{5.20}$$

Re, Im, real and imaginary parts of the complex measuring signal, respectively.

In broadband systems working in the frequency domain, it is possible to get the time signal by applying an inverse Fourier transform. This allows focusing of the measuring system on the object under interest by windowing, ie, by eliminating parts of the time signal which originate from disturbing reflectors and scattering objects (clutter). This can, for example, be the transmitting antenna itself which produces disturbing impulses at its own edges if it is used in monostatic mode, ie, if transmitting and receiving antennas are identical.

Commonly used schemes of microwave measurement are shown in Fig. 5.8. In the bistatic transmission mode (Fig. 5.8(a)) the MMT signals from the generator are sent by the transmitting antenna. One part of the MMT waves is transmitted through the OUT and is captured by the receiving antenna. The detector processes (mixes) this signal with another signal coming directly from the generator. Depending on the detector type the resulting signal consists of the complex signal scattered at the OUT or it consists of the magnitude and phase. In the monostatic reflection mode (Fig. 5.8(b)) the signal which is scattered back at the OUT is captured by one antenna, which at the same time transmits the MMT waves. A directivity coupler or circulator is

(a)

(b)

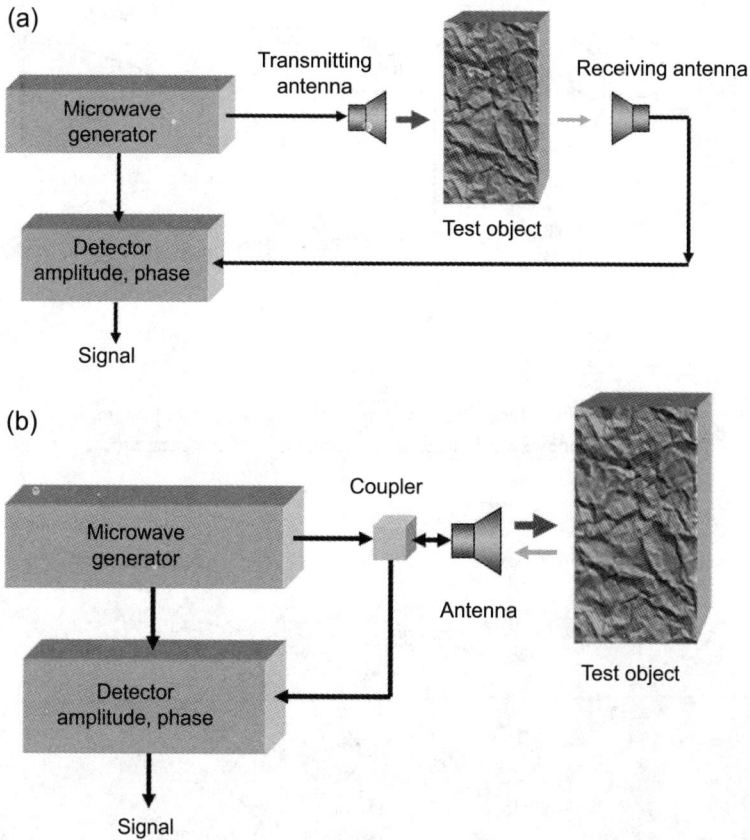

Figure 5.8 (a) Bistatic measuring arrangement in MMT domain; (b) Monostatic measuring arrangement in MMT domain.

then directing this signal to the detector, which generates the measuring quantities similar to the transmission arrangement.

These measuring setups can be implemented for noncontact as well as for contact mode. No coupling agent like in standard ultrasound technique is needed.

5.2.4.3 Narrowband systems

Homodyne and heterodyne systems

In homodyne microwave or millimeter wave systems the EMW are excited by a generator, mostly based on semiconductors (eg, diode), and transmitted by an antenna or wave guide. The EMW scattered by the OUT are either received by the same antenna (monostatic mode) or by another antenna (bistatic mode). In monostatic devices the received signal is carried to a mixer (multiplier) with the help of a circulator. This signal is mixed with a part of the generator signal (Fig. 5.9). In a fixed-frequency device, this results in a DC voltage signal at the mixer output.

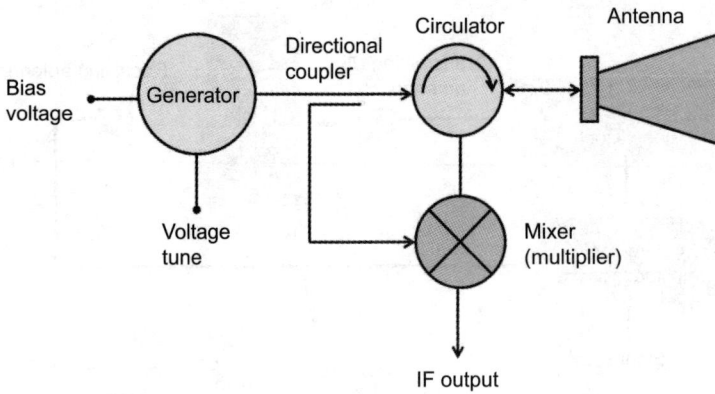

Figure 5.9 Scheme of a homodyne measuring system. In a wideband system, it is possible in contrast to the fixed-frequency system to tune the frequency of the generator.

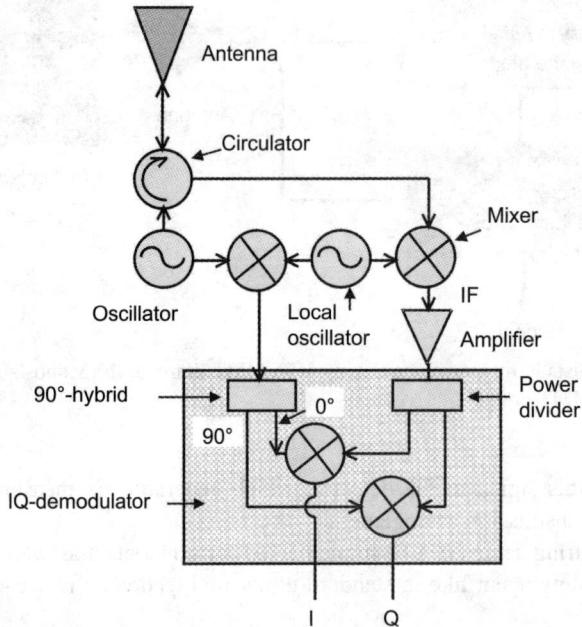

Figure 5.10 Scheme of a heterodyne measuring system.

In the heterodyne measuring device the microwaves generated by the oscillator are conducted over the circulator and are radiated by the antenna (Fig. 5.10). They are scattered by the OUT and in the layout shown here they are received by the same antenna and are conducted to the mixer again by the circulator. The local oscillator (LO) also generates oscillations in the MMT range which are shifted in their frequency (mostly some MHz) compared to the first oscillator. In the mixer the transmitted and received signals are then downmixed with the LO signal to the base band resulting in a low frequency signal (intermediate or difference frequency (IF)). In an in-phase and quadrature (IQ)-demodulator the signals are split, partly shifted by 90° and conducted to additional mixers resulting in the real (I) and imaginary (Q) part of the received signal. The following relationships allow to determinate the signal amplitude and phase:

$$A = \sqrt{I^2 + Q^2}$$
[5.21]

and

$$\varphi = \arctan\,(Q/I)$$
[5.22]

where A is the amplitude and φ is the phase angle. $\tan \varphi$ is also called loss tangent or dissipation factor. If A and φ are given, I and Q can be calculated by

$$I = A \cos \varphi$$
[5.23]

$$Q = A \sin \varphi$$
[5.24]

If only the real part of a non-DC signal is available the complex signal, which contains the imaginary part (original real sequence with a 90° phase shift), can be calculated with the help of the Hilbert transform.

By downmixing the high-frequency signal scattered at the OUT to a lower frequency, it can be processed more easily with conventional low-cost electronic devices. Homodyne systems are simpler and cheaper than the heterodyne ones since they need only one oscillator and mixer. However, their dynamic range is usually smaller.

Resonator

The measuring principle of the resonator sensor, which is often used for materials characterization, consists in a parallel circuit with an inductance and a capacitance which are part of an electromagnetic resonator. The OUT can be inserted into the capacitor. When there is no test object the resonator generates a sharp peak at a defined frequency. As can be seen in Eq. [5.21] the resonance frequency f_r is shifted in dependence of the permittivity of the OUT (Nyfors and Vainikainen, 1989):

$$f_r = \frac{1}{2\pi\sqrt{LC_0\varepsilon_r'}}$$
[5.25]

with L, complex inductivity of the resonator and C_0, capacity, depending on the dimensions of the resonator. The quality factor of the peak, too, is altered:

$$Q = \frac{1}{\tan \delta_L + \tan \delta_C} = \frac{1}{\tan \delta_L + \varepsilon_r''/\varepsilon_r'}$$
[5.26]

where $\tan \delta_L$ and $\tan \delta_C$ are the loss tangent of the inductive and capacitive components of the resonator. The resonance frequency and the quality factor of the air-filled resonator ($\varepsilon' = 1$, $\varepsilon'' \cong 0$) are given by (Trinks and Kupfer, 1999):

$$f_r^* = \frac{1}{2\pi\sqrt{LC_0}} \qquad\qquad [5.27]$$

and

$$Q^* = 1/\tan \delta_L \qquad\qquad [5.28]$$

The dependencies of the resonance frequency and quality factor of a resonator filled with a dielectric material are then given by

$$f_r = \frac{f_r^*}{\sqrt{\varepsilon_r'}} \qquad\qquad [5.29]$$

and

$$Q = \frac{Q^*}{1 + Q^* \cdot \varepsilon_r''/\varepsilon_r'} \qquad\qquad [5.30]$$

In its nondestructive version the resonator sensor is an open resonator whose interior contains the electromagnetic field (Fig. 5.11; Daschner and Knöchel, 2001). The sensor is put on the OUT, ie, it normally works in a contacting but nondestructive manner. The field leaks through an aperture in the cavity into the test object. In most cases the frequency is set to 2.45 GHz which can be used royalty-free. There exist several commercial producers of resonator sensors for moisture measurement.

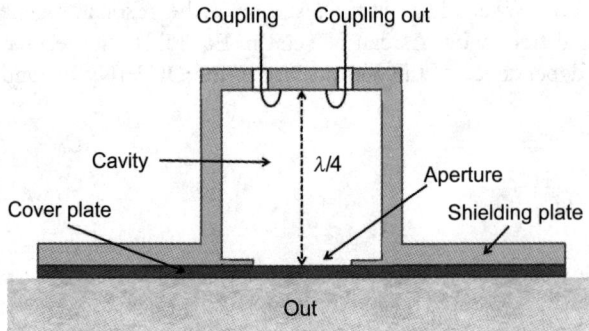

Figure 5.11 Scheme of a resonator sensor.

5.2.4.4 Wideband systems

Principles

In the MMT range, wideband systems can be realized both by sweeping the frequency over a wide frequency range or by generating short-time pulses. Sweeping can be done in a stepwise way (frequency-stepped continuous wave technique) or by continuous frequency modulation (frequency-modulated continuous wave [FMCW] technique).

Network analyzer

Vector network analyzers (VNA) use the heterodyne principle and are mainly applied for characterization of electronic parts in the high-frequency range. The term "vector" means that they handle both the real and imaginary parts of the signal. They can also be applied to nondestructive materials characterization in free field and in contact mode from low microwave range up to the low THz range. They work with continuous waves in the frequency domain, while synthetic time signals can be generated by a Fourier transform. A very broad frequency band can be produced whose signal amplitudes can be constant for all used frequencies thanks to highly developed calibration procedures. VNA are mainly applied for research purposes and rarely in the industrial environment since they are expensive and relatively slow.

Frequency-modulated continuous wave (FMCW) radar

The FMCW radar (radio detection and ranging) uses frequency modulated continuous waves instead of the pulses of the better known pulse radar (see later in this section). The continuous microwave signals are generated by a voltage controlled oscillator or generator, the frequency of which can be modulated by an external function generator (Fig. 5.9). Normally the shape of the modulation function is a triangle or a sawtooth. The microwaves are then conducted to the antenna which emits them to the OUT. The scattered microwaves are received by the same or another antenna and are mixed with one part of the emitted signal. Due to the frequency modulation there arises a frequency difference between an emitted and received signal, which is proportional to the distance between radar device and test object. The sinusoidal signal resulting at the mixer is called IF signal. Its frequency is given by

$$f_{if} = \frac{2dB}{Tc} \qquad [5.31]$$

with d, distance between sensor and OUT; B, frequency band width of the generator; T, duration of period of the modulation function, eg, ramp; and c, velocity of propagation. By an inverse Fourier transform of the IF signal it is possible to get echo signals in the time domain which are equivalent to pulse radar signals. The FMCW radar possesses some advantages compared to pulse radar, eg, simpler electronics (Yamaguchi et al., 1996). The complex scattering and reflection properties of the OUT are mirrored by amplitude and phase of the IF signal.

Single-chip FMCW monolithic microwave integrated circuit (MMIC) radar at the center frequency 94 GHz has been developed which contains all high-frequency components on an area of a few mm^2 (Tessmann et al., 2002; Zech et al., 2013). It can be used as core of a variety of measuring and testing systems.

Pulsed systems

Besides the above-described systems which use continuous waves, there exist systems which generate and receive pulses in a more or less wide frequency range, mainly in the microwave range. The most important NDT application of these systems can be found in the field of GPR which nearly all work in pulsed mode in the frequency range from much less than 1 GHz to about 3 GHz.

Radar systems usually apply microwave pulses which are used to detect objects and to measure their distance (and in some cases velocity) by means of the pulse time of flight. For antenna-object (target) distances, which are much bigger than the wavelength, the received power of a radar system is given by the radar equation

$$P_r = P_t \frac{G_t G_r \lambda^2 \sigma}{(4\pi)^3 R_t^2 R_r^2}$$
[5.32]

with P_t, transmitted power; G_t, gain of transmitting antenna; G_r, gain of receiving antenna; λ, wavelength; σ, radar cross-section, a measure of how an object is detectable with a radar; R_t, distance between the transmitting antenna and the target; R_r, distance between the receiving antenna and the target.

The GPR is normally equipped with a time-varying gain in order to amplify the weaker signals from deeper reflectors. One big advantage of GPR is its ability to directly determine and display the depths of reflectors according to Eq. [5.14] by measuring the time-of-flight of the signals (Daniels, 1996) if the permittivity of the OUT is known.

Time-domain reflectometry (TDR)

The TDR is most often applied to detect and locate defects in metallic cables. A short pulse or a step with short rise time is transmitted along the conductor. This pulse is partly or completely reflected at the defect while the location of the defect can be determined with the help of the pulse time-of-flight (travel time). In earth and agricultural sciences, TDR can be used to determine the content of moisture. Normally a transmission line technique is used: the sensor consists of two or more parallel metal rods embedded in soil or sediment. In a variant of the method the sensor has a flat but flexible surface and is put in a noninvasive manner on the surface of the OUT resulting in a direct contact. No coupling agent is needed.

Fig. 5.12 shows an example of a measurement of the gravimetric moisture (referred to wet material) in green sanitary ceramics (Sklarczyk et al., 2013). The curve can be approximated by a polynomial of the second order and thus can easily be linearized. With this calibration a measuring device can be implemented which directly displays the moisture content of the material. If there is a slight bending of the OUT surface it can be compensated by a flexible sensor surface.

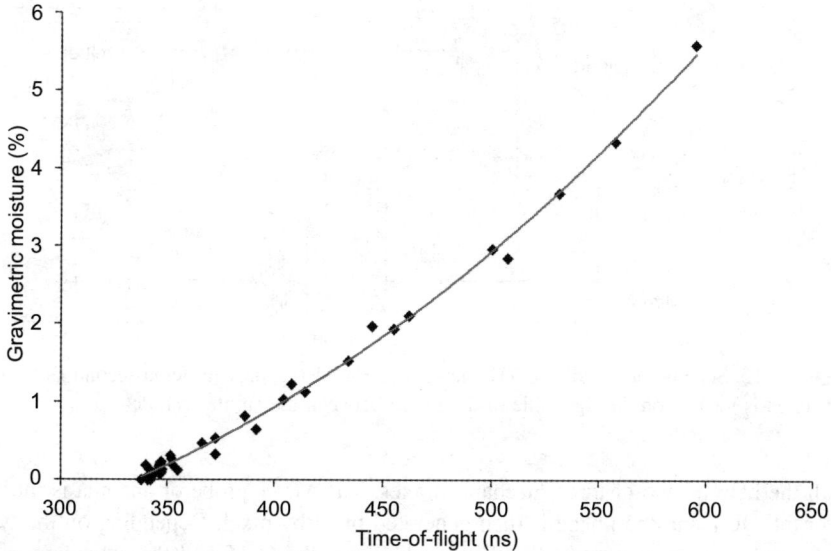

Figure 5.12 Correlation of the gravimetrical moisture in green sanitary ceramics with the time-of-flight of a TDR device which is put nondestructively on the OUT.

THz systems

In the past the frequency range between the millimeter wave range and the infrared range has rarely been used due to the lack of high-power frequency sources (so-called THz gap). This gap has been filled with a variety of THz generators like the backward wave oscillator, the far infrared laser, Schottky diode multipliers, quantum cascade laser, the free electron laser, synchrotron light sources or photomixing sources.

Especially in NDT area and for research purposes, one of the most often used THz schemes consists of an infrared femtosecond (fs) laser, the pulses of which are focused on a photoconductive switch which generates the THz radiation and acts as an antenna. The THz radiation is then focused at the specimen by lenses or concave mirrors (Fig. 5.13). The reception of the THz pulses is performed in a similar but reverse way. With the help of a delay line the optical path of the THz pulse can be changed and the pulse can be sampled in a time domain (stroboscope effect), resulting in a pulse with a bandwidth of up to several THz. There exist many types of THz systems varying in this scheme. Both reflection and transmission setups are feasible. Fiber-coupled THz emitters and detectors allow to change the measuring setup as required by the application. Due to the very high bandwidth time-domain, spectroscopy is feasible.

5.2.5 Imaging techniques with mono- and multichannel systems

In the area of NDT, imaging is used in the MMT technique mainly to detect and characterize defects. Usually imaging is performed with multiaxial scanning systems, ie,

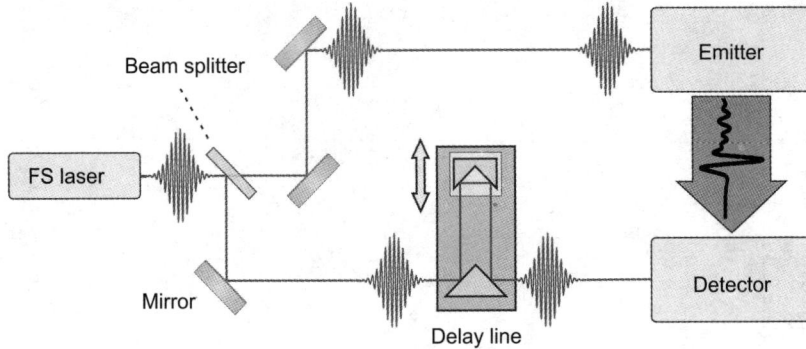

Figure 5.13 Scheme of broadband THz generation and detection with femtosecond laser. The specimen is located on the right side of the figure between the emitter and detector.

with the help of two or three mechanical axes. The MMT probe or antenna is moved over the OUT and the image is then generated pixel by pixel. Depending on the type and purpose of measurement the distance between the OUT and the antenna can be very small, even zero (touching mode) or very big (far field). As an extreme example the investigation of other planets such as Venus, Mercury or Jupiter's moons can be taken with Earth-based radar. Defects in the OUT are usually found by contrast differences compared to the nondefective parts of the OUT. For a well-characterized system, ie, with known lift-off or distance, antenna characteristics and incidence angle and with an OUT which preferably exhibits big planar surfaces it is possible to determine the permittivity of the material of the OUT from the measuring quantities (see Section 5.2).

The lateral (cross) resolution in monostatic mode is given by (Klausing and Holpp, 2000):

$$\delta_l = \theta \cdot R_z \approx 0.64 \frac{\lambda}{d_a} R_z \qquad\qquad [5.33]$$

where θ, width of the main lobe of the antenna; d_a, real aperture antenna; R_z, distance between antenna and OUT.

To optimize and to increase the lateral resolution, the so-called synthetic aperture radar (SAR) can be used. The coherent radar signal is measured at different positions on a path perpendicular to the image plane (Ahmed et al., 2012). The lateral resolution of the SAR is given by

$$\delta_{SAR} = 0.88 \frac{\lambda}{2L_{SAR}} R_z \qquad\qquad [5.34]$$

where L_{SAR}, length of synthetic aperture.

For optimal geometrical parameters

$$\delta_{SAR} = d_a/2, \qquad\qquad [5.35]$$

ie, for SAR the lateral resolution is independent from wavelength and distance. The axial (in-depth) resolution of an MMT system is given by

$$\delta_r = c/2B \qquad\qquad [5.36]$$

δ_r, axial resolution; B, frequency bandwidth.

Aside from these considerations the resolution is limited to about half a wavelength in the OUT (Rayleigh criterion).

As the image is built up pixel by pixel, it may take a long time to complete it if only one probe is used. Therefore, several attempts have been undertaken to develop a real-time imaging system in the MMT range by applying many probes and measuring channels which operate simultaneously. Most of those systems are applied in the homeland security domain as personnel scanners. One system uses a combination of mechanical scanning and a linear sensor array which rotates around the person to be scanned (McMakin et al., 2007). The image is generated within a few seconds, where part of this time is needed by the computer system to automatically separate the suspicious object (metallic and nonmetallic weapon and/or explosives) from its surroundings (human body and clothing) and to identify it as innocent or dangerous/illegal.

Another system works in the mm-wave range (70–80 GHz) and uses a planar array with more than 6000 channels and with no need of mechanical movement of any component (Ahmed et al., 2013). The 3-D image is generated by a sophisticated procedure of transmitting and receiving antennas (multiple input–multiple output) in connection with a sparse array. This type of array needs much less antenna elements compared to a fully populated array with a maximum element distance of half the wavelength in both x- and y-axis. With the help of mathematical methods, which apply the algorithms of synthetic aperture and pattern recognition, dangerous and illegal metallic and nonmetallic objects like weapons and explosives concealed under clothing are detected from distances of about 1 m. Here image processing is performed within about 2 s. This imaging system could basically be used for nondestructive purposes as well. However, up to now no application of this system in the NDT domain has become known.

The prototype of a portable real-time "microwave camera" has been developed for a frequency of 24 GHz (Ghasr et al., 2012). This system is based on a 2-D array (576 elements) of switchable slot antennas and on the modulated scatterer technique. In the future, a further developed version of this camera is planned to be used for NDT applications.

5.3 Applications

5.3.1 Determination of thicknesses in layered systems

5.3.1.1 Single and multiple dielectric layers

With the help of Eq. [5.14] it is possible to directly determine the thickness of dielectric layers or covers from the temporal positions of the echoes at the interfaces (Fig. 5.14).

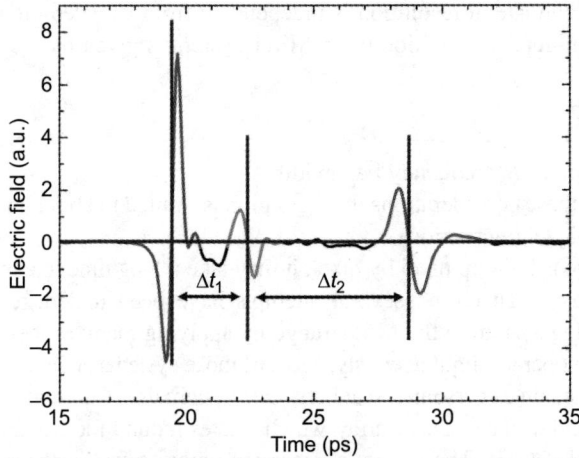

Figure 5.14 Determination of layer thickness in reflection mode (two lacquers on metallic substrate) with broadband THz time domain spectroscopy using time differences. The first echo is from interface air, first lacquer; the second echo from interface first lacquer, second lacquer; and the third echo from interface second lacquer, metal (also see Fig. 5.1).
Source: Jonuscheit, J., 2014b. Zerstörungsfreie Analyse. QZ 59 (5), 94−96.

The preconditions are that (a) the bandwidth of the MMT measuring system is sufficiently high in relation to the layer thickness to get a good axial resolution (Eq. [5.35]) and (b) the permittivities of the layers are known.

In the THz frequency domain, where very high-frequency bandwidths can be realized, it is possible to resolve systems of multilayers with individual layer thicknesses in the range between 10 μm and 500 mm. The axial resolution can be in the range of about ±1 μm (Jonuscheit, 2014a,b; Su et al., 2014).

In those cases where the bandwidth is not sufficient or the layer thickness is too small, multiple reflections occur between the interfaces resulting in a regular sequence of maxima and minima of the magnitude as function of frequency. In the example given in Fig. 5.15 the applied frequency band was between 75 and 100 GHz. The single-layer ceramic specimen was positioned in free field according to the experimental arrangement as given in Fig. 5.8. From the distance between two frequency maxima or minima Δf the product from specimen thickness d and square root of permittivity can be obtained according to

$$d\sqrt{\varepsilon_r} = \frac{c_0}{2\Delta f} \qquad [5.37]$$

If the thickness is known the permittivity can be determined, or vice versa (Zoughi, 1990; Sklarczyk et al., 1998). For multilayer specimens, Eq. [5.36] cannot be used anymore. For that, multilayer approaches have been developed which are based on fundamental equations like Eqs. [5.8]−[5.12]. The reflection and transmission coefficients as function of frequency can be calculated by inserting the material and

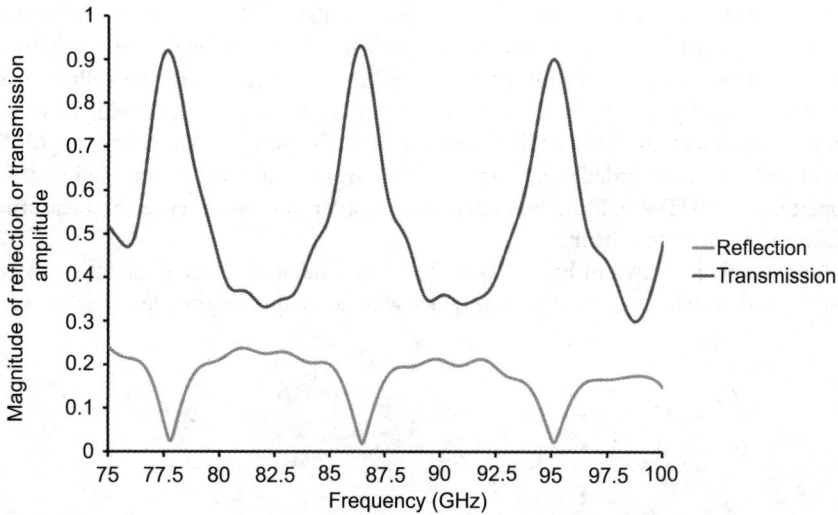

Figure 5.15 Magnitude of reflection and transmission amplitude for an aluminum sheet with thickness 3 mm. The measurement arrangement was performed in free field according to Fig. 5.8.

geometric parameters like dielectric permittivity, electric conductivity and thickness into the equation systems (Orfanidis, 2014; Sklarczyk et al., 1998).

The calculation of the layer parameters from the experimental values, however, is at the most possible for special cases. Here the method of parametric optimization can be applied, which is based on minimization of the target function

$$J = \sum_{i=1}^{N} \left\{ |S(b_j,f_i)|_c - |S(b_j,f_i)|_m \right\}^2 \qquad [5.38]$$

$|S(b_j,f_i)|_c$ and $|S(b_j,f_i)|_m$ are the theoretically calculated, respectively, measured reflection or transmission values for the frequencies f_i ($i = 1, ..., N$) within a relatively broadband, while b_j stand for the estimated values of the parameters (Friedsam and Biebl, 1997).

This procedure to determine the thickness and/or complex permittivity in multilayer structures can be supported by means of genetic algorithms and sequential quadratic programming as described in Elhawil et al. (2009).

Principally, even with only one frequency the layer thickness can be determined. Since the magnitude (or the real or imaginary part) of the MMT signal exhibits a periodical behavior with maxima and minima as function of thickness the range of thicknesses to be expected should be roughly known to avoid ambiguities. By using one or more additional frequencies these ambiguities can be eliminated. Of course, if only one or a few frequencies are available the statistical errors are bigger compared to the measuring procedure, which involves many frequencies.

If the MMT sensor is positioned very close to the OUT, near-field effects arise which hardly can be described mathematically. Then the determination of geometrical and material parameters of the OUT becomes difficult or even impossible with the help of formulae like Eq. [5.13]. Nevertheless the layer can be characterized with a narrow-band measurement in the near-field range if a calibration is done. Here the MMT measuring value is recorded as a function of the layer thickness or the quested material property. The OUT should be homogeneous in order not to influence the measuring values in an undefined manner.

An example is shown in Fig. 5.16. Here the specimen consists of an anticorrosive coating made of some sublayers of polymer with some quartz granules attached on a

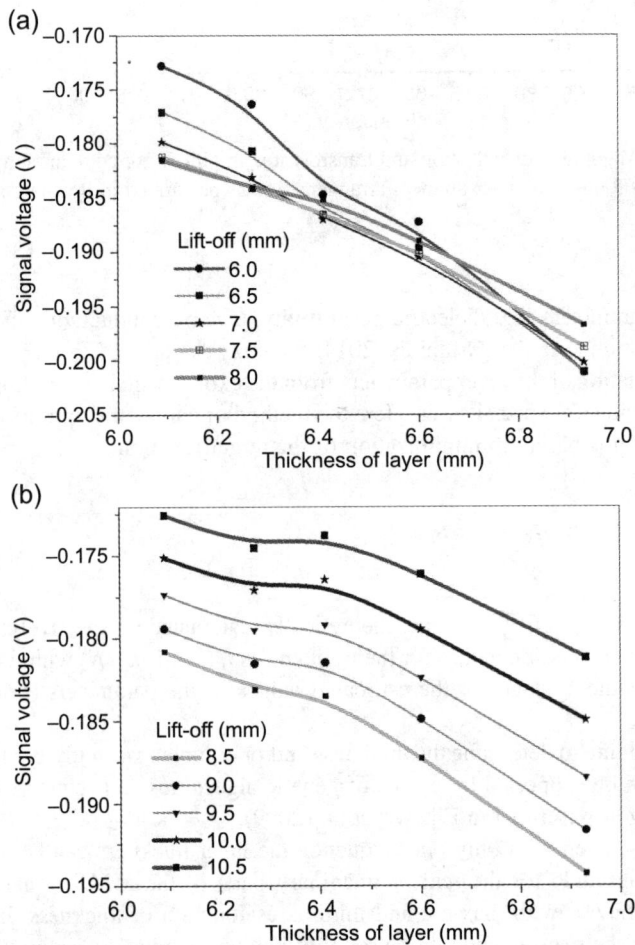

Figure 5.16 (a) Narrowband signal voltage at 9.5 GHz for an anticorrosive coating on concrete substrate for different lift-offs. The lift-off-range is optimized; (b) Narrowband signal voltage at about 9.5 GHz for an anticorrosive coating on concrete substrate for different lift-offs. The lift-off-range is not optimized.

substrate made of concrete. The measured voltage decreases as function of the layer thickness. For better manageability and to avoid wear of the sensor (fixed-frequency 9.5 GHz), a lift-off of some millimeters is used. However, the lift-off acts as an additional layer which influences the measured voltage. Therefore the lift-off has been optimized in such a way that its variation generates only a slight change of the measured voltage, whereas a variation of the anticorrosion layer thickness results in a much stronger change of the measured voltage. Fig. 5.16 shows that the measured voltage decreases with increasing coating thickness. In the optimized lift-off range the curves are relatively close to each other (Fig. 5.16(a)) although the influence of lift-off cannot be removed completely. However, in the nonoptimized range (Fig. 5.16(b)) the scattering of the curves is much higher.

In order to explain and illustrate the effects of lift-off theoretical calculations using a multilayer model (Sklarczyk et al., 1998; Orfanidis, 2014) have been performed as an example. The model takes into account all reflections and transmissions of EMW impinging on a system of four or five dielectric layers. The calculations show a complex behavior of both real and imaginary parts of the reflection coefficient for a variation of lift-offs and coating thicknesses. In Fig. 5.17 the real part of the reflection coefficient is displayed, consisting of an ensemble of maxima and minima and giving a qualitative explanation for the experimental findings in Fig. 5.16 with the narrowband 9.5 GHz sensor. It is to be considered that in Fig. 5.16 the lift-off is given by the distance between the antenna leading edge and the surface of the specimen. The true

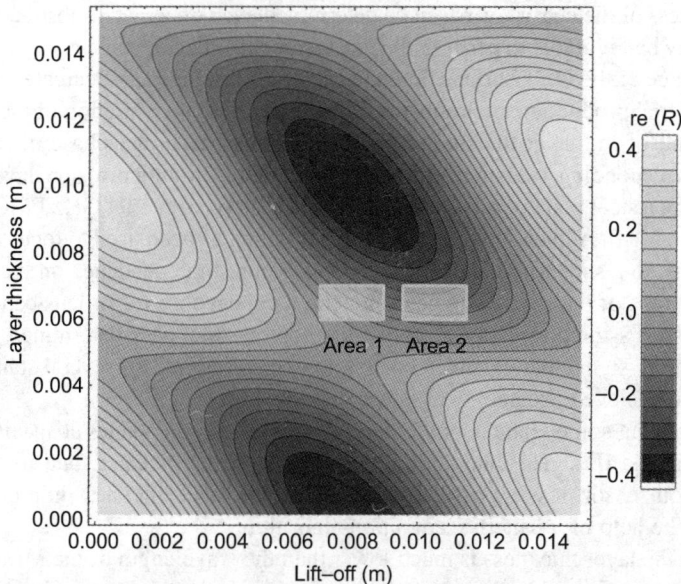

Figure 5.17 Calculated reflection coefficients (real part) as function of true lift-off and layer thickness at 9.5 GHz (planar waves with normal incidence). Area 1 and 2 approximately correspond to the variation of layer thickness and lift-off as given in Fig. 5.16(a) and (b), respectively. The permittivity of the layer was assumed to be 2.5.

lift-off, however, is bigger by about a 3 mm since the distance between the leading edge of the antenna and its reference point, which lies inside the wave guide, must be added.

As shown in Fig. 5.17 there are some domains with low sensitivity with regard to a change in lift-off and with a big sensitivity with regard to a change of the layer thickness, eg, near to $(x, y) = (0.005$ m, 0.003 m$)$ or $(0.014$ m, 0.003 m$)$. In doing so, different polarities of the slope can be found, eg, positive slope for lift-off 0.005 m and negative slope for lift-off 0.014 m at layer thickness 0.004 m. On the other hand, some other areas show big sensitivities with regard to both lift-off and layer thickness, eg, $(0.01$ m, 0.002 m$)$. No area can be found where the sensitivity with regard to the lift-off is zero while the sensitivity with regard to the coating thickness is maximal. But there are some lift-off areas which are more favorable than others for the sake of determination of coating thickness. The calculation in Fig. 5.17 cannot completely fit the experimental data since not all real physical influencing factors could be considered, eg, the shape of the wave fronts. But it can explain the principal effects.

In the publication of Hinken and Beller a plurality of parallel curves representing the real and imaginary parts of the reflection signal (11 GHz and adjacent frequencies) is used to measure the thickness of glass containers even for a variety of lift-offs of the microwave probe (Hinken and Beller, 2006). For this end it is necessary to perform many measurements at different thicknesses and lift-offs and to construct the curves by interpolation. The method is working for other insulating materials as well.

5.3.1.2 Dielectric layer on metal substrate

The thickness of dielectric layers can be determined even on a metal substrate. In some cases it may be favorable to perform the measurements in a noncontact way in order to avoid damage of the OUT surface. Then the standoff distance can fluctuate, especially in an industrial production environment. Here again, both the variation of the standoff distance and of the layer thickness influence the amplitude and phase of the MMT signal. Corresponding measurements have been done on anticorrosion layers made of plastic on a steel plate or pipe (Sklarczyk et al., 2006) with a 94 GHz FMCW radar sensor (Fig. 5.18). A quasi-optical experimental setup has been used to focus the millimeter waves to a spot with a diameter of about 5 mm. In dependence on the geometrical conditions of the measuring environment, a concave mirror or lens can be used. Fig. 5.19 shows the correlation between the coating thickness determined from the phase of the IF signal and the given coating thickness (target) for several standoff distances. The achieved accuracy was better than 0.1 mm.

By extracting some features of the radar IF signal (voltage values at specified locations within the IF signal) and combining them with the distance data from a laser triangulation sensor, it was possible to determine the layer thickness in a noncontact way with the help of a pattern recognition algorithm.

Even if the layer thickness is much lower than the wavelength of the MMT waves, in some cases it is possible to measure it. As an example the thickness of lacquer layers on a substrate of carbon fiber-reinforced polymer (CFRP) could be determined by means of microwaves using a resonance sensor (Hinken and Richter, 2012). The method can be applied for both flat and slightly curved specimens. It is not dependent

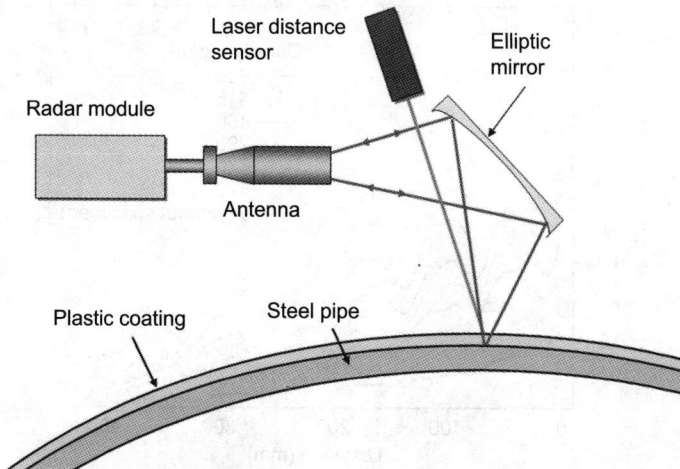

Figure 5.18 Noncontact measurement of the thickness of an anticorrosion coating on a steel plate or pipe with a 94 GHz-FMCW-radar.

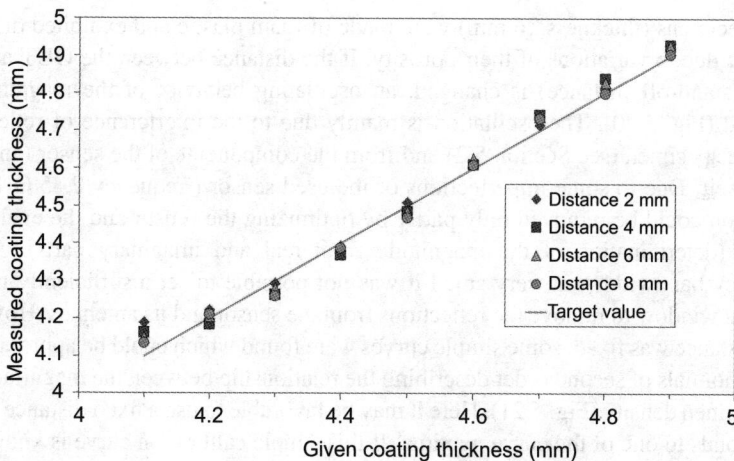

Figure 5.19 Measured versus given coating thickness. The measurements have been carried out with a compact 94 GHz-FMCW-radar with different lift-off distances.

on the permittivity of the paint but only on the distance (lift-off) between the sensor and the surface of the conductive CFRP. This distance is identical to the thickness of the lacquer layer since the resonance sensor is working in contact mode.

5.3.2   Determination of material properties

5.3.2.1   Density and porosity

Below, an example is given for the noncontact determination of the density and porosity of test objects with a fixed-frequency microwave sensor (Sklarczyk et al., 2006). The

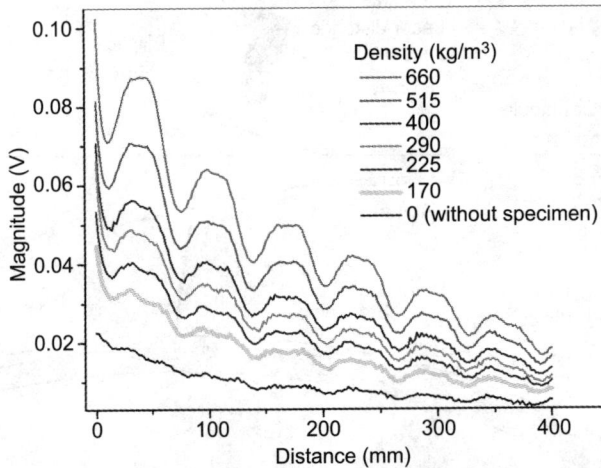

Figure 5.20 Magnitude as function of distance for foam plastic specimens with different densities.

used specimens (thickness 26 mm) were made of foam plastic and exhibited different densities due to variations of their porosity. If the distance between the OUT and the sensor (stand-off distance) is changed, an oscillating behavior of the magnitude is observed (Fig. 5.20). The oscillation is mainly due to the interference of reflections from the specimen (see Section 5.2) and from the components of the sensor (antenna, lines) itself. Due to some imperfections of the used sensor (frequency 2.45 GHz) the oscillation could be removed only partly by optimizing the sensor and the evaluation method (determination of the magnitude from real and imaginary parts). As the frequency bandwidth was very small it was not possible to set a sufficiently narrow temporal window to remove the reflections from the sensor and its antennas. However, if the distance was fixed, some simple curves were found which could be approximated by polynomials of second order describing the relationship between the magnitude and the specimen density (Fig. 5.21). Here it may be favorable to use a fixed distance which corresponds to one of the curve maxima. If this simple calibration curve is known the unknown density can be easily found only by measuring the magnitude provided that the measuring arrangement and the specimen dimensions are unchanged.

Even if the standoff distance changes, it is possible to determine the density of the specimens with the help of pattern recognition algorithms (Sklarczyk et al., 2006) or a neural network in a similar way as already mentioned above. Therefore, it is necessary to generate a sufficiently big training data set with well-characterized specimens and well-known geometrical conditions. The standoff distance has to be measured with a second sensor, eg, a laser triangulation sensor.

5.3.2.2 Area resistance

The penetration depth of MMT waves in metals is very low. However, if the metallic layer is sufficiently thin, it can be penetrated and the layer can be characterized with

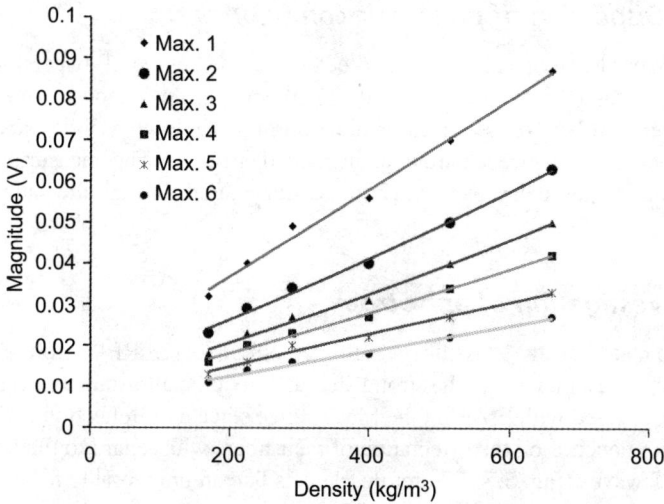

Figure 5.21 Magnitude as function of material density for the maxima in Fig. 5.20.

regard to its properties like the area resistance. In Fig. 5.22, paper sheets which were metallized with aluminum (layer thickness few μm) resulting in a linear relation between the measuring voltage and the area resistance. The measurements have been performed with a simple fixed-frequency sensor at 9.4 GHz in noncontact transmission mode. Similar results have been found on glass plates metallized with silver whose thickness amounted to few tens of nm (Sklarczyk, 2002).

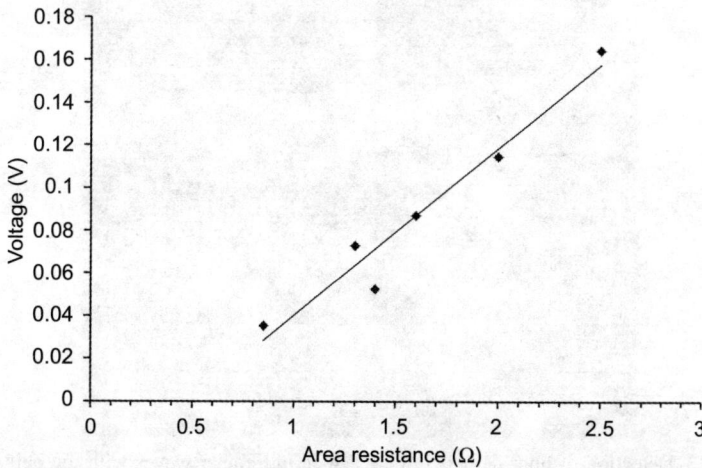

Figure 5.22 Measuring voltage as function of the area resistance in metallized paper sheets.

5.3.2.3 Dispersion of polymeric compounds

Terahertz spectroscopy has been used to determine the degree of dispersion of polymeric compounds (Krumbholz et al., 2011). Wood plastic composites (WPC) with poorly dispersed fillers or additives exhibited a bigger refractive index and a higher absorption than the same substrate with good dispersion. The measuring quantity used for that was the delay of the THz-pulse transmitted through the sample and its amplitude.

5.3.3 Investigation of anisotropy

Anisotropic materials like glass fiber-reinforced polymers (GFRP) are often not easily testable with ultrasound due to the strong attenuation of the ultrasound waves. But they can be characterized with EMW in the MMT range since their reflection and transmission behavior depends on the orientation of their fibers with regard to the polarization of the MMT waves (Fig. 5.7). Therefore, defects like an unfavorable fiber orientation or condition (eg, swirl; Dobmann et al., 2012) as well as diverse other types of defects (cracks, voids, delaminations, defective bonds and joints, foreign material, defective distribution and/or inadequately cured adhesive) are detectable. The example shown in Fig. 5.23 displays the area scan of a fiber swirl in a leaf spring made of GFRP at a frequency of about 94 GHz (noncontact mode). The fiber orientation in selected regions can be quantified with the help of a dedicated software. In Fig. 5.24 there are clear peaks of the distribution of fiber orientation (Regions 1 and 2) while the defective fiber distribution (swirl) in Region 3 does not exhibit any peak. The penetration depth of the millimeter waves used here is higher than in the optical wavelength region.

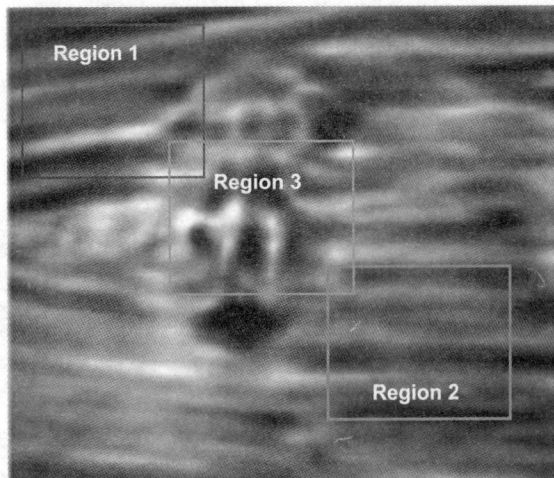

Figure 5.23 Detection of fiber defects in GFRP with millimeter waves with the help of an area scan.

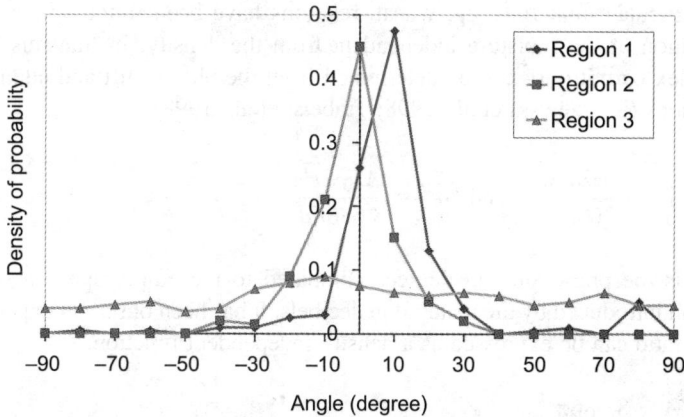

Figure 5.24 Distribution of fiber orientation in selected regions in Fig. 5.23. Region 1 represents a distinct but tilted orientation of the fibers in comparison to the main fiber orientation. Region 2 is the strongly defective fiber swirl while Region 3 is an undisturbed domain.

The measuring frequency is not critical for that type of inspection. In Beller et al. (2007) a similar image of the fiber orientation of GFRP could be obtained with a 10 GHz sensor in contact mode.

Defects under insulating protective covers which are undetectable with optical means can also be investigated. Defects like disbonds under the leading layer can be found since they result in a change of the standoff distance between the sensor and the OUT.

The fibers in CFRP are electrically conductive and thus almost totally reflect linearly polarized MMT waves if their orientation is parallel to their electric field vector. If the orientation is perpendicular, some penetration of the EMW is possible. Therefore, it depends on the type of layers and their orientation if interior defects can be investigated by MMT waves or not. By combining different polarizations in one single sensor at least the uppermost layers in CFRP composites are testable (Kharkovsky et al., 2008). On the other hand, disbonds may be found due to the change of the standoff distance (Akuthota et al., 2004) resulting in a change of the measured voltage.

5.3.4 Measurement of moisture

Measurement of moisture content is one of the most important applications of MMT technique as the permittivity of water is much higher than for most other dielectric materials (Nyfors and Vainikainen, 1989) in a broad frequency band. This physical peculiarity of water results in a higher reflectivity of moist materials compared to dry ones. Often, resonant moisture sensors are used, the resonant frequency of which is shifted and the quality factor of which is altered in dependence on the amount of moisture (see Section 5.2). The sensors can be equipped with various types of antennas which exhibit different penetration depths. This allows them to gain insight into the moisture content in different domains of depth (Göller, 1999).

For particulate materials, eg, wheat, relations have been found which allow the determination of the moisture independent from the density. In transmission mode the complex permittivity can be determined from the phase shift and attenuation by the equations (Kraszewski et al., 1998; Trabelsi et al., 1998):

$$\varepsilon' \approx \left(1 + \frac{\phi\lambda_0}{360d}\right)^2 \quad \text{and} \quad \varepsilon'' \approx \frac{A\lambda_0\sqrt{\varepsilon'}}{8.686\pi d} \qquad [5.39]$$

where ϕ is the phase shift in degrees compared to the empty space and A is the attenuation introduced by the material in decibels. It has been observed experimentally that their ratio can be expressed as a density-independent function:

$$X = \frac{A}{\phi} = \frac{8.686\pi}{360} \frac{\varepsilon''}{\sqrt{\varepsilon'\left(\sqrt{\varepsilon'} - 1\right)}} \qquad [5.40]$$

The calibration equation for moisture content M determination can have the following form:

$$M = \sqrt{aX + c} + b \qquad [5.41]$$

where a, b, and c are constants which have to be determined experimentally with the help of a set of specimens with known moisture content. Depending on the material to be tested, other types of equations for X and M can appear (Kraszewski et al., 1998; Meyer and Schilz, 1981).

In dependence on the respective application the measuring quantities attenuation and phase shift can be measured with horn antennas arranged in transmission mode similar to Fig. 5.8 (Trabelsi et al., 1998).

In many cases these methods can be applied using resonant sensors. However, for more complex geometries, eg, if the material to be tested is transported on a conveyor belt, the density-independent measuring methods may no more be feasible. Then the application of a multilayer feed forward artificial neural network can help to measure the moisture with high accuracy (Daschner et al., 2001). But the calibration effort is significant because the number of samples for the training of the network has to be large.

Another example for the application of the pattern recognition in the field of NDT with MMT waves is the determination of the water-to-cement (w/c) ratio in the concrete industry with an artificial neural network which has been used in the frequency range between 8.2 and 12.5 GHz (Hasar et al., 2009).

One possibility to estimate the moisture content in soils and buildings is to evaluate the reflection echoes from the interfaces by using a broadband device like a GPR. Whenever the moisture changes, the propagation velocity of the EMW changes, too, resulting in a temporal shift of the echo. There exist some techniques for the estimation of simultaneously permittivity and depth by using the echoes from reflective

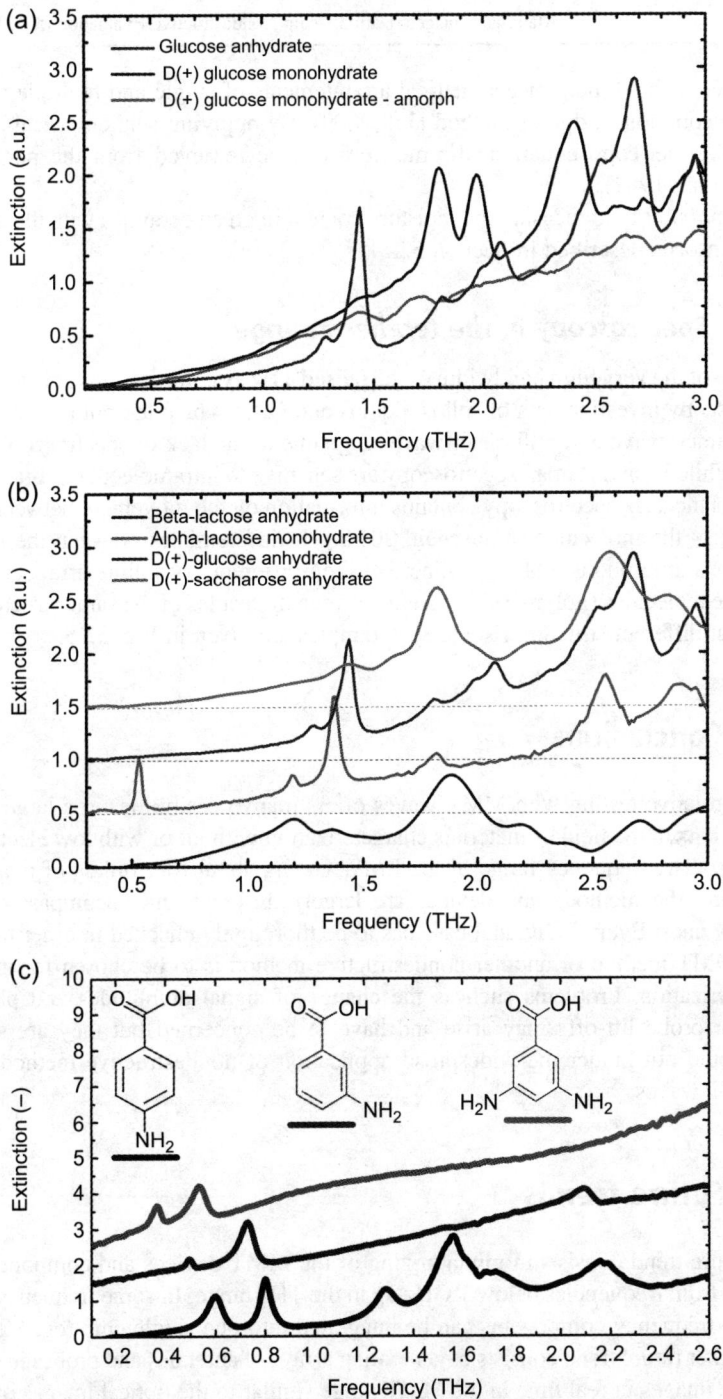

Figure 5.25 (a) Absorption spectra of amorphous and crystalline glucose; (b) Selected absorption spectra of diverse sugar types; (c) Extinction spectra of the three isomers PABA (black), MABA (light gray) and DiABA (gray).
Source: Jonuscheit, J., 2014a. Strukturanalyse mittels Terahertz. GIT Labor-Fachzeitschrift 5/2014, 27−29.

interfaces with the help of geometrical arrangements of single and multiple antennas like the common midpoint method (Jol, 2009). By applying semiempirical relationships like the Topp equation, the moisture can be retrieved from the permittivity (Topp et al., 1982).

The possibility to measure the moisture content in green ceramics with the help of a TDR sensor is described in Section 5.2.

5.3.5 *Spectroscopy in the terahertz range*

Because of the very high bandwidth of the pulsed THz systems it is possible to perform spectroscopy investigations of solids, liquids and gases which are not possible in that way in microwave and millimeter wave range due to the lack of spectroscopic signatures. While IR and Raman spectroscopy are sensitive to intramolecular vibrations and rotations the THz spectroscopy contains information on the movement between (intermolecular) the molecules (Jonuscheit, 2014a). It differentiates between the physical conditions amorphous and crystalline, between different crystalline arrangements of the same molecules (polymorphism) and between molecules of the same structural formula but different structure (isomers). Examples are given in Fig. 5.25.

5.4 Conclusions

Nondestructive testing with MMT waves offers many possibilities in a large field of applications in the field of materials characterization without or with low electric conducuvity. The frequency range of the EMW covers about four orders of magnitude. Therefore, the methods and devices are largely different and encompass classical and new ones. Every individual case has to be thoroughly checked in order to decide if the MMT method or another nondestructive method is to be chosen for materials characterization. Problems such as the change of signal amplitudes and phases at different probe lift-offs may arise and have to be concerned but they are solvable and should not hinder the widespread application of nondestructive methods based on MMT waves.

5.5 Future trends

One future trend consists in miniaturizing of the MMT devices and components and extends from frequencies below 1 GHz up to the THz range. In some frequency ranges all high-frequency components can be integrated into one single chip (eg, MMIC).

Another future trend consists in generating arrays of antennas and probes in order to produce images in real time in the MMT range similar to the optical images of digital cameras. This tendency may move away from the field of security, especially personnel screening and scanning, into the area of NDT. It is to be expected that the costs of these systems will considerably drop in future.

In this context the usage of computers and dedicated software is imperative, eg, for generation of 2-D and 3-D images with the help of synthetic aperture.

The application of sensors and probes which respond to other physical effects and their fusion into the nondestructive measuring system will enhance the capabilities of nondestructive materials characterization as well.

References

Ahmed, S.S., Schiessl, A., Gumbmann, F., Tiebout, M., Methfessel, S., Schmidt, L.-P., 2012. Advanced microwave imaging. IEEE Microwave Magazine 13 (6), 26–43.

Ahmed, S.S., Genghammer, A., Schiessl, A., Schmidt, L.-P., 2013. Fully electronic-band personnel imager of 2 m aperture based on a multistatic architecture. IEEE Transactions on Microwave Theory and Techniques 61 (1), 651–657.

Akuthota, B., Hughes, D., Zoughi, R., Myers, J., Nanni, A., 2004. Near-field microwave detection of disbond in carbon fiber reinforced polymer used for strengthening cement-based structures and disbond repair verification. Journal of Materials in Civil Engineering 16 (6), 540–546.

Al-Qadi, I.L., Jiang, K., Lahouar, S., January 22–26, 2006. Analysis tool for determining flexible pavement layer thickness at highway speed. In: Transportation Research Board 85th Annual Meeting. Washington, D.C., TRB 2006 Annual Meeting CD-ROM, paper no. 06–1923.

von Aschen, H., Gumbmann, F., Schmidt, L.-P., 2011. High resolution permittivity reconstruction of one dimensional stratified media from broadband measurement data in the W-band. In: Proceedings of the 8th European Radar Conference, Manchester, UK, October 12–14, 2011, pp. 45–48.

Beller, T., Hinken, J., Voigt, M., 2007. Hochauflösende Mikrowellen-Defektoskopie, DGZfP-Jahrestagung 2007-Vortrag 53.

Born, M., Wolf, E., 1986. Principles of Optics, sixth ed. Pergamon Press (Chapter 1).

Daniels, D.J., 1996. Surface-penetrating radar. Electronics & Communication Engineering Journal 8 (4), 165–182.

Daschner, F., Knöchel, R., 2001. Microwave moisture sensor with nonlinear data processing using artificial neural networks. In: SIcon'01 Sensors for Industry Conference, Rosemont, Illinois, USA, November 5–7, 2001, pp. 246–250.

Daschner, F., Knöchel, R., Kupfer, K., May 13–16, 2001. Resonator based microwave moisture meter with digital phase signal processing. In: 4th International Conference on Electromagnetic Wave Interaction with Water and Moist Substances, Weimar, Germany, pp. 125–131.

Dobmann, G., Altpeter, I., Sklarczyk, C., Pinchuk, R., 2012. Non-destructive Testing with Micro- and mm-Waves - Where We Are and Where We Go, Welding in the World 56, No. 1 and 2.

Elhawil, A., Koers, G., Zhang, L., Stiens, J., Vounckx, R., 2009. Comparison between two optimization algorithms to compute the complex permittivity of dielectric multilayer structures using a free-space quasi-optical method in W-band. IET Science, Measurement & Technology 3 (1), 13–21.

Friedsam, G.L., Biebl, E.M., 1997. Precision free-space measurements of complex permittivity of polymers in the W-band. In: IEEE MTT-S - International Microwave Symposium Digest, New York, NY, USA, vol. 3, pp. 1351–1354.

Ganchev, S.I., Qaddoumi, N., Bakhtiari, S., Zoughi, R., 1995. Calibration and measurement of dielectric properties of finite thickness composite sheets with open-ended coaxial sensors. IEEE Transactions on Instrumentation and Measurement 44 (6), 1023–1029.

Ghasr, M.T., Abou-Khousa, M.A., Kharkovsky, S., Zoughi, R., Pommerenke, D., 2012. Portable real-time microwave camera at 24 GHz. IEEE Transactions on Antennas and Propagation 60 (2), 1114−1125.

Göller, A., 1999. Moisture mapping - Flächen- und tiefenaufgelöste Feuchtemessung mit dem Moist -Verfahren, Feuchtetag '99. In: Umwelt Meßverfahren Anwendungen, Oktober 7/8, 1999. BAM, Berlin. DGZfP-Berichtsband BB 69-CD, Poster 10.

Hasar, U.C., Akkaya, G., Aktan, M., Gozu, C., Aydin, A.C., 2009. Water-to-cement ratio prediction using ANNS from non-destructive and contactless microwave measurements. Progress in Electromagnetics Research, PIER 94, 311−325.

Hinken, J., Beller, T., 2006. Contactless thickness measurements of glass walls by using micro- wave reflections. In: 9th European Conference on NDT Berlin, September 25−29, 2006.

Hinken, J., Richter, M., 2012. Non-destructive measurement of paint thickness on curved CFRP surfaces. In: 4th International Symposium on NDT in Aerospace 2012 − Tu.2.B.2.

Ida, N., 1992. Microwave NDT. In: Developments in Electromagnetic Theory and Applications. Kluwer Academic Publishers, p. 10.

Jol, M. (Ed.), 2009. Ground Penetrating Radar: Theory and Applications. Elsevier (Chapter 7).

Jonuscheit, J., 2014a. Strukturanalyse mittels Terahertz. GIT Labor-Fachzeitschrift 5/2014, 27−29.

Jonuscheit, J., 2014b. Zerstörungsfreie Analyse. QZ 59 (5), 94−96.

Kharkovsky, S., Stephen, V., Ryley, A.C., Robbins, J.T., Zoughi, R., 2008. Preservation of Missouri Transportation Infrastructures: Life-cycle Inspection and Monitoring of FRP-strengthened Concrete Structures Using Near-field Microwave Nondestructive Testing Methods. Report No. OR 09−008. Center for Infrastructure Engineering Studies, Missouri University of Science & Technology.

Klausing, H., Holpp, W. (Eds.), 2000. Radar mit realer und synthetischer Apertur. Oldenbourg (Chapter 4).

Kraszewski, A.W., Trabelsi, S., Nelson, S.O., 1998. Comparison of density-independent ex- pressions for moisture content determination in wheat at microwave frequencies. Journal of Agricultural Engineering Research 71, 227−237.

Krumbholz, N., Hochrein, T., Vieweg, N., Radovanovic, I., Pupeza, I., Schubert, M., Kretschmer, K., Koch, M., 2011. Degree of dispersion of polymeric compounds deter- mined with terahertz time-domain spectroscopy. Polymer Engineering and Science 51 (1), 109−116.

Kupfer, K. (Ed.), 2005. Electromagnetic Aquametry, Springer (Chapter 6).

Marion, J.B., Heald, M.A., 1980. Classical Electromagnetic Radiation, second ed. Academic Press.

McMakin, D.L., Sheen, D.M., Hall, T.E., Kennedy, M.O., Foote, H.P., 2007. Biometric iden- tification using holographic radar imaging techniques, sensors, and command, control, communications, and intelligence (C3I) technologies for Homeland security and Homeland Defense VI. In: Carapezza, E.M. (Ed.), Proceedings of SPIE, vol. 6538, 65380C.

McMaster, R.C., McIntire, P. (Eds.), 1986. Nondestructive Testing Handbook. Electromagnetic Testing, ASNT, Section 18, Microwave Methods and Applications in Nondestructive Testing, vol. 4.

Meyer, W., Schilz, W.M., 1981. Feasibility study of density-independent moisture measurement with microwaves. IEEE Transactions on Microwave Theory and Techniques, MTT 29 (7), 732−739.

Nyfors, E., Vainikainen, P., 1989. Industrial Microwave Sensors. Artech House (Chapter 3).

Orfanidis, S.J., 2014. Electromagnetic Waves and Antenna (online), Available: http://www.ece. rutgers.edu/~orfanidi/ewa/.

Shangguan, P., Al-Qadi, I.L., Lahuar, S., 2014. Pattern recognition algorithms for density estimation of asphalt pavement during compaction: a simulation study. Journal of Applied Geophysics 107, 8−15.

Sklarczyk, C., Ehlen, F., Netzelmann, U., 1998. Schichten charakterisieren. Materialprüfung 40, 149−153.

Sklarczyk, C., Pinchuk, R., Melev, V., 2006. Nondestructive and contactless materials characterization with the help of microwave sensors. In: Proceedings of ECNDT 2006, Berlin, September 25−29, 2006, Tu.1.8.3.

Sklarczyk, C., Tschuncky, R., Melev, V., Boller, C., 2013. Mikrowellensensoren zur zerstörungsfreien Materialuntersuchung. In: Berichtsband zur 7th CMM Tagung, Weimar, Germany, September 24, 2013, pp. 85−96.

Sklarczyk, C., 2002. Contactless characterization of coatings with a microwave radar sensor, nondestructive characterization of materials XI. In: Proceedings of the 11th International Symposium, Berlin Germany, June 24−28, 2002. Springer, pp. 743−748.

Su, K., Shen, Y.-C., Zeitler, J.A., 2014. Terahertz sensor for non-contact thickness and quality measurement of automobile paints of varying complexity. IEEE Transactions on Terahertz Science and Technology 4 (4), 432−439.

Tessmann, A., Kudszus, S., Feltgen, T., Riessle, M., Sklarczyk, C., Haydl, W.H., 2002. Compact single-chip W-band FMCW radar modules for commercial high-resolution sensor applications. IEEE Transactions on Microwave Theory and Techniques 50 (12), 2995−3001.

Topp, G.C., Davis, J.L., Annan, A.P., 1982. Electromagnetic determination of soil water content using TDR: I. Applications to wetting fronts and steep gradients. Soil Science Society of America Journal 46, 672−678.

Trabelsi, S., Kraszewski, A.W., Nelson, S.O., 1998. A microwave method for on-line determination of bulk density and moisture content of particulate materials. IEEE Transactions on Instrumentation and Measurement 47 (1), 127−132.

Trinks, E., Kupfer, K., Oktober 7/8, 1999. Erfassen dielektrischer Stoffeigenschaften mit Resonatoren. Feuchtetag '99. BAM, Berlin, DGZFP Berichtsband BB 69-CD, Vortrag M3.

Venkatesh, M.S., Raghavan, G.S.V., 2005. An overview of dielectric properties measuring techniques. Canadian Biosystems Engineering 47, 7.15−7.30.

Yamaguchi, Y., Sengoku, M., Motooka, S., 1996. Using a van-mounted FM-CW radar to detect corner-reflector road-boundary markers. IEEE Transactions on Instrumentation and Measurement 45, 793−799.

Zech, C., Hülsmann, A., Weber, R., Tessmann, A., Wagner, S., Schlechtweg, M., Leuther, A., Ambacher, O., October 7−10, 2013. A compact 94 GHz FMCW radar MMIC based on 100 nm InGaAs mHEMT technology with integrated transmission signal conditioning. In: Proceedings of the 43rd European Microwave Conference, Nuremberg, Germany, pp. 1407−1410.

Zhang, Y., Okamura, S., 2006. A density-independent method for high moisture content measurement using a microstrip transmission line. Journal of Microwave Power & Electromagnetic Energy 40 (2), 110−118.

Zoughi, R., 1990. Microwave nondestructive testing: theories and applications. International Advances in NDT 15, 255−288. Gordon and Breach.

Acoustic microscopy for materials characterization

6

R.Gr. Maev
University of Windsor, Windsor, ON, Canada

6.1 Introduction

It is well known that one picture is in fact worse than 1000 words. Imaging technology is largely based on manipulating optical waves, but since optics does not provide all of the information we need, at the end of twentieth century, scientists turned to other alternative solutions; one of them was acoustical imaging. Today, acoustic imaging is an integral and important part of the continuing effort to extend our ability to "see." Ultrasound renders real-time images where the operator can dynamically select the most important sections for documenting the changes in microstructure of an object as a result of the impact of physical or chemical factors. Breakthrough developments in high resolution acoustic imaging technology in the early 1970s facilitated advanced acoustic imaging methods for the examination of the internal microstructure of non-transparent solids and the monitoring of internal stress. In addition to measuring elastic properties, this technique is also used to examine adhesion in multilayered structures and has many other applications. Acoustic microscopy has become not only a new imaging method extensively used in many areas of physics, biology, and industrial technology but also a new and efficient tool of quantitative characterization of the microstructure of various materials (Maev, 2008).

The role of high-resolution ultrasonic imaging in academic condensed matter study and various applications for microstructural materials characterization in solid-state physics, material sciences and technology is rapidly increasing. The whole spectrum of original physical and methodological approaches to ultrasonic imaging results brought a significant improvement to the quality of the developed technology. New generations of ultrasonic imaging systems continue to decrease in size and will soon enter the realm of pocket-sized dimensions. New transducer materials, including advanced composites, recent MEMS applications, powerful custom microchips, etc.—all of these factors contribute to substantial changes in the design and efficiency of high-resolution ultrasonic imaging systems.

Novel physical solutions, including new results in the field of adaptive methods and inventive approaches to inverse problems, original concepts based on high harmonic imaging algorithms, impressive concept of vibroacoustic imaging and vibromodulation technique, etc., were successfully introduced and verified in numerous studies of industrial materials and biomaterials in the past few years. Together with the above-mentioned traditional academic and practical avenues in ultrasonic imaging research, intriguing

Materials Characterization Using Nondestructive Evaluation (NDE) Methods
http://dx.doi.org/10.1016/B978-0-08-100040-3.00006-7

scientific discussions have surfaced in various fields and will hopefully continue to bear fruit in the future (Maev, 2013).

Another important achievement is that over the past few decades, the ultrasonic technique has become a common instrument for the industrial nondestructive evaluation (NDE) of materials, parts, and assemblies. However, huge demand still exists for nondestructive methods in quality monitoring tasks. Today's technological and industrial advancements bring new standards and requirements for NDE inspection. Almost each area of manufacturing requires the monitoring of properties and composition of materials as well as structural integrity (Bray and Stanley, 1996; Maev, 2007). Microchip manufacturing, emerging nanotechnologies and even conventional joining methods benefit from cost-saving NDE and quality control. A wide variety of advanced materials characterization methods are used to obtain specific information about the quality of materials (Bray and Stanley, 1996). Eddy currents, ultrasonic, X-rays, magnetic resonance and other techniques are widely used. Each has its own advantages and drawbacks. Ultrasonic methods benefit from simplicity, safety, and cost efficiency. The line of ultrasonic devices starts from simple pulse-echo flaw detectors and includes complex multichannel systems and, obviously, scanning acoustic microscope with sophisticated transformation of ultrasonic waves and received signals; each particular task requires optimized configuration of the corresponding acoustic microscopy instrument. In particular, requirements for high spatial resolution can be met by using a high frequency and focused acoustical beam. Data acquisition during the motion of the ultrasonic beam adds imaging capability, especially important for fast interpretation of the test results. More complex and sophisticated algorithms of signal processing provide additional information about viscoelastic or even anisotropic or nonlinear properties of materials.

This chapter offers a number of new results that show aspects of the development of novel physical principles, new methods or implementation of modern technological solutions into current acoustic imaging devices and new applications of high-resolution imaging systems. The goal of this chapter is to provide an overview of the recent advances in high-resolution ultrasonic imaging techniques and their applications to materials evaluation and industrial materials. In this chapter, a collection of novel results and techniques in SAM are brought together that were developed by leading research groups worldwide.

6.2   Basic principles

SAM hardware combines a mechanical scanning system and an ultrasonic emitting–receiving system. A short pulse of ultrasonic wave generated by the piezoelectric transducer penetrates into the tested sample. Variations in the acoustical impedance at the boundaries of the sample structure cause reflection and scattering of the sound. These scattered waves are received and used to produce the output scan. With the motion of the transducer from point to point along the surface, it is possible to register the spatial distribution of the ultrasonic characteristics of the specimen represented in the obtained images. Spherical focusing of acoustic beam by use of specifically designed transducers (acoustic lens) greatly increases sensitivity and provides spatial resolution compatible to sound wavelength. Acoustic contact between the lens and the specimen is achieved by immersing them in liquid (usually water).

The reflection mode is the most common and appropriate method to perform most of the testing. The same acoustic transducer is used to send and receive sound waves. The received echoes are converted to electrical signals, amplified and digitized. The digital waveforms are sent to the computer for further processing and can be presented to the user as raw waveforms (A-scans) or processed to provide an acoustical image. The two most widely used types of acoustical images assembled from the sequence of A-scans as the transducer is moving along the surface of interest are:

- B-scan is the 2-D image formed during a single linear motion of the transducer along the sample surface. Each vertical line of pixels is an individual A-scan in amplitude-color representation. The position of this line in a horizontal direction corresponds to the position of the acoustical lens over the object. Actually the B-scan mode is the cross-sectional display showing the ultrasonic wave reflections at different interfaces through the thickness of the object. It shows the location of interfaces in the material at various depths along the chosen direction.
- C-scan is the 2-D image where the color of each pixel represents the amplitude of signal within a specific gated depth range as the transducer scans the area of interest in a raster pattern. Usually, C-scan represents the cross-section of the material in the plane perpendicular to the direction of the A-scan, or nearly parallel to the surface of the specimen. It shows 2-D microstructure details of the material at selected depths.

The acoustic microscope is the device which very often can bring a lot of advantages while having a relatively simple design and low cost. Its resolution can reach submicron ranges; it is fast and easy to use, is safe for personnel and is nonhazardous for medical and biological specimens. While the principles of operation are still the same since the first acoustic microscope was introduced by Cal Quate (Stanford University, 1974), the design and characteristics of particular devices are widely varied (Maev, 2008; Bray and Stanley, 1996; Maev, 2007). Different requirements and application areas are reflected in corresponding technical solutions. Several years ago, Tessonics, Inc. (Windsor, ON, Canada), a manufacturer of equipment for industrial ultrasonic inspection, introduced to the market its own pulse-echo acoustic microscope, combining a universal approach that is highly adaptive to specific applications.

6.3 Resolution and contrast mechanisms

The contrast of an acoustical C-scan comes from variations of the reflective coefficient, ie, variations of specific acoustic impedance across the sample. However, advanced numerical processing allows working with thin samples and extracting important information in case of interleaving signals. Their interference is determined by the material's acoustic properties: interference in frequency is determined by the thickness and sound speed of the sample, interference in intensity by the surface reflection coefficient and sound attenuation in the material. Quantitative measurement of the sound speed in this case is based on analysis of the interference of frequency-dependent characteristics and it can be obtained either by serial measurements with varying frequencies or by fast Fourier transform of a single broadband pulse.

Lateral resolution of ultrasonic imaging is determined by the width of the acoustic beam. Spherical focusing enables reducing it down to the theoretical limit, which is roughly equal to the wavelength multiplied by the angular aperture. High resolution requires waves with short wavelength and corresponding high acoustic frequency. As a rule of thumb, the size of the visualized objects entirely determines the minimal operational frequency of the ultrasonic imaging system. Attenuation of ultrasound waves in tissue sharply increases with frequency, which limits the imaging depth for ultrasonic devices. A proper compromise between the opposing demands for a large penetration depth and good spatial resolution has to be found for each particular application.

Classical acoustic microscopy expects achievement of maximal resolution only at a certain depth, determined by the focal distance of the acoustic lens. Imaging the entire volume of thick samples requires repeated scanning with the focus located at different depths in order to maintain resolution along the axial direction. The final image should be composed from subsequently acquired short B-scans. Synthetic aperture focusing technique (SAFT) is another approach for depth-invariant imaging, which involves numerical processing of the B-scans from the aggregate of acquired data. According to this technology, the focal point is modeled as a point source of acoustic far-field with the transducers opening angle (Maev, 2008)—each individual point of the image is calculated by summing up echo signals with a corresponding delay. Further, 2-D scanning in both lateral and elevation directions allows obtaining images with constant high resolution along both coordinates. As an additional benefit, this method allows for significantly extended range of the system beyond the focal depth of the transducer. Taking into account diffraction effects provides further improvement of acoustical images, eg, a noticeable increase in contrast and signal-to-noise ratio (SNR) (Maev, 2013).

A variation of this method is the limited angle spatial compounding (LASC) technique. It includes multidirectional echo measurement and incoherent superposition of B-mode images from different directions. Instead of linear scanning in the elevation direction, the focused transducer is tilted in the plane of the B-scan. The sequence of SAFT-processed B-mode images obtained by lateral scanning at various tilting angles should then be combined into one output image. The compound B-mode image is obtained by the superposition of scan-converted B-mode images (Maev, 2008). Advantages of this approach are significant improvements in contrast, suppression of speckle and electronic noise, and reduction of image artifacts.

6.4   Tessonics AM1103 acoustical microscope

The design of the microscope is aimed at nondestructive material evaluation and laboratory investigations of the internal structure for both industrial and biomedical samples. The device has a compact desktop design and offers wide possibilities for 3-D data analysis (see Fig. 6.1). It provides precise scanning and positioning of the acoustical lens in a 3-D volume $100 \times 150 \times 50$ mm with repeatability not worse

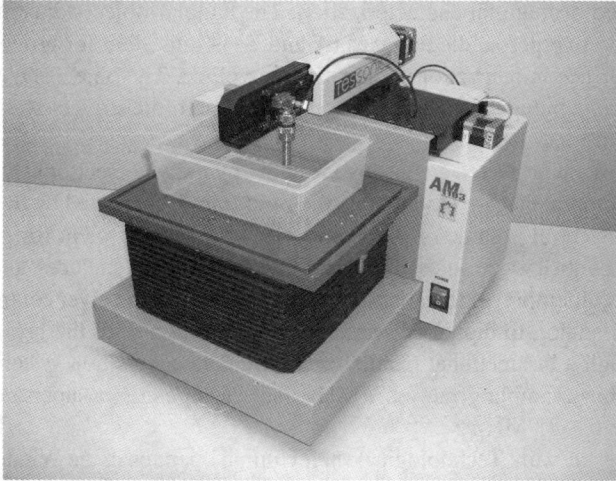

Figure 6.1 Tessonics AM1103 acoustical microscope.

than 2 μm in the X and Y directions and not worse than 0.2 μm in the Z direction. Typically, the microscope is operated with spherically focused transducers in the frequency range of 5−400 MHz.

The device can operate in two modes. The standard operation is a pulse-echo regime when the same transducer is used for the irradiation of the acoustic wave and for receiving a backscattered signal. An additional possible pitch-catch mode requires two separate transducers with corresponding mounting hardware. The standard commercially available immersion transducers (lens) with weight up to 200 g can be fastened by an axial UFH male connector. This feature provides the possibility of an electrical connection both through the main connector and through the side-mounted Microdot connector.

The acoustical lens is driven by the pulse generator, which provides a short negative square pulse with an amplitude of 120 V at 50 Ω load. Pulse duration can be electronically adjusted for proper transducer matching. Receiver input low-noise preamplifier provides an overall gain up to 50 dB in frequency range 1−400 MHz which is digitally adjustable to deliver optimal signal amplitude for digitizer circuitry.

Two-dimensional motion and positioning of the lens in the plane normal to the axis of the acoustic beam (XY plane) is provided by the main 2-D mechanical scanner. The fast X-axis was designed in outrigger form to provide maximum convenience for the operator and give three-side access to the object. The lens has a velocity up to 20 cm/s in this direction. The bigger and more rigid stage is used for "slow" Y-axes. The motion smoothness for both coordinates is provided by using microstepper-driven leadscrews and linear ball bearing.

The important part of the device conception is the motorized object stage, which provides vertical movement of the specimen (Z-axis) and angular adjustment of its surface. The small vertical step size makes possible not only the focusing of the ultrasound beam on the prescribed depth inside the sample but also the realization of V(z) method of material characterization. The angular object adjustment ± 10 degree relative to two perpendicular axes (A and B) is important for proper leveling of the sample surface in respect of the lens moving plane. The object stage is designed as a functionally independent module including electronic drivers and can be used separately if needed.

All five axes are protected against overrunning by the optoelectronic limit switches. Mechanical components are mounted on the rigid frame, supported by the four adjustable levelers. Central volume of the frame forms the compartment for power supply modules and control electronics. The main controller board includes a digital signal processor, flash memory, external bus for motion and pulse/receiver control, computer interface, and analog-to-digital converter. The connection with the host computer is realized through a bidirectional parallel port. The data acquisition system has a resolution of 10 bits, sampling rate 66.7 MHz and can operate in supersampling mode with a rate up to 533 MHz.

Controlling software Technological shell Unimic operates in the Windows environment. The embedded program module loads to the controller flash memory during board initialization. The main program has a user friendly and intuitively understandable interface system. Basic regimes of device operation include manual control of lens position, A-scan, when the oscillogram window in live mode is used for control of the relative position of reflected signals; B-scan with data acquisition during the lens movement along X- or Y- axes; and C-scan regime, in which 2-D raster scanning of a rectangular area is realized (see Fig. 6.2). The last mode is most useful because

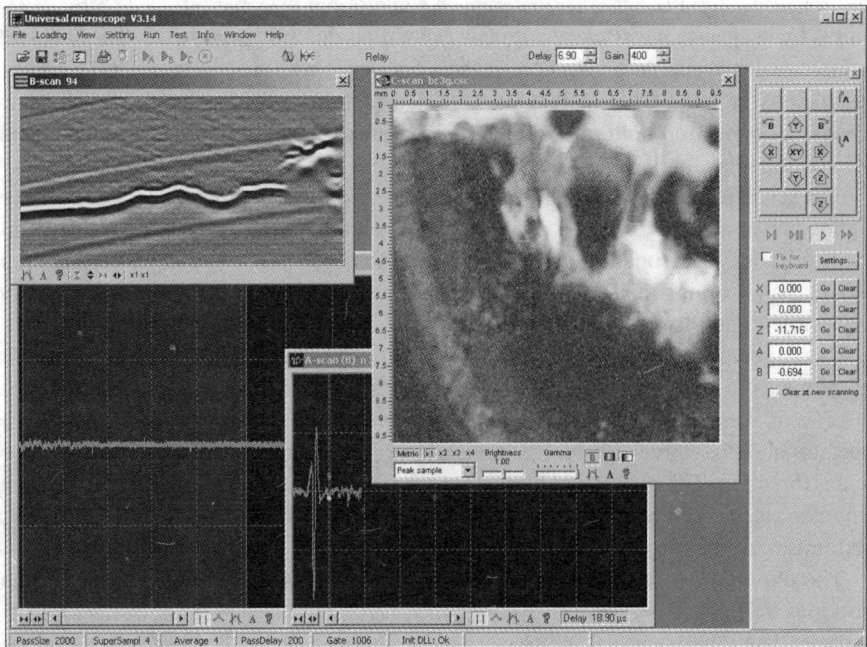

Figure 6.2 Screenshot of Unimic software.

the full information about the sample volume under the scanning area is collected and stored. The unique feature of this device is the differential mode when the difference between real-time signal and previously stored waveform is registered. This makes it possible to remove interfering responses and display the tiny elements of the sample structure. The sets of pop-up menu and control panels allow for fine adjustments of the hardware and program settings.

The whole device is simple in operation and does not require sophisticated maintenance and service.

Most of the commercially available SAM software is capable to present only B-scan or C-scan but not both simultaneously due to the large amount of data collected. In fact, scanning provides a 3-D cube of data, and the Tessonics AM1103 microscope has the capability to save all this data on the computer hard drive. The advantage of storing all three dimensions of data is that they can be processed later, without physical access to the specimen. This processing may be done by a separate software and can be as sophisticated as necessary.

6.5   Materials characterization for industrial applications

In most cases, material evaluation is performed for two kinds of materials essentially different in their ultrasonic properties: metals and plastics. In both cases, the object of inspection may be volumetric integrity (detection of porosity or heterogeneous inclusions) or quality of joints. Exact inspection parameters (frequency, aperture, focal distance, etc.) are determined by the nature of the expected defects. A frequency range 5−50 MHz is suitable for most of the metal samples. The resolution of the obtained acoustical images may reach a value of 0.1 mm (50 MHz near sample surface) and noticeably decreases with focusing depth. Surface roughness also may drastically reduce the resolution and detection thresholds. In practice, standard acoustic lenses with aperture 3−12 mm allow to confidently inspect metal samples with thicknesses up to 40−50 mm. As an example, Fig. 6.3 shows acoustical images (C-scan and

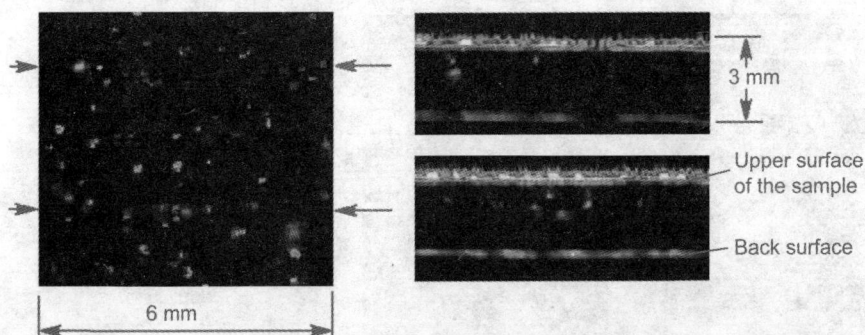

Figure 6.3 C-scan and two B-scans of aluminum casting sample. White dots represent porosities.

two B-scans) of porosities in aluminum castings at 50 MHz. Pores of size 0.1—0.5 mm are clearly visible; their position and distribution can be precisely determined.

6.6   Scanning acoustic microscopy inspection of spot welds

The first results related to the development of inspection methods for resistance spot welding based on high-resolution acoustic microscopy were first published in 1996 (Maev, 2007; Maev et al., 1996). Further studies demonstrated the great potential of this approach, and significant progress in the development of quality evaluation algorithms based on SAM images has been made (Lee et al., 2003; Pthchelinsev and Maev, 2001; Chertov et al., 2007; Chertov and Maev, 2005).

Strong reflection of the ultrasonic wave at the interface between welded metal sheets allows a clear distinction between the weld and no-weld zones. The absence or presence of the reflected pulse at that specific depth corresponds to the absence or presence of the physical discontinuity. Fig. 6.4 presents typical B-scans obtained outside and across the weld. Here the grayscale color of each pixel represents the amplitude of the A-scan. White corresponds to positive values; black corresponds to negative values of amplitude. Obviously, the visible disappearance of the reflected signal in the area of a weld nugget indicates metal integrity here.

In order to have a clear C-scan image of the weld, the data gate should be positioned at a specific depth, corresponding to the thickness of the first sheet. The typical C-scan shown in Fig. 6.4 was obtained for the good, sound weld. The black area indicates the absence of the reflection at the level of the metal—metal interface. The size measurement can be done using the built-in software instruments of the microscope as shown in Fig. 6.5 for the C-scan and central B-scan, respectively. The acoustical measurement of 4.5 mm is very close to the physical size of 4.6 mm obtained after the peel test. This is an important result which could be used for other applications as well.

These C- and B-scans were both obtained with 20 μm resolution for better image representation. However, 100 μm is enough for size and shape measurements and it

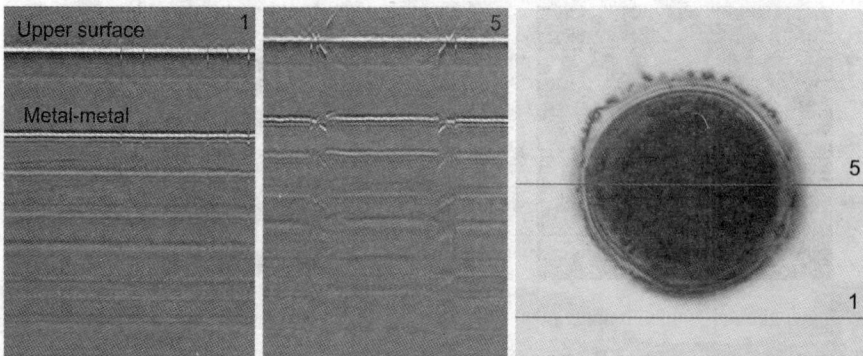

Figure 6.4 Spot weld: B-scans obtained outside the weld (1); across the weld (5); and C-scan.

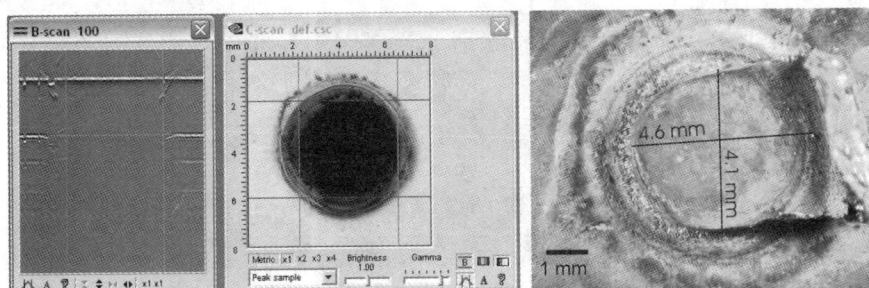

Figure 6.5 Measuring the nugget on the C-scan, on the B-scan and during destruction testing.

requires much less time. Extensive tests were performed to estimate the difference between the microscope and the peel test results. A series of welds with different diameters was produced using mild galvanneal steel, 1.2 mm thick. The SAM measurements were compared with the destructive test measurements. It was found that the smaller weld produces a bigger relative error but the actual difference rarely exceeds 0.25 mm, indicating the precision of the method.

The shape of the weld is also an important issue as this affects the weld integrity and strength. Improper electrode alignment, electrode degradation and/or contamination, dirty or oily plates often cause an odd shaped weld. The total area of the nugget could be big enough but the nugget could be unacceptable. Examples of such irregular shaped welds and their acoustic images are shown in Figs. 6.6 and 6.7. The shape of the weld can be easily revealed with the scanning acoustic microscope.

Another important feature of acoustic microscopy is its capability to reveal internal discontinuities and voids inside the volume of the weld. Voids, cracks, and bubbles contribute to the mechanical degradation and possible failure of the weld under stress. For example, the oval-shaped weld in Fig. 6.7 was produced with deliberately contaminated electrodes. The uneven redistribution of the current density led to the irregular weld shape and formation of the void in the bulk volume of the weld nugget. The acoustic microscope can easily show such cavities in the volume as the white spot inside the

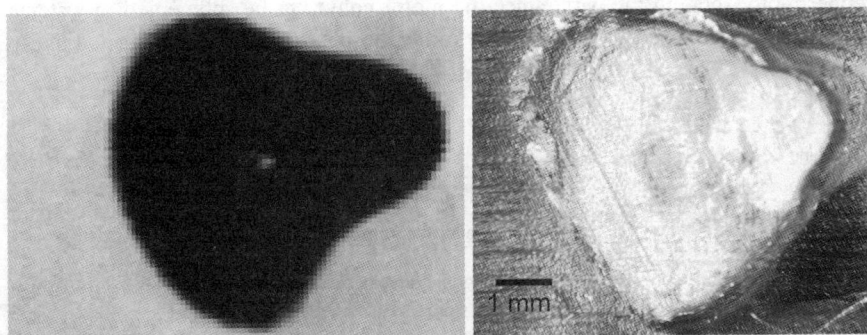

Figure 6.6 Odd shaped weld. Acoustic image, 7 × 10 mm (left), and real image (right) of the weld.

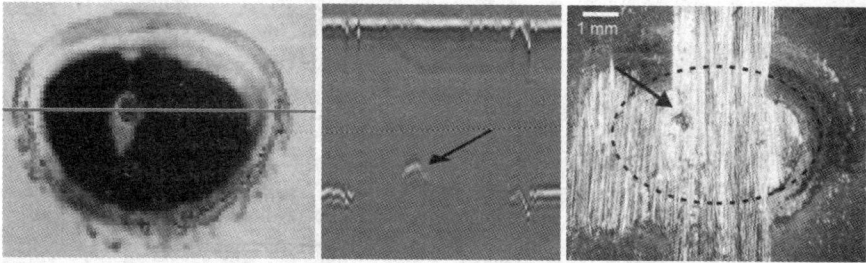

Figure 6.7 Oval-shaped spot weld with the void inside. C-scan, B-scan, and machined-off nugget.

black oval of the weld nugget. The standard destructive peel test tells nothing about possible voids inside the metal. Only gradual removal of the weld material made it possible to find the small void exactly at the predicted location. The acoustic microscope reveals such imperfections easily and quickly (within 20–25 s of scanning).

6.7 Real-time imaging of seam welds

Linear welding of metal plates, especially butt welding, is a more challenging configuration for acoustic microscopy. The bond interface is normally oriented to the outer surface and in general requires more complex microscope setup with two inclined acoustical lenses operating in pitch-catch mode. In some particular cases the weld seam may be inspected directly in reflection mode. For example, the specific shape of some samples (powertrain parts) allows side access to the weld seam and B-scan in axial direction and easily reveals its depth (Fig. 6.8).

The cylindrical nature of this part necessitates rotating rather than moving the scanning arm across the sloping surface in order to get 2-D scanning. As a result, a special rotating fixture was fabricated and installed on the Tessonics AM1103 instead of the regular Z-stage (Fig. 6.9). The scanner was then rewired so that the Y-control circuitry, which would normally have moved the scanning arm laterally, would instead drive the stepper motor rotating the part. Since the motor could not be submerged in water, a pump system was used to deliver water into the area of acoustic contact. Finally,

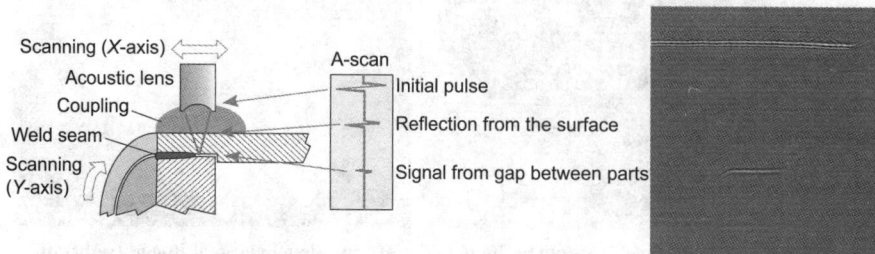

Figure 6.8 Side assess to the weld seam and obtained B-scan.

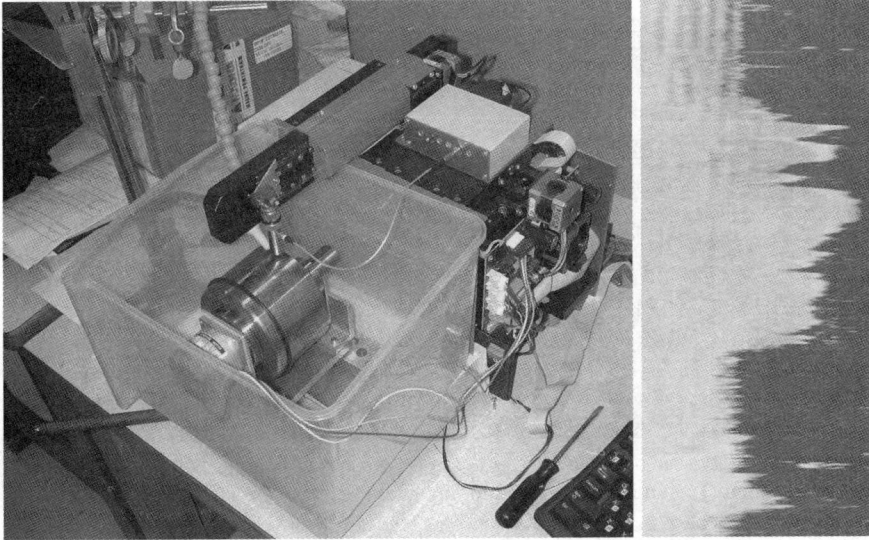

Figure 6.9 Microscope adaptation for cylindrical parts.

the C-scans that showed welding details all around the sample were obtained. As usual, dark areas on the scans represent properly welded regions, whereas light areas mean completely unbounded regions (Fig. 6.9).

6.8 Imaging of adhesive bonding between various metals

Adhesive bonding becomes an increasingly important part of industrial joining technology due to its lightness, improved load distribution, reduced corrosion, extended durability, and insulation. The use of adhesive bonding in combination with traditional joints (spot welds and rivets) has significant advantages for automotive production (Maev et al., 2012). The problem of nondestructive quality assessment of thin metal sheets joined using a layer of polymer adhesive is complicated by several factors. First, there is a high acoustic mismatch between the layers: 4.3×10^7 kg/(m^{2}s) for steel versus 0.3×10^7 kg/(m^{2}s) for epoxy-based structural adhesive. Second, there is high attenuation inside the adhesive layer itself. The attenuation coefficient is almost 0.76 mm^{-1} at 7.5 MHz, and even worse for higher frequencies (Hagemaier, 1996). As a result, a weak signal reflected from the lower interface is totally covered by strong reverberations in the upper sheet (Maeva et al., 2004). Proper interpretation of these complicated and cluttered A-scans requires careful analysis (Fig. 6.10).

The first and second pulses represent reflections from the upper and lower metal sheet surfaces. The operational frequency should be high enough for their separation and the optimal frequency range for metal sheets with thickness of 0.7—2.0 mm is 20—50 MHz. Mapping of the second pulse amplitude (C-scan) gives images of

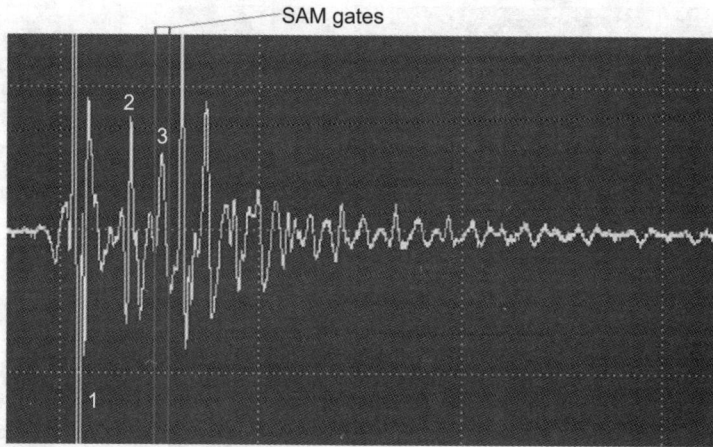

Figure 6.10 The A-scan obtained on steel-epoxy-steel adhesive bonding with 25 MHz spherical lens. Focal point (12 mm in water) tuned to first interface.

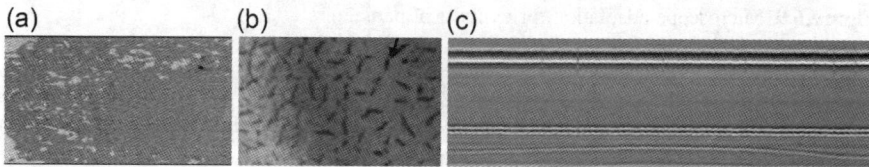

Figure 6.11 (a) The C-scan images representing water penetration into the epoxy—metal interface, first interface image; (b) crack appearance in the adhesive layer after bonded sample was exposed to forming operation, second interface image; (c) B-scan image demonstrates the adhesive layer thickness variations of cross-section through the sample.

interface conditions and voids (Fig. 6.11). Significant variations in adhesive parameters (curing state, density) can also be detected (Maeva et al., 2007). The contrast of images (the difference in reflected amplitude between areas with good bonding and bad quality bonding) is approximately 10%. The third pulse, which corresponds to reflection from the bottom of the adhesive layer, usually is very small in amplitude and often masked by strong reverberation oscillations. This pulse can be recognized on the B-scan of the sample with variable adhesive thickness as an inclined line, crossing the parallel lines of stable sheet reflections (Fig. 6.11). Some special methods of filtration reduce masking of the second interface signal (Challis et al., 1996); however, uncertainties of adhesive parameters complicate the problem.

6.9   Adhesive bonding of composite and biocomposite materials

The inspected specimen was a three-layer adhesive bond structure (Fig. 6.12). The acoustic impedance of the middle adhesive layer is close to the impedance of the

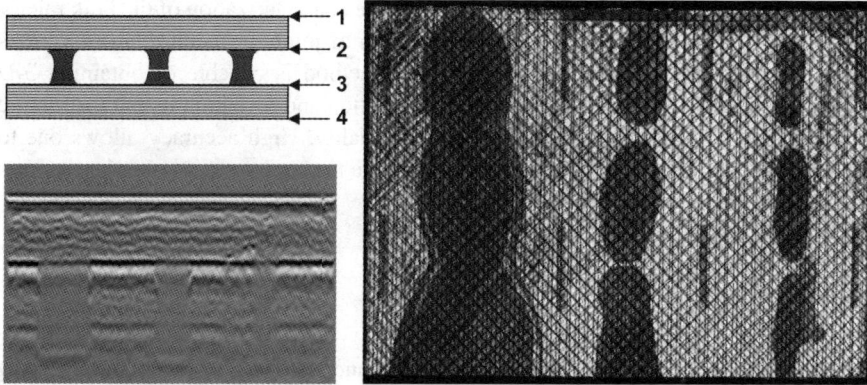

Figure 6.12 Schematic structure, B- and C-scans of the composite sample; 15 MHz focused transducer. 278 × 200 mm scanning area.

composite on either side, which makes the accurate visualization of the interface difficult. The visualization is further obscured by the fact that the reflected signals are hidden among the numerous echoes from the fibers in the upper sheet of the composite. These facts very much complicate the SAM use for testing of these samples. However, correct choice of frequency and aperture of acoustic lens highlights certain features of the sample and existing defects, like impact damage (Maeva et al., 2011).

Fig. 6.12 represents the acoustic images of the adhesive joint of the specimen taken at the composite/adhesive interface. Scanning was performed with a 15 MHz focused transducer. The regions with adhesive at the interface are easily detectable on the C-scan of the sample. The easily recognizable fiber reinforcement pattern covers all the scanning area.

B-scans visualize the sample cross-section and estimate the real width and depth of the adhesive joint. Slight variation of the joint depth is noticeable on the B-scan. It is difficult to determine the adhesive thickness variations precisely as the speed of sound in the adhesive is not known. However, some estimation can be made: based on the measurement of time-of-flight in the adhesive material and in water in two close points, the longitudinal sound velocity has a value of 1660 ± 20 m/s. The measured variations in the joint depth are $1.7-1.4$ µs. Taking into account the estimated value of the speed of sound in the adhesive, the depth of the adhesive joint varies within $1.5-2.2$ mm.

6.10 Conclusions

Acoustic microscopy is a relatively new but cost-effective method in material characterization. Its advantages such as sensitivity, high resolution and reliability have been demonstrated by different research groups over the world as a strong and efficient instrument for material evaluation. Using an acoustic microscope provides a way to visualize small-scale cavities, failures, disbonding, and other defects at different depths. Acoustic microscopy makes it possible to inspect the fine details of internal structures of joints. It also provides an opportunity for reliable characterization of the bulk microstructure of different materials as well as the area of joints between them of various nature. In addition to size measurements, this method is capable of obtaining 3-D high-resolution images of the area under investigation and revealing its microstructure with possible imperfections affecting the joint quality. High accuracy allows one to consider acoustic microscopy results as an ultimate measure for a whole range of materials characterization and NDE tasks.

References

Bray, D.E., Stanley, R.K., 1996. Nondestructive Evaluation: A Tool in Design, Manufacturing and Service. CRC Press LLC, 586 p.

Challis, R.E., Freemantle, R.G., Wilkinson, G.P., White, J.D.H., 1996. Compression wave NDE of adherent lap joints: uncertainties and echo feature extraction. Ultrasonics 34, 315−319.

Chertov, A., Maev, RGr., 2005. One-dimensional model of acoustic wave propagation in the multilayered structure of the spot weld. IEEE Trans Ultrason Ferroelectr Freq Control 52 (10), 1783−1790.

Chertov, A., Maev, RGr., Severin, F., 2007. Acoustic microscopy of internal structure of resistance spot welds. IEEE Trans Ultrason Ferroelectr Freq Control 54 (8), 1521−1529.

Hagemaier, D.J., 1996. Adhesive-bonded joints. In: ASM Handbook Nondestructive Evaluation and Quality Control, ninth ed., vol. 17, USA, pp. 610−640.

Lee, H.-T., Wang, M., Maev, R., Maeva, E., 2003. A study on using scanning acoustic microscopy and neural network techniques to evaluate the quality of resistance spot welding. Int J Adv Manuf Technol 22, 727−732.

Maev, RGr., Watt, D.F., Levin, V.M., Maslov, K.I., Pan, R., 1996. Development of high resolution ultrasonic inspection for welding microdefectoscopy. In: Acoustical Imaging, vol. 22. Plenum Press, pp. 779−784.

Maev, RGr., Titov, S., Bogachenkov, A., Ghaffari, B., Lazarz, K., Ondrus, D. Method for Assessing of Quality of Adhesively Bonded Joints Using Ultrasonic Waves, US Provisional Application No. 61/623838, Filed: April 13, 2012.

Maev, RGr., (guest editor), August 2007. Special issue on high resolution ultrasonic imaging in industry, materials and biomaterials applications. IEEE Trans Ultrason Ferroelectr Freq Control 54 (8).

Maev, RGr., August 2008. Acoustic Microscopy. Fundamentals and Applications. John Willey and Sons − VCH, New York, London, Berlin, approx. 450 p.

Maev, RGr., (editor & contributor), November 2013. Advanced in acoustic microscopy and high resolution imaging. From Principles to Applications, 14 Chapters. Wenham: John Wiley and Sons − VCH.

Maeva, E., Severina, I., Bondarenko, S., Chapman, G., O'Neill, B., Severin, F., et al., 2004. Acoustical methods for investigation of adhesively bonded structures: a review. Can J Phys 82, 981−1025.

Maeva, E.Y., Severina, I., Chapman, G.B., 2007. Analysis of the degree of cure and cohesive properties of the adhesive in a bond joint by ultrasonic techniques. Research Nondestr Eval 18, 121−138.

Maeva, E.Y., Severina, I., Wehbe, H., Erlewein, J., 2011. Ultrasonic imaging techniques to evaluate quality of fiber reinforced composite materials and their adhesive joints. In: Proceedings of Fifth Pan American Conference for NDT Cancun, Mexico, October 2011.

Pthchelinsev, A., Maev, R.Gr., 2001. Monitoring of pulsed ultrasonic waves' interaction with metal continuously heated to the melting point. In: Review of progress in quantitative nondestructive evaluation, vol. 20B. Plenum Press, pp. 1509–1517.

Ultrasonic techniques for materials characterization

7

G. Hübschen
Formerly Fraunhofer Institute for Nondestrutive Testing (IZFP), Saarbrücken, Germany

7.1 Introduction

Nondestructive evaluation (NDE) by ultrasound offers a wide variety of techniques for materials characterization. Being volumetric in nature, ultrasonic examination can give an idea about bulk material properties. Ultrasonic parameters are significantly affected by changes in microstructural or mechanical material properties. Ultrasonic material characterization has also been used to qualify various processing treatments like precipitation hardening, case hardening, rolling texture, etc. and to assess the damage of metallic components due to various degradation mechanisms like fatigue, creep, corrosion and embrittlement. Ultrasonic techniques for material characterization cover a very wide field of applications in metallic and nonmetallic solid materials, gases and fluids. This chapter deals with applications on metal alloys.

7.2 Principles of ultrasonic technique

This section describes in a condensed form the basic properties of ultrasonic waves propagating in solids and the probes used to generate and receive ultrasonic waves. Parameters like wave velocity, attenuation, scattering and the propagation as bulk and guided ultrasonic waves are covered. In many cases, eg, the determination of texture, residual stress, hardness and fatigue damage by ultrasound highly precise time-of-flight (TOF) measurements are carried out. For the measurement of grain size and hardening depth, the scattering of ultrasonic waves is evaluated. Furthermore, ultrasonic attenuation measurements are also applied, eg, to determine grain size, hardness and yield strength of metallic alloys.

7.2.1 Wave propagation in solids

Ultrasonic waves can exist in different media including gas-like media, fluids and solids. The possible forms of the propagating waves vary depending on the kind of the medium. In gas-like and fluid media only compression (longitudinal) waves exist. In solid media, besides compression waves, shear (transversal) waves can propagate. For shear waves many different types of polarization are possible. For nondestructive

Materials Characterization Using Nondestructive Evaluation (NDE) Methods
http://dx.doi.org/10.1016/B978-0-08-100040-3.00007-9

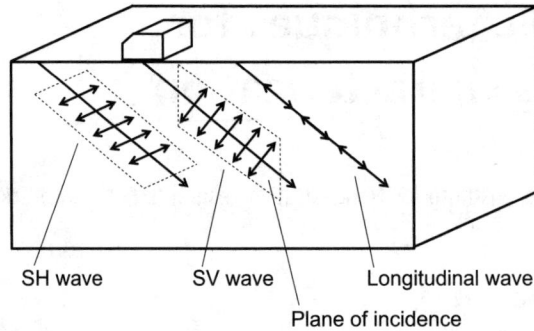

Figure 7.1 Polarization of bulk waves.

evaluation mostly linear polarized shear waves are applied. Using piezoelectric transducers, which are widely used in practical applications, exclusively shear vertical (SV)
waves are excited. This wave type is linearly polarized in the plane of incidence (given
by the surface normal and the propagation direction). A second pronounced linearly
polarized shear wave is the shear horizontal (SH) wave. Here the polarization direction
is oriented perpendicularly to the plane of incidence, which is in most cases tangential
or horizontal to the material surface (see Fig. 7.1).

During interactions with boundaries, compressional and SV waves are coupled
together and mode converted because in most situations the boundary conditions
can only be fulfilled by the existence of the two wave types.

Besides the differentiation concerning the displacement direction as compression or
shear wave with different polarization directions, there are two other classifications for
elastic waves in solids, namely their appearance as bulk waves in thick-walled components or as guided waves in thin-walled components and on boundaries between two
media. On a traction-free surface of a solid, a so-called surface wave (Rayleigh wave)
can propagate (Viktorov, 1967). The particle displacements of Rayleigh waves are
elliptically polarized and the penetration depth at the material surface is in the order
of one to two wavelengths (see Fig. 7.2).

For ferritic steel the propagation velocity of a Rayleigh wave v_R on plane surfaces is
given by:

$$v_R = 0.92\, v_T \qquad\qquad\qquad [7.1]$$

v_T is the sound velocity of the transverse wave.

Other forms of guided waves, eg, Stoneley waves, so-called interface waves, can
exist at the (plane) boundary between two solid media. In case of waves in thin plates,
different modes are possible in the form of Lamb modes, elliptically polarized in the
plane of incidence, or SH modes, linearly polarized perpendicular to the plane of incidence and parallel to the plate surface (Fig. 7.3). For more details about the particle
displacement distribution over the thickness of the plate (see, eg, Achenbach, 1973;
Rose, 1999).

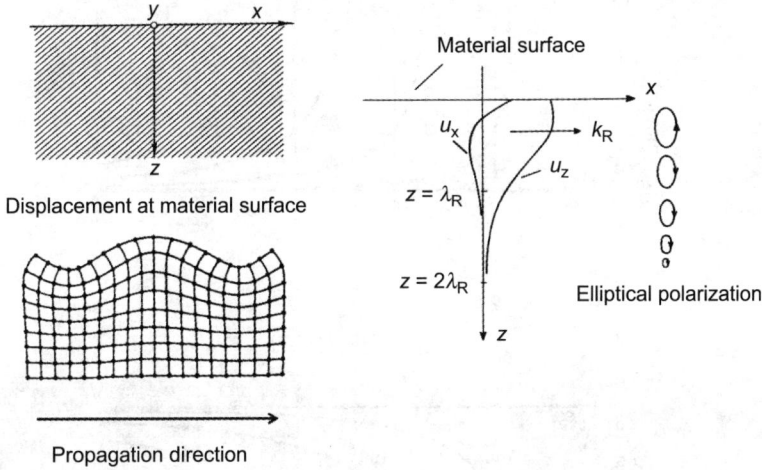

Figure 7.2 Rayleigh waves: particle displacements (tangential component u_x and normal component u_z), polarization and penetration depth $\approx 1\lambda_R$ to $2\lambda_R$ (after Victorov, 1967; Langenberg, 2009).

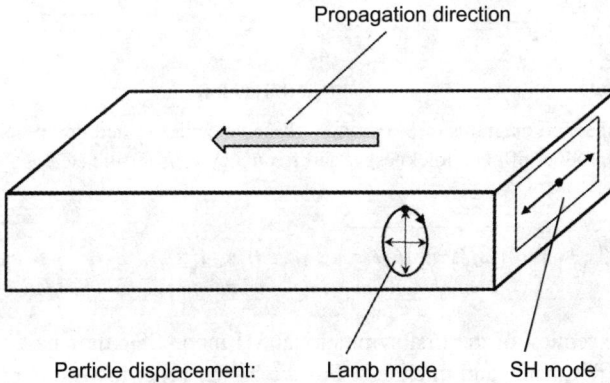

Figure 7.3 Schematic sketch of polarization of Lamb modes and SH modes.

With the exception of the lowest SH mode SS_0, all these plate wave modes are dispersive, ie, the phase and group velocities depend on the frequency and the plate thickness. This is shown for SH modes in Fig. 7.4 (upper diagram: phase velocities c_p, lower diagram: group velocity c_g). Here the phase and group velocities are presented as a function of the product $f \cdot d$ (f, frequency; d, thickness of the plate). These diagrams are numerical evaluations of the following equations:

$$c_p = c_T \bigg/ \sqrt{\left(1 - (nc_T)^2 \big/ 4(df)^2\right)} \quad n = 0, 1, 2, 3\ldots \qquad [7.2]$$

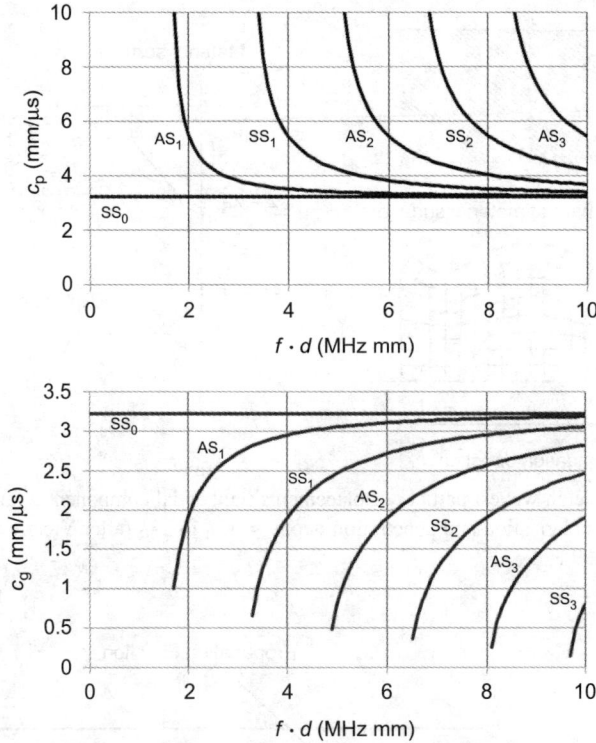

Figure 7.4 Dispersion diagrams of SH modes, phase velocities c_p and group velocities c_g as a function of the product of plate thickness d and frequency f for ferritic steel.

$$c_g = c_T \left/ \sqrt{\left(1 - (nc_T)^2 \middle/ 4(df)^2\right)} \right. \quad n = 0, 1, 2, 3\ldots \qquad [7.3]$$

With the exception of the first symmetrical SH mode SS_0, the phase velocities are dependent on frequency and thickness. For the higher order modes ($n = 1,2,3\ldots$) frequency limits exist where the phase velocity values tend to infinity. At smaller values of $f \cdot d$ the corresponding modes are no more capable to propagate. The group velocity of the first symmetrical SH mode also does not depend on the frequency f and the plate thickness d and has the same values as the phase velocity. These values are identical to the phase velocity of a bulk SH wave under free propagation. The group velocities of all higher modes are smaller than the group velocities of the first mode and depend on $f \cdot d$. The group velocities of the higher modes ascend from the value 0 at the corresponding frequency limit very fast and approximate for large values of $f \cdot d$ the group velocity of the first symmetrical mode SS_0. The single modes are designated as symmetrical (SS) and asymmetrical (AS) concerning their particle displacement distribution over the plate thickness whereby the symmetry plane is the middle plane of the plate.

7.2.2 Sound velocity

In the linear elasticity theory the stress—strain relation is described by Hooke's law. The elastic strain ε is proportional to the stress σ. This is macroscopically described by the following relation:

$$\sigma = E\varepsilon \qquad [7.4]$$

where E is the Young's modulus for tensile or compressive stress. Considering the symmetry of stress and strain microscopically, the proportionality between stress and strain is described by the elasticity tensor c_{ij}:

$$\sigma_i = c_{ij}\varepsilon_j$$

Here the symmetry of the elasticity tensor itself has been taken into account. The indices have to be read in the following way:

Tensor notation	11	22	33	23, 32	13, 31	12, 21
Matrix notation	1	2	3	4	5	6

The c_{ij} are the anisotropic elastic constants with respect to the three spatial directions and the two possible wave modes, namely longitudinal waves L and shear waves (transverse waves) T. For a cubic single crystal three nonvanishing independent elastic constants exist: c_{11}, c_{12} and c_{44}. In a single crystal the sound velocities (phase velocities) and their directional dependence are directly determined by the elastic constants (Truell et al., 1969) as follows:

Longitudinal wave velocity in [100] direction:

$$v_L[100] = \sqrt{c_{11}/\rho} \qquad [7.5]$$

Longitudinal wave velocity in [110] direction:

$$v_L[110] = \sqrt{(c_{11} - A/2)/\rho} \qquad [7.6]$$

Longitudinal wave velocity in [111] direction:

$$v_L[111] = \sqrt{(c_{11} - 2A/3)/\rho} \qquad [7.7]$$

For shear waves the following equations are valid:

$$v_T[100] = \sqrt{c_{44}/\rho} \qquad [7.8]$$

$$v_T[110] = \sqrt{c_{44}/\rho} \text{ polarization in } [001] \text{ direction} \qquad [7.9]$$

$$v_T[110] = \sqrt{(c_{44} + A/2)/\rho} \text{ polarization perpendicular } [110] \text{ direction} \qquad [7.10]$$

$$v_T[111] = \sqrt{(c_{44} + A/3)/\rho} \qquad [7.11]$$

A designates the elastic anisotropy factor and has the form:

$$A = c_{11} - c_{12} - 2c_{44} \qquad [7.12]$$

The quantity A plays an important role in grain size and texture determination, as will be described later.

For an iron single crystal the functional distribution of the phase velocities of the longitudinal wave and the two transverse waves, one with polarization in the cube face and one with polarization direction perpendicular to the cube face, propagating in the x_1 and x_2 are shown in Fig. 7.5 (Musgrave, 1970). Here the orthogonal coordinate system is selected so that the edges of the cube coincide with the x_1, x_2 and x_3 axes. One can see that the phase velocities of the longitudinal wave and the transverse wave with polarization in the x_2–x_3-plane vary with the propagation direction. Only the phase velocity of the transverse wave with polarization in x_3-direction is constant and independent of the propagation direction.

In an isotropic and stress-free solid the wave velocities (in nondispersive media the phase and group velocities are identical) of the different bulk wave types are determined by the elastic properties, described via elastic moduli Young's modulus E, shear modulus G and Poisson ratio v of the solid media, and they are constant in all directions (see Fig. 7.6).

The sound velocities of longitudinal and shear waves are given by the following formulas:

$$v_L = \sqrt{(E(1-v))/(\rho(1+v)(1-2v))} \qquad [7.13]$$

$$v_T = \sqrt{G/\rho} \qquad [7.14]$$

Often instead of E, G and v the so-called Lamé constants λ, μ are used. The relations between both are:

$$\lambda = vE/((1+v)(1-2v)) \qquad [7.15]$$

$$\mu = G \qquad [7.16]$$

The wave velocities are:

$$v_L = \sqrt{(\lambda + 2\mu)/\rho} \qquad [7.17]$$

$$v_T = \sqrt{\mu/\rho} \qquad [7.18]$$

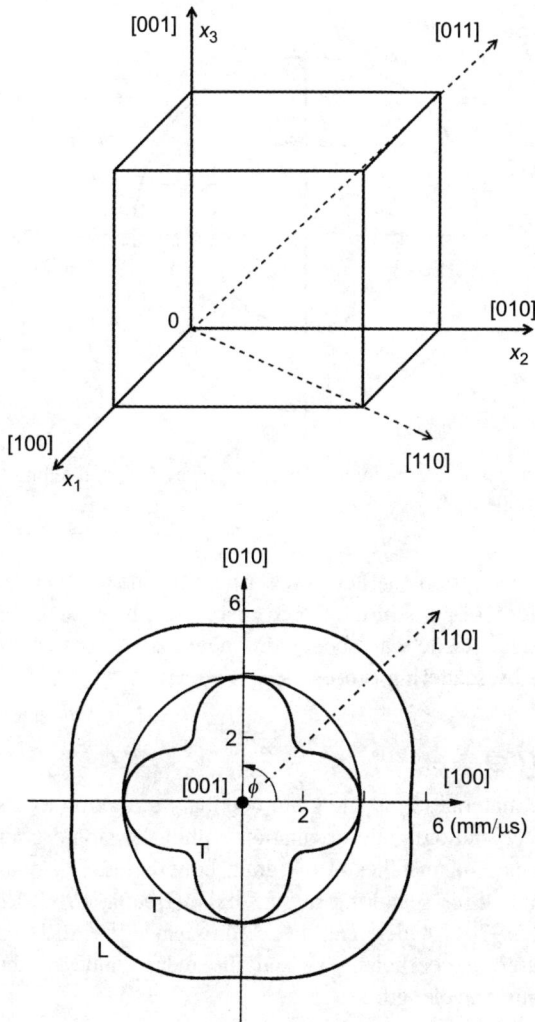

Figure 7.5 Coordinate system crystallographic axes in cubic system, phase velocity distribution in iron single crystal.

7.2.3 Attenuation

The wave velocity describes the elastic interaction between the solid material and the wave. Therefore, the velocity is a direct function of elasticity and the mass density. The attenuation coefficient indicates mainly the energy loss of a propagating wave. Two main mechanisms contribute to ultrasound attenuation, namely absorption and scattering. The attenuation coefficient α can be written as:

$$\alpha = \alpha_A + \alpha_S \tag{7.19}$$

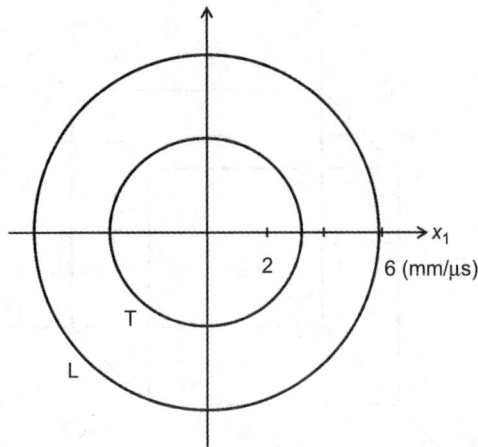

Figure 7.6 Phase velocity distribution of a transverse wave (T) and a longitudinal wave (L) in isotropic polycrystalline iron.

Here α_A is the absorption coefficient describing the inelastic wave behavior characterized by dissipation of parts of ultrasonic energy into heat; α_S is the scattering coefficient describing the elastic wave behavior where parts of the energy are distributed into all directions by scattering processes.

7.2.4   Scattering

In polycrystalline material ultrasonic grain boundary scattering exists due to the grain structure and the crystallographic orientation within the single grains, which causes acoustical impedance mismatches at the grain boundaries. For quasi-isotropic polycrystalline materials three scattering regions exist (Bhatia, 1967; Bhatia and Moore, 1959; Goebbels, 1980), which can be distinguished by different functionalities between the scattering coefficient α_S and the mean value of the grain diameter d and the ultrasonic wavelength λ:

Rayleigh scattering, $d \ll \lambda$ $\alpha_S = S_1 d^3 f^4$ [7.20]

Stochastic scattering, $d \approx \lambda$ $\alpha_S = S_2 d f^2$ [7.21]

Diffuse scattering, $d \gg \lambda$ $\alpha_S = S_3/d$ [7.22]

For the Rayleigh scattering on grains the interaction of the ultrasonic wave takes place with a (theoretical globular) obstacle, which is small in comparison to the ultrasonic wavelength. Hereby longitudinal and shear waves are generated by mode conversion.

In the case of stochastic scattering the grain diameter and the ultrasonic wavelength are in the same dimension so that resonance phenomenon can occur (generation of standing waves in the single grains) and therefore also with quantitatively larger scattering losses compared to the two other scattering mechanisms.

In current nondestructive testing (NDT) applications mainly the Rayleigh scattering case plays an important role. Here the scattering is proportional in a first approximation to the volume of the scattering obstacle and to the fourth power of the frequency f. For longitudinal and shear waves the scattering parameter S_1 is given by the known properties: ρ, mass density; v_L, velocity of longitudinal wave; v_T, velocity of shear wave and the elastic anisotropy factor $A = c_{11} - c_{12} - 2c_{44}$:

$$S_{1L} = \left(8\pi^3/375\right)\left(A/\rho v_L^2\right)^2 \left(1/v_L^4\right)\left(2 + 3(v_L/v_T)^5\right) \qquad [7.23]$$

$$S_{1T} = \left(6\pi^3/375\right)\left(A/\rho v_T^2\right)^2 \left(1/v_T^4\right)\left(3 + 2(v_T/v_L)^5\right) \qquad [7.24]$$

During the scattering process mode conversions occur. For an incident longitudinal wave, scattered longitudinal and shear waves are generated and also an incident shear wave is scattered in both wave types. The terms in the Eqs. [7.23] and [7.24] including (v_L/v_T) and (v_T/v_L) represent the mode conversion contribution.

7.2.5 Ultrasonic transducers

In nearly all applications of materials characterization of steel and aluminum products by ultrasonic techniques, piezoelectric and electromagnetic ultrasonic transducers are applied. Comprehensive treatments of piezoelectric transducers can be found, eg, in Krautkrämer and Krautkrämer (1977) and Nakamura (2012). For texture measurements, residual stress measurements and for the determination of fatigue of materials, different types of Electromagnetic Acoustic Transducers (EMATs) are widely used. The basic principles of piezoelectric and electromagnetic ultrasonic transducers are sketched in Fig. 7.7.

Piezoelectric transducers generate an ultrasonic wave by a crystal inside the transducer housing, and with the help of a coupling medium the wave is transferred into the component that has to be investigated (upper part of Fig. 7.7).

An EMAT consists basically of a magnet (electro- or permanent magnet) and an RF coil. Such a transducer does not need any coupling medium and can generally be driven with an air gap between the housing and the component (lower part of Fig. 7.7). Here ultrasonic sources are induced in the material surface by alternating electromagnetic fields. From these sources an ultrasonic wave is generated inside the material under test.

Note that EMATs can only be applied on electrically conductive materials.

Due to the fact that no coupling material is necessary applying EMATs in the NDT practice for metallic materials, very often such transducers are used for materials characterization. In the following the different types of EMATs used for materials characterization will be described (for more details, see, eg, Hübschen, 2012).

Fig. 7.8 shows the principle of normal beam EMAT for linearly polarized shear waves. These transducer types are applied, eg, for residual stress measurements and texture analysis.

Piezoelectric ultrasonic transducer

EMAT

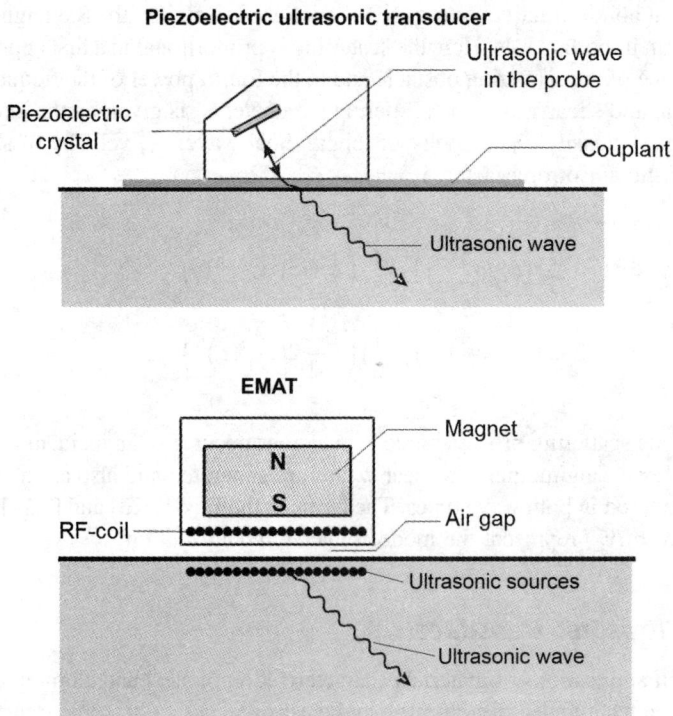

Figure 7.7 Principles of piezoelectric ultrasonic transducer and Electromagnetic Acoustic Transducer (EMAT).

In a cross-section through the transducer and the material the instantaneous direction of RF currents in the RF coils and the resulting vibration direction of the excited ultrasonic wave in the material are indicated. In the case of linearly polarized

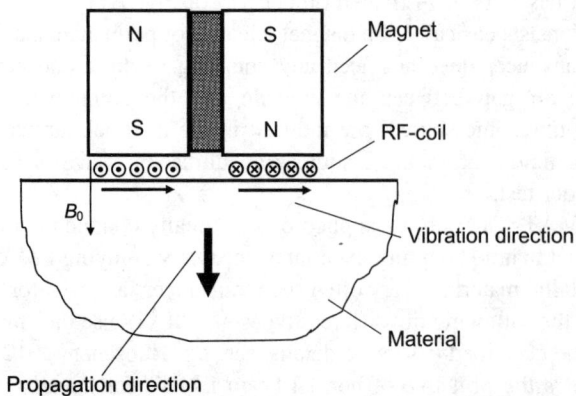

Figure 7.8 Normal beam EMAT for linearly polarized shear wave (Hübschen, 2012).

transverse waves the vibration under the two pole pieces of the two magnets is oriented in the same direction. Here a rectangular RF coil is used.

For the excitation and reception of guided SH waves in rolled sheets to evaluate texture and for stress measurements applying SH waves under grazing incidence, the following two types of angle beam EMATs are widely applied. The first type of transducer consists of an arrangement with periodic permanent magnets (upper part Fig. 7.9) and a rectangular RF coil. This configuration generates ultrasonic sources

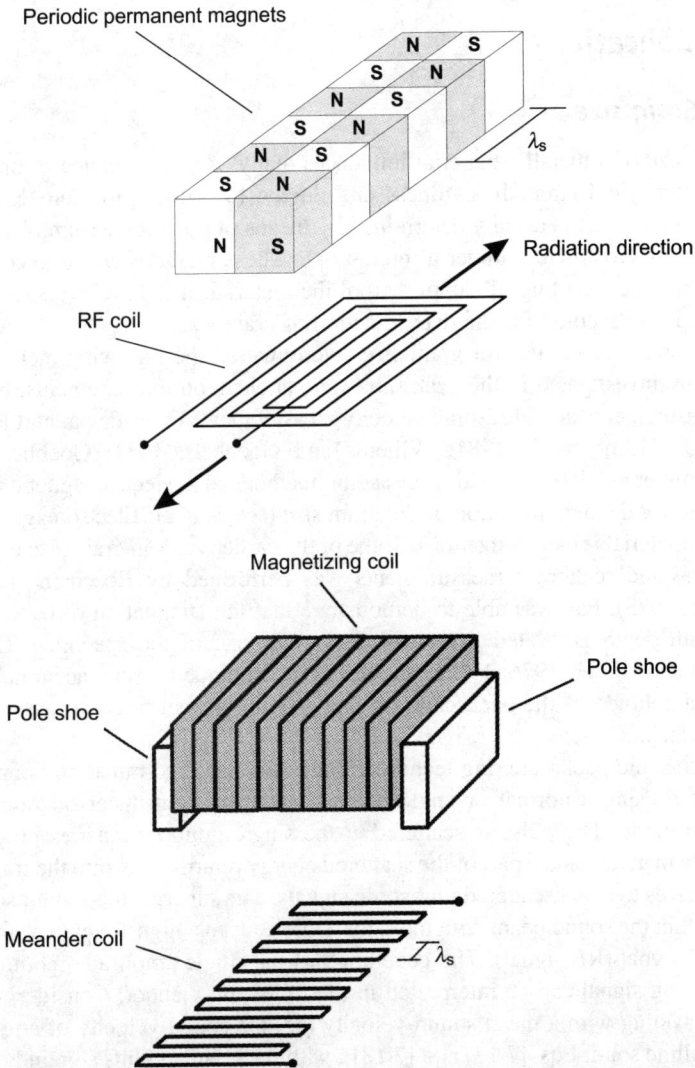

Figure 7.9 Angle beam EMATs for SH waves Lorentz force type (upper part), magnetostriction type (lower part) (Hübschen, 2012).

with alternating vibration directions parallel to the surface necessary to excite SH waves. A characteristic parameter of such a transducer is the period or the spatial wavelength λ_s. The period and the frequency of the RF current feed into the coil determine the beam angle of such a transducer. The transduction mechanism bases mainly on Lorentz forces.

The second type of angle beam EMAT is built up with an electromagnet and a meander-type RF coil. Here the period λ_s of the transducer is determined by the nearest distance between two line elements of the coil with the same current direction.

7.3 Applications

7.3.1 Grain size

The grain size of a metallic material influences many of its mechanical properties as fatigue, creep yield strength, hardness and ductile to brittle transition temperature and so on. Grain size is usually determined by means of optical micrographs or while viewing the microstructure under a microscope. These methods are time consuming and often require a cutting of samples from the material. It is advantageous to apply a nondestructive method for the determination of grain size.

As ultrasonic techniques for grain size determination the following methods have been mainly investigated in the past: ultrasonic attenuation measurements, backscattering measurements and ultrasonic velocity measurements (Vasudevan and Palanichamy, 2002; Hecht et al., 1981; Willems and Goebbels, 1981; Goebbels, 1994; Palanichamy et al., 1995). Besides ultrasonic methods also electromagnetic methods can be used for the determination of the grain size (see chapter: Electromagnetic techniques for materials characterization). Some of the earlier work in grain size estimation using ultrasonic scattering measurements was performed by Beecham (Beecham, 1966; Fay, 1973). Fay was able to demonstrate that the attenuation of backscattered echoes with depth is related to the average grain size of the specimen. Goebbels (Goebbels and Höller, 1976, 1980) refined Fay's technique to more accurately determine the amplitude of the backscattered echoes with respect to depth using various averaging techniques.

The ultrasonic backscattering technique for grain size determination is applied in practice, eg, using a normal or angle beam shear wave transducer operated in the pulse-echo mode. The pulse is scattered at the single grain boundaries propagating through the material, and a part of the scattered energy returns back into the transducer, which receives the backscattered ultrasonic signals. The superposition of all scattering signals within the sound beam with the same TOF leads to a high-frequency amplitude modulated signal (RF signal). The generation of the single amplitude-modulated RF backscattering signal can be interpreted in the following manner: Considering that a wave is traveling with a mean sound velocity (eg, the sound velocity of an isotropic polycrystalline solid, Eqs. [7.17] and [7.18]), within a grain and hits a grain boundary. If the neighboring grain is oriented so that it exhibits the same velocity, no scattering would occur. If the velocity in the grain v_{Gr} is smaller than v_{poly} in the polycrystal, then

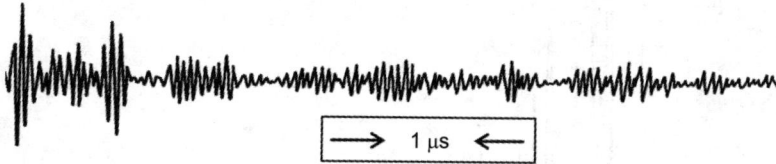

Figure 7.10 RF backscattering signal.

the scattering is excited with an amplitude proportional to the difference $(v_{Gr}-v_{poly})$. Depending on the different single crystal wave propagation and the possible polarization directions, this difference can reach a maximum value.

A typical back scattering RF signal is shown in Fig. 7.10.

Such a signal contains the necessary information, but it cannot be evaluated immediately. The evaluation becomes feasible using different averaging methods to add up rectified backscattered signals from different points of insonification ("spatial averaging"), from different angles of insonification ("directional averaging"), and of different frequencies ("frequency averaging"; Goebbels, 1980) and from the weighted average over a certain time window of amplitudes, moved through the hole backscattering curve ("moving filter averaging"; Pongratz et al., 1983). The backscattering amplitude $A_s(x)$ measured as a function of the sound path x for homogeneous material (neglecting multiple scattering and geometrical propagation conditions) is given by (Fay, 1973):

$$A_s(x) = A_0 \sqrt{\alpha_S} \exp(-\alpha x) \qquad [7.25]$$

where A_0, sound pressure amplitude at the sample surface, $x = 0$; α_S, scattering coefficient; α, attenuation coefficient, $\alpha = \alpha_A + \alpha_S$; α_A, absorption coefficient.

Typical backscattering amplitude curves are shown in Fig. 7.11 (Willems et al., 1987). The upper part is a single A-scan and the lower part an averaged curve of 1024 A-scans.

The scattering coefficient α_S can be calculated (Goebbels, 1980) for shear waves and material with cubic crystals in the Rayleigh scattering region by Eq. [7.26]:

$$\alpha_S = S d^3 f^4 \qquad [7.26]$$

with the frequency f and the scattering parameter S:

$$S = (6\pi^3/375)(A/\rho v_T^2)^2 (1/v_T^4)\left(3 + 2(v_T/v_L)^5\right)$$

Here $A = c_{11} - c_{12} - 2c_{44}$ is the anisotropy factor with the elastic moduli. For ferritic and austenitic steels, the scattering parameter has the value $S \approx 0.0131$ $(mm/\mu s)^{-4}$. If α_S is known the mean grain size diameter can be calculated. If the frequency dependence of α_A (eg, $\alpha_A \sim f$) is known, two backscattering measurements with two different frequencies f_1 and f_2 are sufficient to calculate the absorption

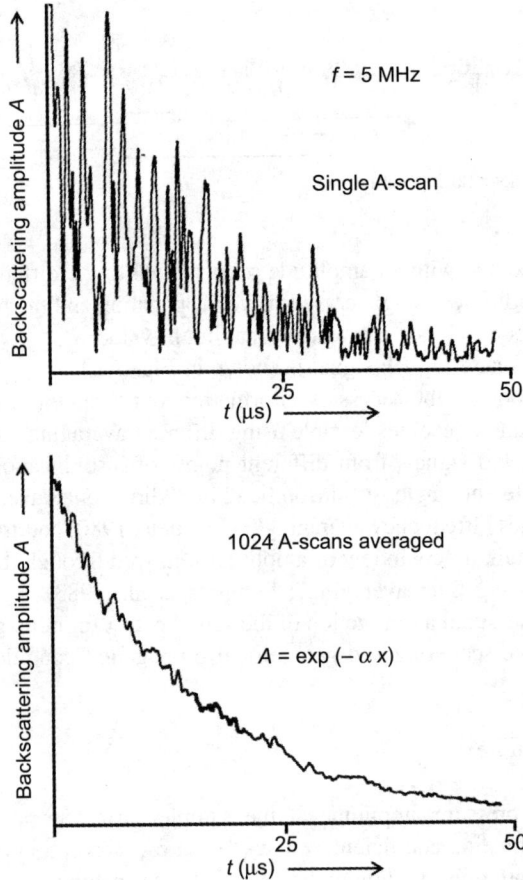

Figure 7.11 Rectified backscattering signals.

coefficient α_A and the scattering coefficient α_S separately (Goebbels and Höller, 1980) (Fig. 7.12).

By logarithmizing the measured backscattered sound pressure amplitude curves, the attenuation coefficients α_1 and α_2 are obtained. For the mean grain size d it follows:

$$d = \left((\alpha_2 - \alpha_1 f_2/f_1) \big/ \left(S f_2 \left(f_2^3 - f_1^3 \right) \right) \right)^{1/3} \qquad [7.27]$$

If the absorption coefficient α_A is the same in two samples with different grain sizes, one backscattering measurement on each of the samples with one frequency leads to α_A and α_S (Goebbels and Höller, 1980) (Fig. 7.13).

Normally the level of backscattering signal amplitudes is about 40–60 dB below typical ultrasonic backwall echo signals. Therefore, a sensitive technique is needed for the quantitative detection of the scattering signals. Goebbels and Höller (1980)

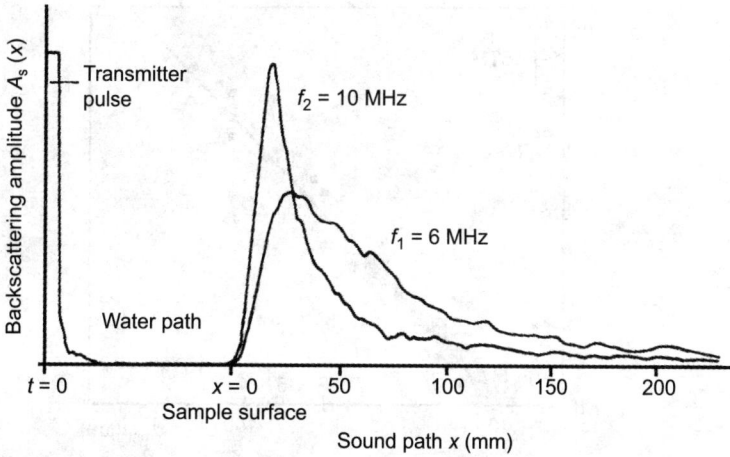

Figure 7.12 Backscattering amplitudes from austenitic steel for two different frequencies.

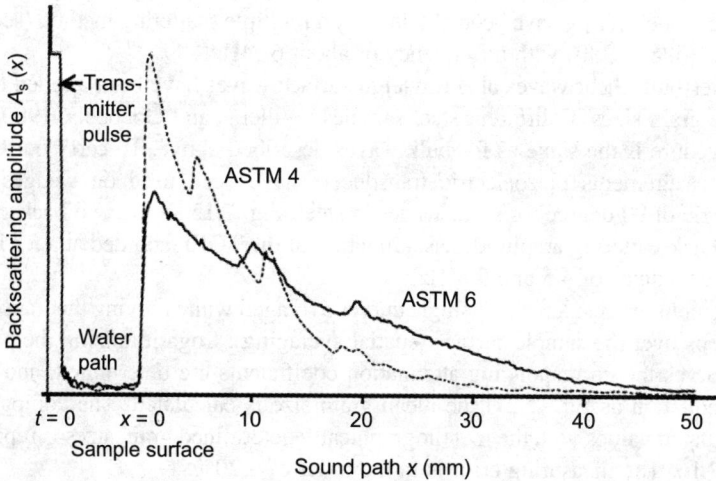

Figure 7.13 Backscattering amplitudes from two different ferritic steels (shear waves, $f = 6$ MHz).

reported that shear wave backscattering measurements in contact as well as in immersion technique with frequencies between 4 and 20 MHz on more than 200 samples of different ferritic and austenitic steels with grain sizes from ASTM 10 to 1 confirmed the quantitative grain size determination by the above-described methods.

Fig. 7.14 shows the grain size values d_{US} determined by the ultrasonic method in comparison to the metallographic determined values d_{met} for about 80 samples of different steels. For the range of $0.05 < d/\lambda < 0.5$, the accuracy of the values d_{US} lies within $\pm 15\%$. Grain sizes between ASTM 10 and 6 and between 6 and 4 have

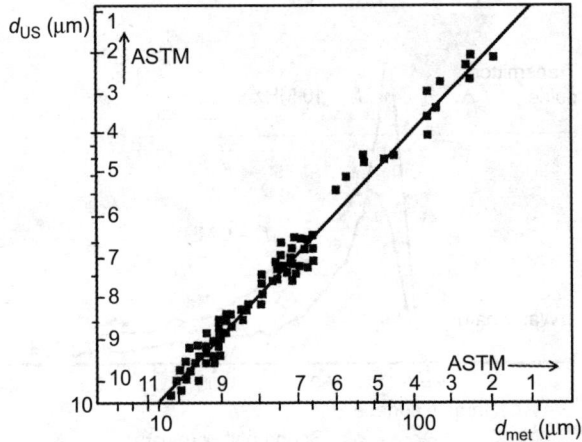

Figure 7.14 Comparison of grain size d_{met} determined by metallographic and grain size d_{US} determined by ultrasonic method.

been analyzed at frequencies from 10 to 20, and 5–10 MHz, respectively. For ASTM 4 to 1 the values of d_{US} have been obtained by a multiple scattering method (see Goebbels and Höller, 1980) with a frequency of about 6 MHz.

Besides bulk shear waves also Rayleigh surface waves have been applied to determine the grain sizes in different steel samples (Willems and Goebbels, 1980, 1981). The procedure is the same as for bulk waves described above. To carry out the scattering measurements piezoelectric transducers have been used on wedges with a beam angle of 90 degree for shear waves in steel. Fig. 7.15 shows two backscattering curves (backscattering amplitudes as a function of the TOF) recorded at a ferritic steel and at frequencies of 4.5 and 9 MHz.

The graphs are results of 512 single curves averaged while moving the transducer in small steps over the sample surface (spatial averaging). Logarithmizing the backscattering curves the corresponding attenuation coefficients are determined, and using a similar equation as Eq. [7.27] the mean grain size is calculated. The comparison of the calculated values with the metallographically determined grain sizes is represented in Fig. 7.16. The measuring error is in the order of ±20%.

7.3.2 Texture

In the crystallography texture describes the entity of the orientations of the crystallites in a polycrystalline solid. From the texture results, in particular, the anisotropy of the mechanical formability of many metallic materials. Examples are the earing at the deep drawing of sheets as well as the anisotropy of the magnetic properties of soft magnetic (eg, grain-oriented transformer sheet) and hard magnetic materials as anisotropic Alnico magnets, Samarium–Cobalt alloys, anisotropic Neodymium–iron–Boron alloys and hard magnetic ferrites. At solidification processes a directional growing of crystallites often happens. Also by forming of a material as cold rolling or drawing, textures are generated. Textures in rolled sheets are developed during manufacturing

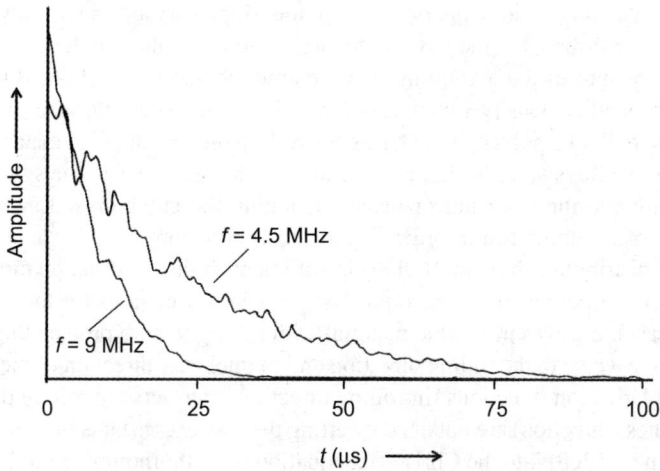

Figure 7.15 Backscattering curves for Rayleigh waves at $f = 4.5$ and 9 MHz.

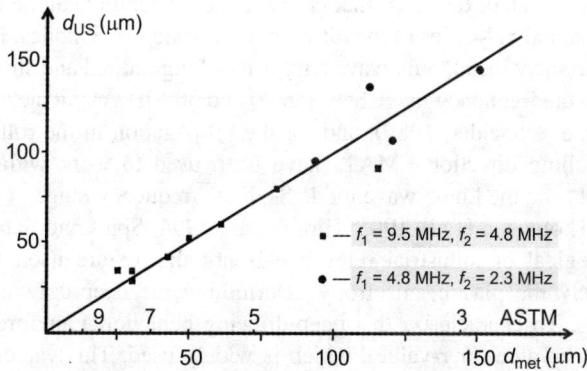

Figure 7.16 Comparison of grain size d_{US} determined by ultrasonic method (Rayleigh waves) and grain size d_{met} determined by metallography.

by deformation processes and heat treatments. Herewith, the elastic, plastic and magnetic behavior of the product becomes directionally dependent. A nondestructive analysis of the texture and their consequences for material properties is of great technical interest. X-ray diffraction techniques are available and used for this purpose (see chapter: X-ray diffraction (XRD) techniques for materials characterization; Kopineck et al., 1989). Further electromagnetic methods have a potential for texture analysis (see chapter: Electromagnetic techniques for materials characterization).

Complementary to these methods ultrasonic techniques have been developed in the past in order to analyze the texture of rolled plates and sheets of steel and aluminum (Thompson et al., 1987, 1989; Sayers and Proudfood, 1986; Clark et al., 1987a,b; Hirao et al., 1994; Murayama et al., 1996; Kawashima, 1990; Spies and Schneider, 1989, 1990).

Texture, the nonrandom distribution of the single crystals in a polycrystalline material, is described by the crystallite orientation distribution function (CODF). The CODF quantifies the texture in the harmonic method (Roe, 1965; Bunge, 1982) and is a probability density function defining how the crystallites are oriented as a whole in the bulk of a polycrystal. During the rolling process of cubic materials as steel and aluminum alloys an orthorhombic texture is generated. Due to the symmetry with respect to three orthogonal mirror planes including the cubic crystal symmetry, the knowledge of the three fourth-order expansion coefficients C_4^{11}, C_4^{12} and C_4^{13} of the orientation distribution function (ODF) is sufficient to describe the texture influence on the elastic properties of the material (Bunge, 1968). Relations for the wave velocities of ultrasonic waves in a cubic material with rolling texture propagating in rolling direction, transverse to the rolling direction and in thickness direction of the sheet with different polarization directions (in rolling direction, transverse to rolling direction, in sheet thickness direction) are obtained inserting the elastic constants of a textured polycrystal (Bunge, 1968) into the Christoffel equation for orthorhombic sample symmetry (Auld, 1973).

The velocities of waves propagating in the plane of a sheet are given in several publications (Thompson et al., 1985; Allen et al., 1985; Sayers and Proudfood, 1986). The expansion coefficients of the ODF that characterize the texture can be determined by measuring the sound velocities of the different ultrasonic wave modes. For the normal incidence in the sheet metal, bulk waves in form of longitudinal and shear waves have been applied in the frequency range between 20 and 30 MHz using piezoelectric transducer (Spies and Schneider, 1989), and for the propagation in the rolling and transverse to the rolling direction EMATs have been used to work with the SH plate wave mode SH_0 and the Lamb wave mode S_0 in the frequency range of a few hundred kHz (see, eg, Thompson et al., 1985; Hirao et al., 1994; Spies and Schneider, 1989).

Of technological or industrial relevance is not the texture itself but its consequences, namely the plastic anisotropy determining the deep-drawing behavior of rolled products. To characterize this deep-drawing behavior, Lankford et al. (1950) introduced the R-value (or r-value), which is widely used. This value is determined in a tensile test. Hereby at a sheet strip the thickness d_0 and width b_0 are measured before and after at least 20% plastic deformation (thickness d_ε, and the width b_ε). The values R and r are defined in the following manner:

$$R = (\ln(b_0/b_\varepsilon))/(\ln(d_0/d_\varepsilon))$$
$$r = R/(R+1)$$

[7.28]

A measure for the direction dependence of the thickness reduction is the quantity Δr, the planar anisotropy and r_m, the normal anisotropy.

$$\Delta r = 0.5(r(0°) - 2r(45°) + r(90°))$$
$$r_m = 0.25(r(0°) + 2r(45°) + r(90°))$$

[7.29]

These values are evaluated after the plastic deformation of strip cuts at 0, 45 and 90 degree to the rolling direction. It has been found that r_m is a measure for the deep draw-ability of a sheet and Δr correlates well with the earing (Vlad, 1972). A correlation of the texture parameters C_4^{11}, C_4^{12} and C_4^{13} (coefficients of the ODF) with the parameters of the plastic anisotropy as, eg, Δr and r_m results from the relations between the R- and r-values and the E modulus found empirically and experimentally (Stickels and Mould, 1970) on one side and between the E modulus and the expansion coefficient C_4^{11}, C_4^{12} and C_4^{13} on the other side (Bunge, 1979). By the determination of the expan-sion coefficients, the ultrasonic method allows the evaluation of the characteristic parameters of the plastic anisotropy of rolled metal products.

Spies (Spies and Schneider, 1990) has reported results concerning the correlation between the expansion coefficients of the ODF determined by TOF measurements with different ultrasonic wave modes and the deep-drawing properties of hot- and cold-rolled ferritic steel samples with thicknesses between 0.75 and 2.53 mm. The coefficients correlate linearly with the r_m- and Δr-values provided by the manufacturer of the sheets. With these correlations between the Δr- and r_m-values and the expansion coefficients C_4^{13} and C_4^{11}, an algorithm has been developed for the evaluation of the Δr (US)- and r_m (US)-values using TOF data of ultrasonic waves. Figs. 7.17 and 7.18 (Spies and Schneider, 1989) show the relation between the deep-drawing parameters given by the manufacturer r_m (mech), Δr (mech) and the parameters determined by the ultrasonic technique Δr (US) and r_m (US).

The correlation shows that ultrasonic techniques have a high potential to charac-terize the deep-drawing behavior of rolled steel sheets.

Besides laboratory investigations on rolled steel samples, possibilities for the online monitoring of rolled sheets using EMATs to determine characteristic properties have also been reported. In Murayama et al. (1996), an online evaluation system

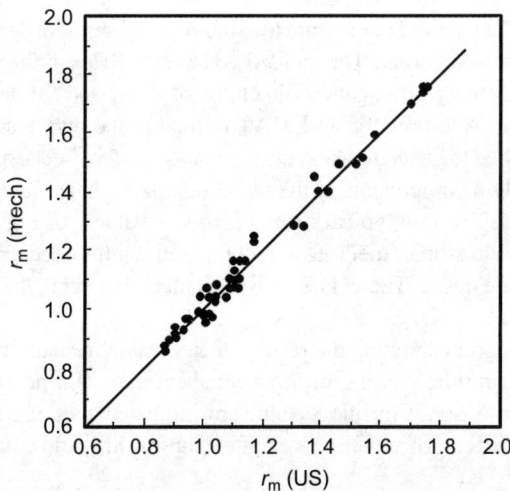

Figure 7.17 Correlation between the deep-drawability parameter r_m evaluated with mechanical and ultrasonic technique.

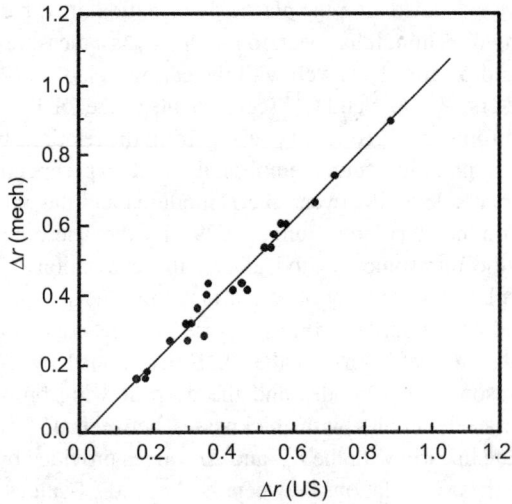

Figure 7.18 Correlation between the deep-drawability parameter Δr evaluated with mechanical and ultrasonic technique.

determining r-values in cold-rolled steel sheet is described, which uses the Lamb wave mode S_0 propagating under 0, 45 and 90 degree to the rolling direction at a frequency of 280 kHz. The generation and reception of the ultrasonic waves has been realized using EMATs with meander-like RF coils (see Fig. 7.9). The capability for measuring r-values has been evaluated using many samples and products in a continuous annealing line. The system could continuously measure r-values in the production line with an accuracy of $1\sigma = 0.07$.

Hyoguchi and Kawashima (1995) describe a method and apparatus for measuring Young's modulus and r-values of cold-rolled thin steel sheets.

Here also EMATs are used to measure the sound velocities of different wave modes in several propagation directions. The guided SH mode SH_0 of the lowest order (see Fig. 7.4) is applied under propagation directions of 0, 45 and 90 degree in relation to the rolling direction. Additionally, an EMAT normal probe with a spiral or a so-called pancake coil is applied to generate thickness resonances in the sheet using the longitudinal wave part and the two components of the radial polarized shear wave part.

Figs. 7.19 and 7.20 show typical results for the r-values. In Fig. 7.19 the r-values measured by the ultrasonic method are compared with the corresponding values measured by tensile tests. There is good agreement between the values measured with both methods.

Fig. 7.20 is a diagram showing the results of an r-value measurement on cold-rolled thin steel sheet during the passing of the steel sheet in an online system. The results measured by the ultrasonic in-line system are indicated with Δ, and the results of measurements by means of tensile tester after cutting the cold-rolled steel sheet are indicated by $\square$.

In addition, here the agreement of the measuring values of both methods is very good.

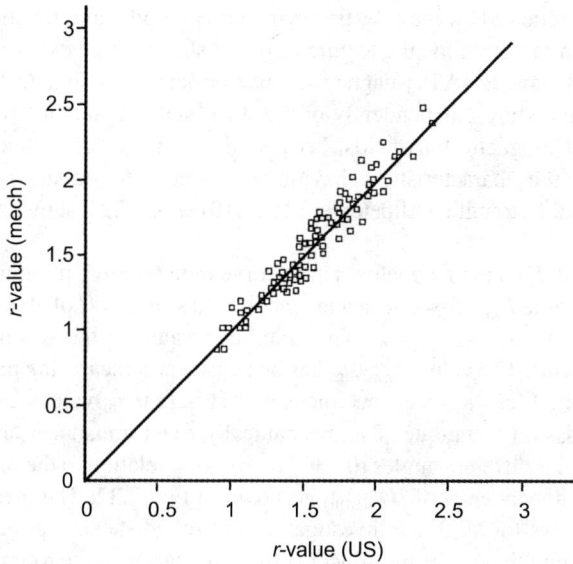

Figure 7.19 Comparison of r-value measured by ultrasound and measured by tensile tests.

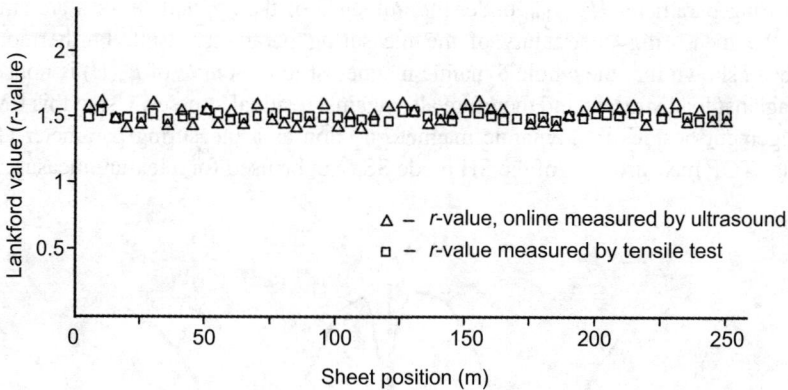

Figure 7.20 Result of online measurement, r-value determined by ultrasound and by tensile test.

Further results of ultrasonic online measurements on hot-dip galvanized steel strips during the production process are reported in Borsutzki et al. (1993). Here, too, a good correlation was found between the r_m- and Δ_r-values determined by the ultrasonic and the destructive technique.

Nearly all known publications dealing with texture analysis of thin-walled rolled steel and aluminum products describe measurements in the laboratory or in production lines carried out using EMATs. In most of the papers SH modes have been applied with transducers working according to the Lorentz force and the magnetostriction principle (see, eg, Thompson, 1979; Hübschen, 2012). Besides the evaluation of the wave velocity of the guided SH wave SS_0 (first symmetrical mode) as a function of the propagation direction to determine the texture of rolled sheets, there exists a second possibility. Using SH wave EMATs (magnetostrictive principle, see Fig. 7.21), it is possible to determine the texture independently of the stress states evaluating the receiver signals at different magnetic bias fields, because the transduction efficiency of such a transducer shows a characteristic behavior of the received signals as a function of the magnetic field strength (Altpeter et al., 2010). Fig. 7.21 shows this functional relation.

This so-called dynamic magnetostriction curve indicates two minima of the excited ultrasonic amplitude E_λ, a first one at a magnetic field strength H of about 0 A/cm and a second one at a value of $H_{E\lambda(Min)}$. The amplitude value of the second minimum is defined as E_λ (Min). The value $H_{E\lambda(Min)}$ has been used as a measuring parameter investigating cold-rolled ferritic sheet specimens with different r_m-values using EMATs in through transmission technique. The mechanically fixed transducer arrangement has been applied under different angles (0–360 degrees) in relation to the rolling direction. The directional dependency of $H_{E\lambda(Min)}$ is shown in Fig. 7.22. The measuring parameter $H_{E\lambda(Min)}$ shows for all of the investigated cold-rolled sheets a pronounced dependency on the orientation of the transducer system in relation to the rolling direction and diminishes in each direction with growing r_m value. Applying load stresses to simulate residual stresses, which can be superimposed to the texture, the changes of the measuring parameter $H_{E\lambda(Min)}$ under the influence of the applied stresses are smaller than the measuring uncertainty of the measuring parameter itself. Furthermore, it has been shown that the whole dynamic magnetostriction curve of $E_\lambda(H)$ is nonsensitive against load stresses and therefore also against residual stresses. Using this EMAT arrangement besides the dynamic magnetostriction as a measuring parameter also a precise TOF measurement of the SH mode SS_0 can be used for a texture measurement

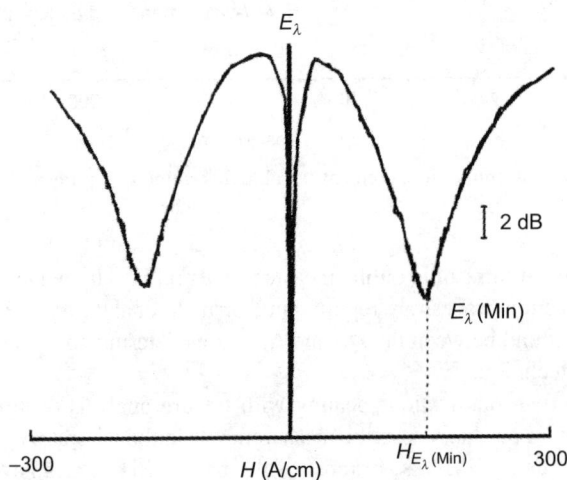

Figure 7.21 Dynamic magnetostriction curve (see also chapter: Electromagnetic techniques for materials characterization).

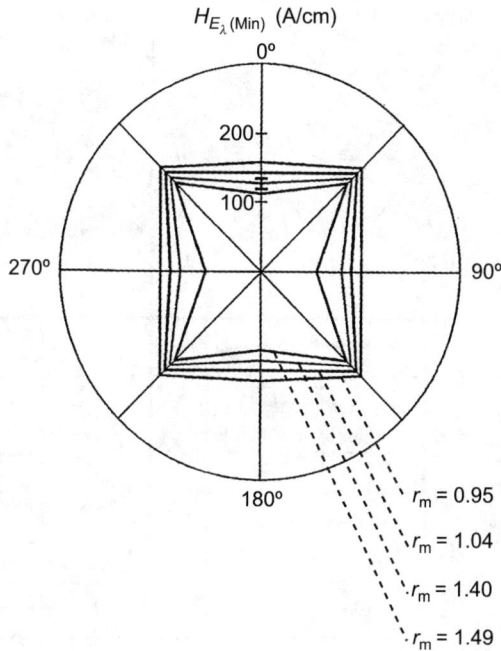

Figure 7.22 Directional dependency of $H_{E\lambda(Min)}$ for ferritic sheets with different r_m-values.

by evaluating in through transmission technique the difference in TOF between two receivers.

7.3.3 Hardening depth

In Section 7.3.1 the backscattering technique has been described in context with the grain size determination in homogeneous steel components. This section deals with a second area to apply the ultrasonic backscattering technique, namely to characterize the thickness of an induction-hardened surface sheet, the so-called hardening depth of steel components.

Hardening improves the wear resistance and enhances the limit of fatigue of dynamically loaded parts. Such a hardened surface layer is characterized by fine grains compared to the coarser grains of the base material. The thickness of such layers is in the order of a few mm. For induction-hardened components, the transition from the fine-grained surface layer to the coarse-grained base material is abrupt. In this case the microstructure changes within a small transition zone in the order of a mm from the martensitic into the ferritic/perlitic state of the parent metal (upper part of Fig. 7.23) (Schneider et al., 2007). Then the hardness distribution curve shows a sharp transition from the surface hardness to the hardness of the parent material (lower part of Fig. 7.23).

Due to the fact that the ultrasonic scattering coefficient (Eq. [7.23]) is proportional to the third power of the grain size d and to the fourth power of the frequency at the

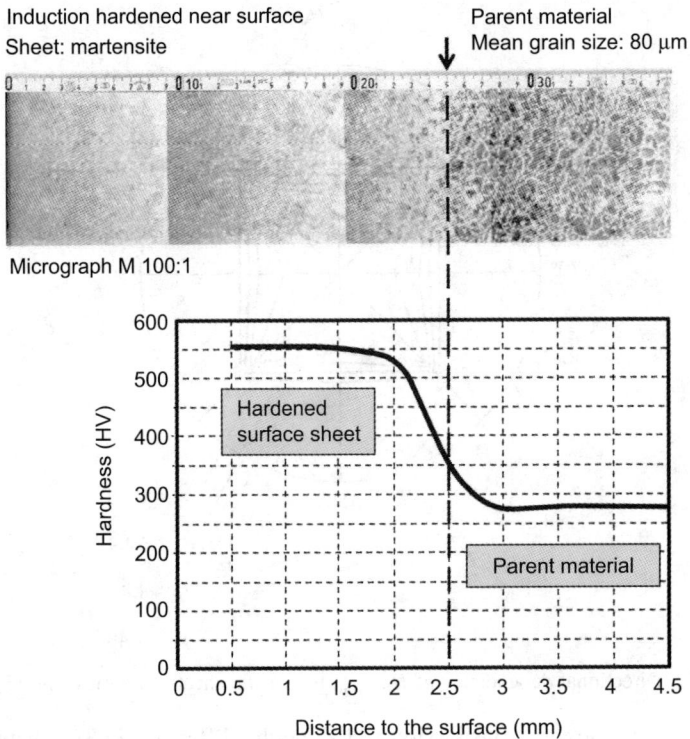

Figure 7.23 Micrographs of the transition zone at induction-hardened component, hardness distribution.

transition between the hardened surface layer and the parent metal, an enhanced scattering occurs.

The ultrasonic frequencies have to be selected in a range above 10 MHz so that sufficient scattering signals are generated. The principal measuring situation is shown in Fig. 7.24 (Willems, 1991). An ultrasonic pulse is coupled into the component in immersion or contact technique in such a way that inside the component a shear wave propagates under beam angles between 40 and 45 degree. Within the hardened layer the wave propagates without any significant scattering.

At the transition zone from the hardened surface layer to the base material, noticeable backscattering appears, which is received in the probe driven in pulse-echo technique. Fig. 7.25 (Willems, 1991) shows the corresponding backscattering signals (A-scan).

Using this technique at induction-hardened parts of steel components at which the extension of the transition zone between a fine-grained martensitic structure of the hardened surface layer to a coarse-grained structure of the tempered base material merges with a steep hardness gradient, the technique is applicable for hardening depth of >1.5 mm (Willems and Neumann, 1996). The ultrasonic backscattering method can be applied reliably at all hardenable steels, if the mean value of the grain size

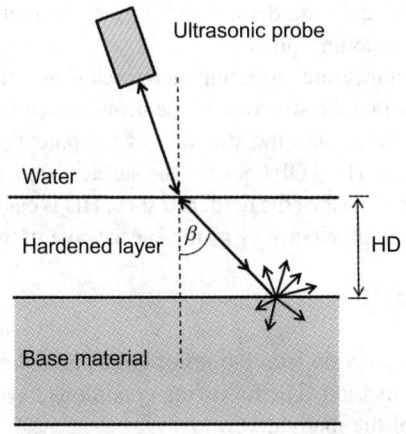

Figure 7.24 Surface hardening depth (HD): schematic sketch of ultrasonic transducer and backscattered signals.

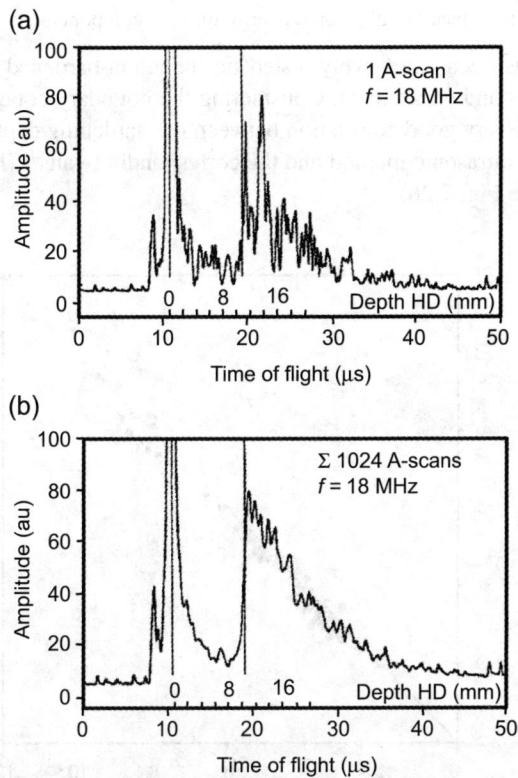

Figure 7.25 Typical backscattering signals at the transition zone from hardened surface layer to the base metal; (a) single A-scan, (b) averaged signal of 1024 A-scans.

is ≥ 20 μm before the induction hardening and if an exclusively martensitic transition takes place during the hardening process.

The frequency of the ultrasonic probe and therefore the wavelength of the shear wave is selected in such a way that the structure of the near-surface-hardened sheet only produces minor scattering effects and the structure of the parent metal, however, creates maximal scattering effects. The TOF between the surface echo and the scattering signal Δt is measured and the thickness of the hardened sheet HD is calculated using the known beam angle β and the sound velocity v_T of the shear wave in the material.

$$HD = (\Delta t \, v_T \cos \beta)/2 \qquad\qquad [7.30]$$

In most cases there exist good correlations of these parameters with the depth determined by the ultrasonic method. The following conditions have to be considered for a successful application of the method:

HD $> 1.5 - 2$ mm
Mean grain size of the parent metal $>20-30$ μm
Hardness gradient $> \sim 35$ HV/0.1 mm
Suitable wedges to induce the ultrasonic energy into the component

The method has been extensively tested on induction-hardened components, eg, roller bearing rings and crankshafts. Considering the boundary conditions mentioned above, there was a very good correlation between the hardening depth (HD$_{US}$) values determined by the ultrasonic method and the corresponding values (HD$_{met}$) gained by metallography (see Fig. 7.26).

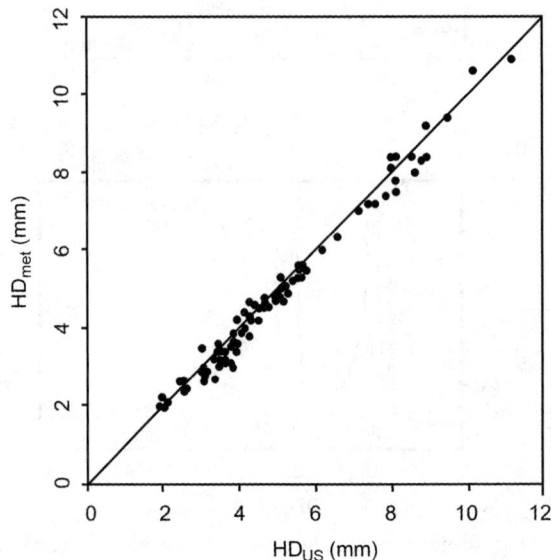

Figure 7.26 Results of hardening depth measurements by ultrasound (HD$_{US}$), compared to results of destructive tests (HD$_{destr}$).

The ultrasonic backscattering method comes to its application limits if surface-hardening depth smaller than ~ 1.5 mm should be evaluated. For surface-hardening depth less than 1.5 mm electromagnetic methods show a high potential (see chapters: Electromagnetic techniques for materials characterization and Hybrid methods for materials characterization). The ultrasonic method is not applicable either if the steel develops an intermediate microstructure due to cooling conditions as well as for steel quenched from the forging head, annealed and that has undergone a surface-hardening process and if the base metal shows a very fine-grained structure (grain size smaller than ~ 20 μm).

The described method has been a laboratory-tested method for a long time. Today there exist a large number of industrial applications in the metal working industry. On the market one can find at least three manufacturers offering different ultrasonic hardware systems (IZFP Rht testing device; Krautkrämer USLT, 2000; Metalscan TRAPUS).

7.3.4 Hardness, fracture toughness and yield strength

7.3.4.1 Hardness

Standard destructive hardness tests require taking samples, which is expensive and time consuming in a production line. These kinds of testing instruments are generally bulky and heavy. A nondestructive solution is to estimate the hardness level based on the correlation between ultrasound wave velocity and attenuation with the real hardness values determined using traditional methods. It needs to be mentioned that a determination of the hardness is also possible with electromagnetic methods (see chapters: Electromagnetic techniques for materials characterization and Hybrid methods for materials characterization). Results for hardness determination in steel evaluating ultrasonic velocity and attenuation are reported in Papadakis (1976), Briks and Green (1991) and Bouda et al. (2000). In each of these papers the results were obtained only on specific types of material. Various materials showed different relationships between hardness and the ultrasonic velocity. In Lukomski and Stepinski (2010) it is shown that ultrasonic velocity measurements can be used to determine the hardness of different types of martensitic steel without the need to adjust the fitting curve in each case. Ultrasonic TOF measurements with longitudinal and shear waves propagating in thickness direction were performed on 17 specimens that represent five types of high-quality martensitic steels (Steel types: AR400, AR450, AR500, AR550, AR600). For shear waves EMATs were used. The time measurement could be performed with a resolution of 0.01 ns by time averaging of 10 echoes, using the zero-crossing method and interpolation of the signal. The hardness was varied by tempering at temperatures of 200, 300, 400, 500 and 600°C between 585 (HB) and 255 (HB). Both chemical composition and thermal treatment of the specimens have been strictly controlled. All specimens showed a decreasing sound velocity with increasing hardness values. The hardness prediction applying multiple regression taking into account velocity, tempering temperature and 14 chemical elements is shown for the shear wave in Fig. 7.27.

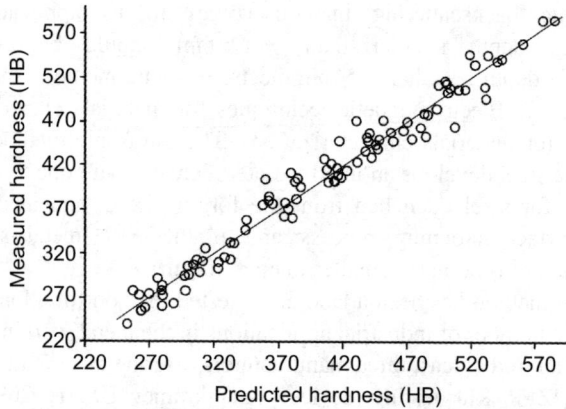

Figure 7.27 Hardness predicted by ultrasonic velocity applying multiple regression (Steel types: AR400, AR450, AR500, AR550, AR600).

These results demonstrate that nondestructive ultrasonic methods can be a very useful tool for continuous monitoring of steel hardening during the production process.

7.3.4.2 Fracture toughness and yield strength

Fracture toughness is an intrinsic material property characterizing the fracture behavior of a material. Fracture toughness is correlated to critical stress intensity factor K_c at which a thin crack in the material begins to grow. For plane stress conditions it is related to the tensile modulus (Young's modulus) E and the critical strain-energy release factor G_c (Vary, 1976) by the following relation:

$$K_c = \sqrt{EG_c}$$ [7.31]

G_c is a measure of the force required to extend the perimeter of a crack by a unit length. This quantity is related to the yield strength σ_y by $G_c = n\sigma_y\delta_c$. Here δ_c is a critical crack opening displacement and n is a numerical coefficient describing strain and associated factors.

Viewing for the ultrasonic assessment of K_c it is important that the quantity E in Eq. [7.31] is related to the ultrasonic velocity of a longitudinal wave v_L (see Eq. [7.13]). Otherwise, K_c is related to the microstructure of a material and because the attenuation of ultrasonic waves is related to the microstructure, K_c should be related to ultrasonic attenuation properties. In practice, K_c can be taken as the plane strain fracture toughness K_{1c}, and therefore a correlation is expected between fracture toughness and ultrasonic propagation properties via ultrasonic velocity and attenuation factors.

Vary (1976) could measure the ultrasonic attenuation as a function of the frequency (frequency range: 10–50 MHz) and found some empirical relations between the fracture toughness and attenuation factors related to grain sizes in two maraging steels and titanium alloys. The materials used for these measurements were specimens of three different materials for which fracture toughness values were mechanically determined. Three different sets of materials consisting of 200-grade, 250-grade maraging steel specimens and a set of titanium alloy specimens with the composition Ti-8Mo-8V-2Fe-3Al have been aged to produce different toughness and yield strength values. Fracture toughness measurements were carried out in conformance with plane strain criteria defined in ASTM E 399-70T, 4.2 (Brown, 1970). The ultrasonic measurements were made using piezoelectric crystal transducer and cylindrical fused quartz buffers. The nominal center frequencies of the broadband transducers were 10, 20, 30 and 50 MHz. For all these specimens the attenuation coefficients α have been measured as a function of the frequency. The slope of these curves $\beta_{\alpha 1} = d\alpha/df$ were evaluated at a frequency were α is 1 Np/cm.

Fig. 7.28 is a plot of $\beta_{\alpha 1}$ versus K_{1c} for the 200-grade maraging steel and the titanium alloy specimens.

It can be seen that the data can be grouped according to yield strength σ_y as well for the data points for the steel as for the data points for titanium. The data pairs can be connected with straight lines that are parallel to the lines for the 250-grade maraging steel. At each line the approximate yield strength shared by the data points is indicated. Thus the $\beta_{\alpha 1}$ versus K_{1c} plot for each material appears to consist of a set of lines that are isometric with respect to σ_y. The magnitude $\beta_{\alpha 1}$ can be written as:

$$\beta_{\alpha 1} = a + bK_{1c} \hspace{6cm} [7.32]$$

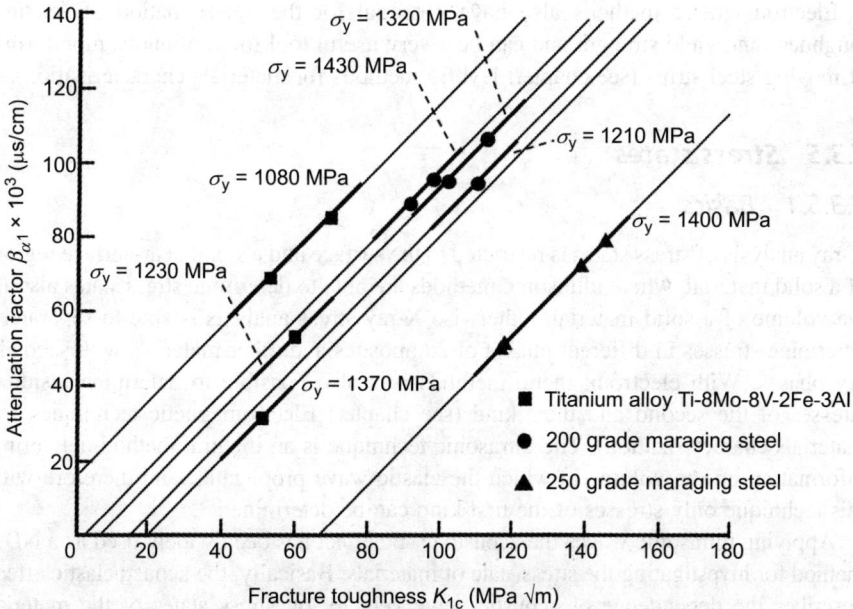

Figure 7.28 Ultrasonic attenuation factor $\beta_{\alpha 1}$ as a function of the fracture toughness K_{1c}.

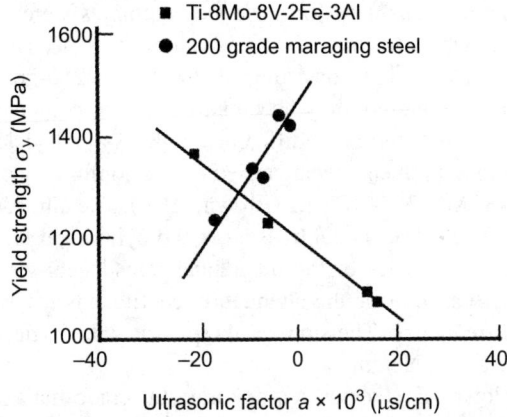

Figure 7.29 Yield strength σ_y as a function of the ultrasonic factor ($a = \beta_{\alpha 1} - bK_{1c}$).

where a is a function of σ_y and b is a conversion factor equal to 1×10^{-3} μs/(cm MPa$\sqrt{\text{m}}$). Further, all the lines of Fig. 7.28 share a common slope of unity. Under the assumption that the linear isometric grouping is correct, an equation for each line can be determined. This gives a set of values for a ($a = \beta_{\alpha 1} - bK_{1c}$) in the relation Eq. [7.32] for specimens with the same yield strength. Fig. 7.29 shows a plot of the yield strength as a function of the ultrasonic factor a for the two specimen materials with sufficient data for this plot. The results presented in Figs. 7.28 and 7.29 show that fracture toughness, yield strength and associated material properties are directly connected with ultrasonic propagation properties.

Electromagnetic methods also have potential for the determination of fracture toughness and yield strength and can be a very useful tool for continuous monitoring of moving steel strips (see chapter: Hybrid methods for materials characterization).

7.3.5 Stress states

7.3.5.1 Basics

X-ray analysis of stress states is restricted to the surface and a small near-surface region of a solid material, where ultrasonic methods are able to determine stress states also in the volume of a solid material. Otherwise X-ray stress analysis is able to separately determine stresses in different phases of composites and other materials with secondary phases. With electromagnetic methods it is also possible to determine residual stresses of the second and third kind (see chapter: Electromagnetic techniques for materials characterization). The ultrasonic technique is an integral method delivering information on the volume in which the elastic wave propagates, and therefore with this technique only stresses of the first kind can be determined.

Applying ultrasonic waves the acoustoelastic effect has been widely used as a NDT method for investigating the stress state of materials. Basically, the acoustoelastic effect describes the dependence of acoustic wave velocity on stress states of the material through which the wave travels. A body's density and elasticity change under stresses, and therefore this change results in a variation in acoustic wave velocity. Applying mechanical stress states to an isotropic solid, the elastic properties are changed and the isotropy is disturbed. The velocities of the elastic waves are influenced in such a way that their values depend on the magnitude and the direction of the stresses. Hughes and Kelly (1953) formulated expressions for the wave velocities in a stressed solid using Munaghan's theory of finite deformations (Munaghan, 1951) and third-order terms in the strain-energy expression. They showed that in the general case of a homogeneous triaxial strain field the wave velocities in an isotropic solid of cubic structure are given by:

$$\rho v_{11}^2 = \lambda + 2\mu + (2l + \lambda)\theta + (4m + 4\lambda + 10\mu)\varepsilon_1 \qquad [7.33]$$

$$\rho v_{12}^2 = \mu + (\lambda + m)\theta + 4\mu\varepsilon_1 + 2\mu\varepsilon_2 - 0.5n\varepsilon_3 \qquad [7.34]$$

$$\rho v_{13}^2 = \mu + (\lambda + m)\theta + 4\mu\varepsilon_1 + 2\mu\varepsilon_3 - 0.5n\varepsilon_2 \qquad [7.35]$$

where ρ, initial density; v_{11}, v_{12}, v_{13}, velocity of waves propagating in 1-direction with particle displacement in 1, 2, 3-direction, respectively; λ, μ, Lamé or second-order elastic constants; l, m, n, Munaghan's third-order elastic constants; ε_1, ε_2, ε_3, components of the homogeneous triaxial principal stains in the 1, 2, 3-direction.

$\theta = \varepsilon_1 + \varepsilon_2 + \varepsilon_3$

1, 2 and 3 are the axes of a Cartesian coordinate system. The first index on the velocity values gives the propagation direction and the second index the polarization or particle displacement direction. v_{11} is the velocity of a longitudinal wave propagating in the 1-direction. v_{12} and v_{13} are the velocities of two shear waves traveling in 1-direction and polarized in 2- and 3-directions perpendicular to each other.

The Eqs. [7.33]–[7.35] describe the acoustoelastic effect and are the fundamental equations for the ultrasonic evaluation of stain or stress states. These equations can only be applied if the elastic wave propagates and vibrates along principal axes. The above equations have been modified by Schneider (1999) to calculate the propagation velocities of elastic waves in cubic polycrystals with load or residual stresses. By replacing the factors $\lambda + 2\mu$ by ρv_{L}^2 and μ by ρv_{T}^2 and with $\rho(v_{11}^2 - v_{L}^2) = \rho(v_{11} - v_{L})(v_{11} + v_{L})$ using the approximation $v_{11} + v_{L} = 2v_{L}$ and using a similar approximation for shear waves, the following basic relations can be developed. Here the principal strains are expressed corresponding Hook's law by the elastic parameters and the principal stresses:

$$(v_{11} - v_{L})/v_{L} = A/C\sigma_1 + B/C(\sigma_2 + \sigma_3) \qquad [7.36]$$

$$(v_{12} - v_{T})/v_{T} = D/K\sigma_1 + H/K\sigma_2 + F/K\sigma_3 \qquad [7.37]$$

$$(v_{13} - v_{T})/v_{T} = D/K\sigma_1 + F/K\sigma_2 + H/K\sigma_3 \qquad [7.38]$$

Here the A, B, C, D, H, F and K are combinations of the elastic constants of the second and third order.

$$A = 2(\lambda + \mu)(4m + 5\lambda + 10\mu + 2l) - 2\lambda(2l + \lambda)$$

$$B = 2(2l + \lambda)(\lambda + \mu) - \lambda(2l + \lambda) - \lambda(4m + 5\lambda + 10\mu + 2l)$$

$$C = 4\mu(\lambda + 2\mu)(3\lambda + 2\mu)$$

$$D = 2(\lambda + \mu)(\lambda + m + 4\mu) - \lambda(2\lambda + 2m + 2\mu - 0.5n)$$

$$H = 2(\lambda + \mu)(\lambda + m + 2\mu) - \lambda(2\lambda + 2m + 4\mu - 0.5n)$$

$$F = 2(\lambda + \mu)(\lambda + m - 0.5n) - \lambda(2\lambda + 2m + 6\mu)$$

$$K = 4\mu^2(3\lambda + 2\mu)$$

The Eqs. [7.36]–[7.38] describe the influence of the principal stresses on the relative velocity changes of longitudinal and shear waves. The corresponding prefactors A/C, B/C, D/K, H/K and F/K weight the influences of the principal stress. Schneider has also formulated an equation to evaluate stress differences in two orthogonal principal axes using the Eqs. [7.33] and [7.34]. Normalizing the difference $v_{12} - v_{13}$ by one of the velocities, eg, by v_{13} the relationship $(v_{12} - v_{13})/v_{13} = ((4\mu + n)/4\mu)/(\varepsilon_2 - \varepsilon_3)$ results with the approximation $(v_{12} + v_{13})/v_{13} = 2v_T^2$. If the strains again are replaced by stress values (Hooke's law) and the velocities are expressed by time of flights and path lengths being identical for both shear waves, the following expressions have been formulated:

$$(t_{13} - t_{12})/t_{12} = (4\mu + n)(\sigma_2 - \sigma_3)/8\mu^2 \qquad [7.39]$$

Eq. [7.39] describes the ultrasonic birefringence effect, which allows the characterization of the stress states in terms of difference of the two principal stresses. This method offers two advantages for stress measurements using the ultrasonic technique namely the independence of the ultrasonic path length and the fact that only the third-order elastic constant n has to be considered. This third-order elastic constant is less influenced by changes in the microstructure compared to l and m. Further relations between stresses and ultrasonic velocities for skimming longitudinal waves, SH waves under grazing incidence and Rayleigh surface waves can be found in Schneider (1999). It should be mentioned that the application of the method is limited to materials with cubic single-crystal geometry.

A quantitative evaluation of stress states by ultrasonic techniques requires knowledge of the material-dependent elastic constants. The second-order constants λ and μ or Young's and shear moduli can be determined according to Eqs. [7.17] and [7.18] by measuring the longitudinal and shear wave velocity and the density of the material. The five possible different changes in wave velocities for longitudinal and shear waves have been measured by Eagle and Bray (1976) for the first time in a tensile test experiment. Fig. 7.30 shows the results of the relative changes of the wave velocities as a function of the elastic strain.

The strongest dependency on the elastic strain or stress shows the longitudinal wave propagating in the direction of the load (v_{11}). The velocity of a shear wave propagating perpendicular to the load direction (in thickness direction) with polarization in load

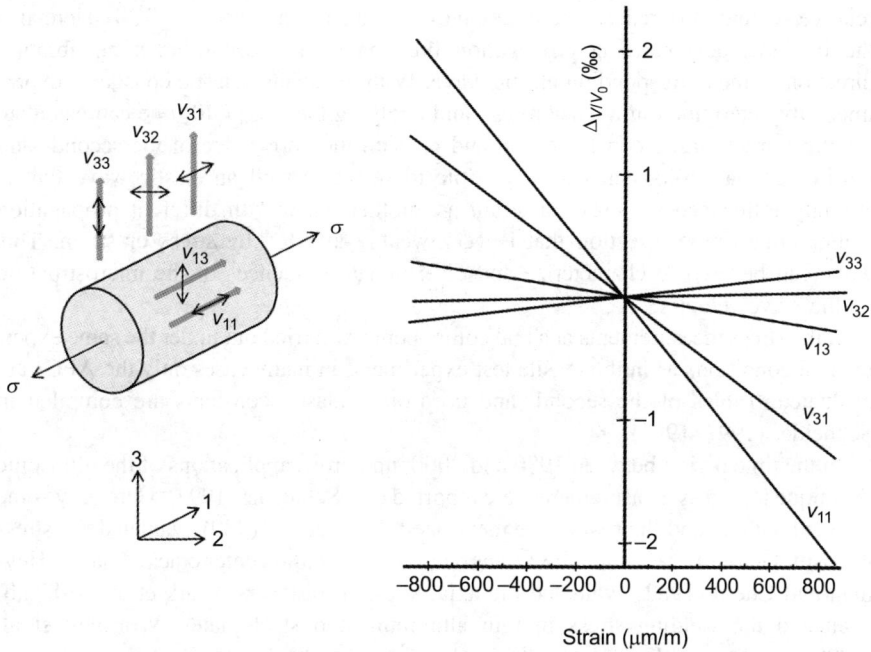

Figure 7.30 Relative changes of wave velocities $\Delta v/v_0$ (‰) as a function of strain.

direction (v_{31}) decreases significantly and increases slightly if the polarization direction is oriented perpendicularly to the load direction. The influence of stress or strain on the shear wave is stronger if the polarization is parallel to the load direction. Usually, the propagation directions of a longitudinal and a shear wave are oriented in the through-thickness direction perpendicular to the load. The particle displacements of the linearly polarized shear waves are parallel (v_{31}) and perpendicular to the load direction (v_{32}). For the described situation with the load in 1-direction and the propagation of the ultrasonic wave in 3-direction, the third-order elastic constants l, m and n are given by Schneider (1999):

$$l = \frac{\lambda(\lambda + \mu)}{\mu} \left\{ \frac{\lambda + 2\mu}{\lambda} AEC_{33} + \frac{4(\lambda + \mu)}{3\lambda + 2\mu} \left[AEC_{31} + \frac{\lambda}{2(\lambda + \mu)} AEC_{32} \right] + \frac{\lambda}{\lambda + \mu} \right\}$$

$$m = \frac{2(\lambda + \mu)}{3\lambda + 2\mu} \left\{ \lambda AEC_{32} + 2(\lambda + \mu)AEC_{31} \right\} - 2\mu$$

$$n = \frac{8\mu(\lambda + \mu)}{3\lambda + 2\mu} \left\{ AEC_{31} - AEC_{32} \right\} - 4\mu$$

$$[7.40]$$

AECs are the acoustoelastic constants, which are a function of the elastic constants of the second and third order. These constants are the slope of the linear change of the relative velocity differences as a function of the elastic strain (see Fig. 7.30). Here also the first index designates the propagation direction and the second index the vibration direction of the corresponding elastic wave. With the acoustoelastic constants, experimentally determined in a tension test, and applying the Eq. [7.40] one can calculate the third-order elastic constants l, m and n. With the knowledge of the second- and third-order elastic constants it is possible to select as well an elastic wave that is strongly influenced by stress or strain as another wave with different propagation and/or polarization direction that is very weakly affected by stress or strain. This wave can be used to characterize and discriminate influences of the microstructure on the wave velocity.

If the stress measurements at a real component are carried out under the same experimental conditions as in the tensile test experiment, in many cases only the AECs are evaluated. Tables of the second- and third-order elastic constants are compiled in Schneider (1997, 1999).

In the time period between 1980 and 2000, numerous applications of the ultrasonic technique for stress measurement were reported (see Schneider, 1997). Here only some selected authors and their work are mentioned. Kino et al. (1980) evaluated the stress intensity factor of cracks and the J integral of an aluminum center cracked panel. Heyman and Chern (1982) evaluated the axial stress in fasteners. Clark et al. (1987a,b) evaluated the welding stress in thin aluminum and steel plates. Wormley et al. (1990) developed a Fourier transform phase slope technique to determine the stress and texture by ultrasonic technique. A commercially available, semi-automatic ultrasonic system for the stress analysis in rails and railroad wheels was built by Deputat and coworkers (Deputat et al., 1989; Deputat, 1993). The first fully automated system for the precise measurement of time of flight and for the stress analysis at components is described in Herzer and Schneider (1989). Schneider et al. (1989) introduced a system for the automated time and locus continuous stress evaluation at surfaces of hardened rolls. Fukuoka et al. (1993) developed a measuring system to analyze stress in thin plates using an ultrasonic resonance method. The widest applications of ultrasonic techniques have been established in the field of stress evaluation in railroad wheels. Deputat introduced the DEBBIE system (Debbie, 2004) on the market. This system is portable and designed for use in the field. Schneider et al. (1989) developed the UER (Ultraschallgerät zur Bestimmung von Eigenspannungen in Radkränzen, engl.: Ultrasonic equipment for the determination of residual stresses in rims) system for fully automated evaluation of stress distribution along the height of wheel rims. These systems are applied in wheel inspection facilities of railroad workshops. In the following three examples, typical applications for the ultrasonic stress measurement technique will be presented.

7.3.5.2   Stresses in screws and bolts

The real load-induced stress in screws and bolts determines the quality of a screwed junction. Using the acoustoelastic effect, it is possible to analyze this stress. Using two different wave types, a longitudinal and a shear wave polarized perpendicular to the screw axis and propagating along the axis, the stress values can be determined

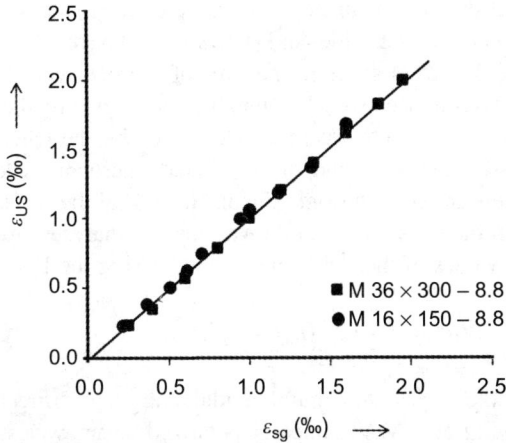

Figure 7.31 Strain measurements in screws, comparison between strain ε_{US} measured by ultrasound and strain ε_{sg} measured by strain gauge.

even if the screw has been in place for a long time period and if the original length of the screw is unknown. According to Fig. 7.30 the velocity of the longitudinal (v_{11}) and the velocity of the shear wave (v_{13}) are changed with growing strain. So the difference of both changes depends also linearly on the strain. Rearranging the Eqs. [7.33] and [7.35] (Schneider, 1999) for the stress value in axial direction (1-direction) the following relations result for the stress:

$$\sigma_1 = (t_{11} - t_{13}Q)/(Dt_{13}Q/K - At_{11}/C) \qquad [7.41]$$

with t_{11}, TOF of longitudinal wave; t_{13}, TOF of shear wave.

$$Q = t_L/t_T = (\mu/(\lambda + 2\mu))^{1/2}$$

Q is the quotient of the TOFs of the two waves in the unloaded screw.

A, C, D and K are the weighting factors, given in the Eqs. [7.36]−[7.38]. A further important fact developing the Eq. [7.41] is the situation that the propagating paths of the two waves are the same as in both the unloaded and the loaded situations.

Fig. 7.31 shows a result of the strains on screws measured with the method described above in comparison to the results determined by strain gauges.

There is very good agreement between the measuring values of both methods.

7.3.5.3   Residual stress measurements in rims of railroad wheels

During braking procedures with shoe-type brakes, rims of railroad wheels undergo heavy thermal loads, which change the residual stress state. Tension residual stresses are generated in circumferential direction. It can happen that small microcracks grow under these residual stresses and in the worst case can lead to the breakoff of the wheel. Schneider et al. (1994) and Schneider and Herzer (1998) have developed a method and system to evaluate the stress states in the rims of railroad wheels. The near-surface circumferential stresses in used wheels cannot be determined by the ultrasonic method due to the changes of the microstructure state caused by the wheel rail contact. The changes in microstructure state and the generated microcracks lead to changes of the elastic parameters and the ultrasonic signals. Instead of the surface stresses circum-ferential changes in the stress states of the volume are analyzed in order to determine maximal tolerable values of these volume stresses. Using the Eq. [7.39] in the form:

$$\sigma_{\text{tan}} - \sigma_{\text{rad}} = \left(\left(8\mu^2 / (4\mu + n) \right) (t_{\text{rad}} - t_{\text{tan}}) \right) / t_{\text{tan}}$$

the difference of the stress in circumferential and radial direction $\sigma_{\text{tan}} - \sigma_{\text{rad}}$ is determined. Applying an EMAT, linearly polarized shear waves (see Fig. 7.8) are insonified at one side of the rim and the transducer is scanned in 1-mm steps while the shear wave is polarized in tangential direction and in a second scan with polarization in radial direction (upper part of Fig. 7.32). The material specific parameter $8\mu^2/(4\mu + n)$ has been determined for different wheel material. Fig. 7.32 shows the change in the residual stress states as a function of systematically enhanced breaking power realized in a breaking test facility.

Because of the fact that stresses acting in radial direction are small and do not change significantly during the braking process, the measured results of the differences $\sigma_{\text{tan}} - \sigma_{\text{rad}}$ show mainly the circumferential stress values.

7.3.5.4   Stresses in rolls

Rolls mainly used in the steel and aluminum industry are hardened with the goal to produce hardness values and hardness depths as high as possible. With respect to the performance lifetime of rolls, the residual stresses in the near-surface zones and their homogeneous distribution are further quality factors of the rolls. To determine the near-surface stress states using the ultrasonic method, transmitter and receiver probes for longitudinal and shear waves are applied at fixed distances. So the measured value is the time of flight of the ultrasonic wave traveling between transmitter and receiver. The relative differences in time of flights for the longitudinal and shear wave as a function of the triaxial stress state are given (Schneider et al., 2009) (see also Eqs. [7.36] and [7.37]):

$$(t_{\text{L}0} - t_{\text{L}})/t_{\text{L}0} = (A/C)\sigma_{\text{axial}} + (B/C)(\sigma_{\text{circ}} + \sigma_{\text{rad}})$$
$$(t_{\text{T}0} - t_{\text{T}})/t_{\text{T}0} = (D/K)\sigma_{\text{axial}} + (H/K)\sigma_{\text{circ}} + (F/K)\sigma_{\text{rad}}$$

[7.42]

A, B, C, D, H, F and K are combinations of the elastic constants of the second-order (E-modulus, shear modulus) and the third-order elastic constants, the so-called acous-toelastic material parameters (see Eqs. [7.36] and [7.37]). As mentioned before these

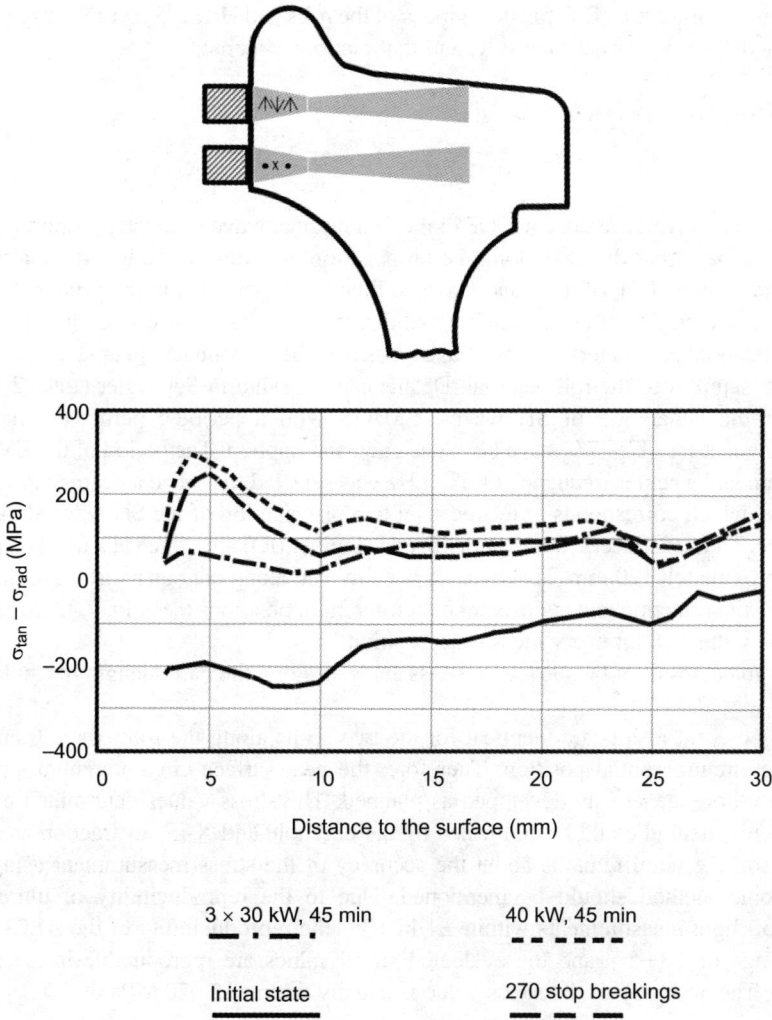

Figure 7.32 Stress states in rims of railroad wheels after braking tests with different braking power and braking procedures.

parameters weight the influences of the corresponding principal stress components on the wave velocities. The TOFs of the longitudinal and shear wave for the stress free material, t_L and t_T are the TOFs of the corresponding waves traveling in axial direction along the length of the roll. The vibration direction of the longitudinal wave is oriented in the axial direction and the shear wave is an SH wave polarized in circumferential direction. The principal stress directions in axial, circumferential and radial direction are σ_{axial}, σ_{circ}, σ_{rad}. The above equations can be simplified because the factor A/C is 20 times larger than B/C for steel grades of the rolls and H/K is 3−5 times larger than D/K and 10 times larger than F/K, and the relations become:

$$(t_{L0} - t_L)/t_{L0} = (A/C)\sigma_{axial}$$

$$(t_{T0} - t_T)/t_{T0} = (H/K)\sigma_{circ}$$
[7.43]

So the relative difference in TOF of the longitudinal wave is directly proportional to the principal stress directed along the propagation direction of the longitudinal wave and the relative TOF of the shear wave is directly proportional to the principal stress in the vibration direction (circumferential direction) of the shear wave. The acoustoelastic material parameters for both wave types have been evaluated in uniaxial pressure tests at samples of the roll material. Details can be found in Schneider et al. (2009).

For the generation of SH waves, EMATs with a periodic permanent magnet arrangement (see Fig. 7.9) have been designed and applied. The period of the EMATs is 5 mm and a center frequency of 620 kHz was selected to realize a beam angle of 90 degree, which corresponds in ferritic steel to a propagation of the SH wave along the surface. The transducers (transmitter (T) and receiver (R)) are driven at a fixed distance of approximately120 mm. This unit has been moved along the outer surface of the roll in the axial direction and at different circumferential positions (see Fig. 7.33) to record precisely the TOF at every measuring position.

A typical result of the ultrasonic stress analysis along the barrel length of a cold roll shows Fig. 7.34.

The present results are identical for the two scans along the roll axis at 0 and 90 degree circumferential position. Therefore, the near-surface circumferential stresses σ_{circ} are homogeneously developed as planned. The stress values determined by the ultrasonic method could be confirmed by the borehole and X-ray diffraction method. Here some general remarks about the accuracy of the stress measurement using the ultrasonic method should be mentioned. Due to the reproducibility of ultrasonic time-of-flight measurements within ± 1 in 10^{-4} the reproducibility of the AECs is in the range of $\pm 3−5\%$ and the evaluated stress values are reproducible in the same range. The accuracy of the stress value is usually within 15−20 MPa $\pm 10\%$.

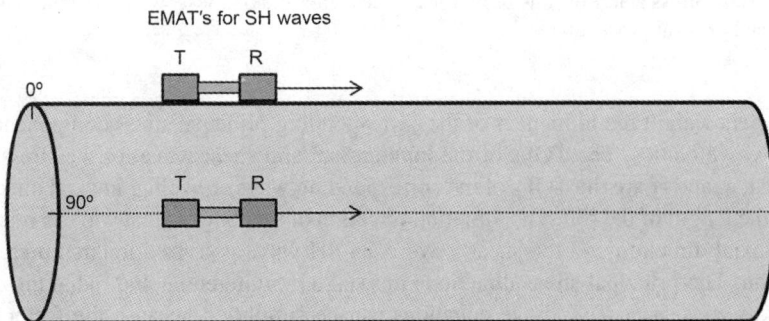

Figure 7.33 Sketch of the transmitter (T) and receiver (R) arrangement for stress measurements at rolls.

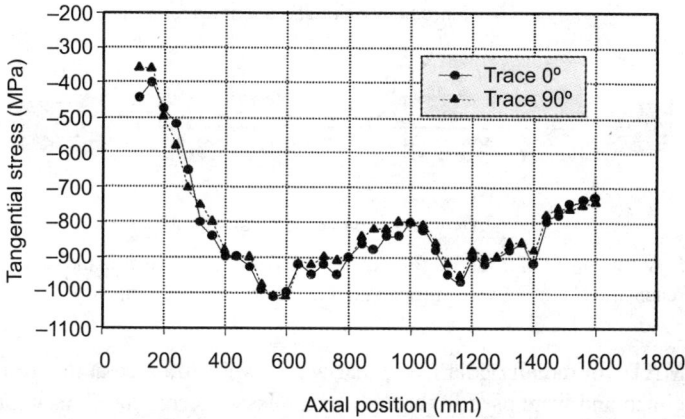

Figure 7.34 Tangential stress values at a roll at two different circumferential positions measured by ultrasonic technique.

7.3.6 Fatigue

Fatigue of materials takes place during plant operation. In power plants and especially in nuclear power plants (NPPs) many components are exposed to thermomechanical load conditions by thermal fluctuations, stratification of cold and hot water, turbulences and vibrations. Austenitic components in the primary circuit are particularly sensitive to thermomechanical loading because of their low thermal conductivity, which can cause high temperature gradients and locally concentrated plastic deformations. These thermomechanical conditions inside the primary circuit generate both high-cycle and low-cycle fatigue and can lead to microstructural changes and microcracks in the material. At low temperatures (room temperature) the plastic deformation can result in a phase transformation from paramagnetic austenite into ferromagnetic α'-martensite. At elevated temperatures, where predominantly the stable austenitic phase exists, ie, changes in dislocation density and arrangement can occur and lead to a further state to intrusion/extrusion and microcracks at the surface. It is of high importance for the safety of NPPs to know the fatigue state of components. Thus, monitoring systems able to give information about the microstructure of components is desirable. It is well known that the microstructure and dislocation density influence the ultrasonic velocity and attenuation (Othani et al., 2006; Zuev et al., 2002; Shankar et al., 2001; Vasudevan and Palanichamy, 2002; Palanichamy et al., 2001; Hirao et al., 2000). Therefore, ultrasonic techniques should offer high potential for monitoring different microstructural changes that occur in a material during fatigue. In NPPs operation temperatures of about 300°C exclude the application of conventional piezoelectric ultrasonic transducers, which require a couplant medium between transducer and the surface of the component under test. Therefore, EMATs have been developed and applied within the frame of a research project with austenitic stainless steel AISI 347 specimens to detect and evaluate fatigue-induced microstructural changes by thermoelastic/plastic and mechanical stress (Altpeter et al., 2011a,b). Fig. 7.35 shows the principal measuring arrangement.

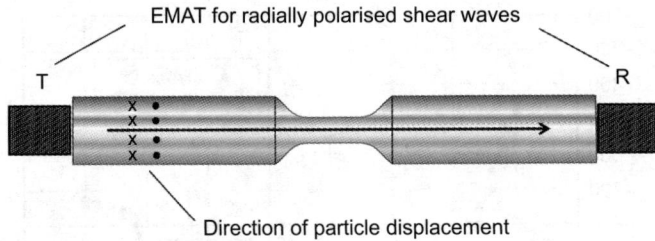

Figure 7.35 Arrangement of EMAT's for radially polarized shear waves at the austenitic fatigue specimen, schematic sketch of wave propagation and particle displacement.

Two EMATs for radially polarized shear waves were arranged at the front surfaces of the specimen and were used in through-transmission technique. This configuration of specimen and transducers was integrated in a servohydraulic testing machine (see Fig. 7.36). The specimens were investigated under different constant total strain amplitude loadings (0.8, 1.0, 1.2 and 1.6%) at room temperature (RT) and at 300°C with a strain ratio $R_\varepsilon = -1$ using triangular load time functions and a frequency of 0.01 Hz.

Besides the nondestructive ultrasonic measurements with EMATs and a Feritscope sensor, the complete stress–strain hysteresis strain measurements have been carried out for the microstructure-related description of the cyclic behavior (Altpeter et al., 2011a,b). At all experiments, especially during the measurements at 300°C, the temperature (T) in the area of the fatigued material was continuously recorded. The heating was induction heating. During the high temperature tests the EMATs integrated in the mechanical clamping devices of the specimen could be cooled, so that the heat flow

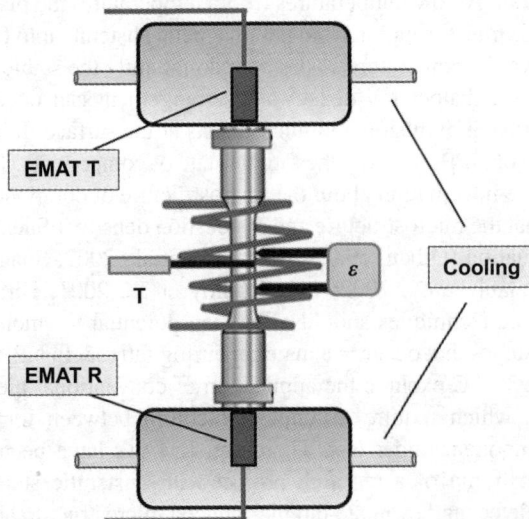

Figure 7.36 Sketch of the measuring setup in a servohydraulic testing machine (Altpeter et al., 2011a,b).

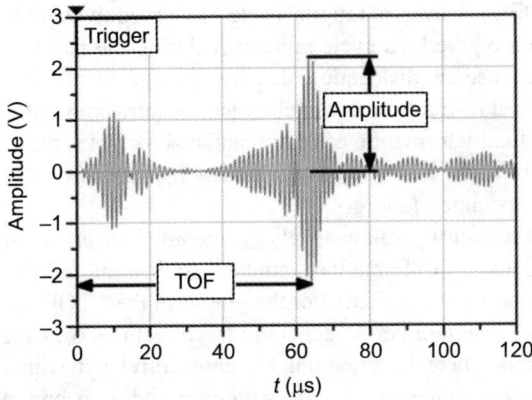

Figure 7.37 EMAT through transmission signal at the fatigue specimen, evaluated time of flight (TOF) and signal amplitude.

from the heated specimen in the clamping mechanic could be reduced. Fig. 7.37 shows a typical through-transmission signal passed through the austenitic fatigue specimen. During the measurements the amplitude and the TOF of the ultrasonic signal was recorded and evaluated. These two parameters carry the highest potential to characterize the fatigue behavior. By connecting the ultrasonic testing system with the servo-hydraulic measuring system and synchronizing both systems with a trigger, reliable data could be recorded during the fatigue tests. So it is guaranteed that during the fatigue tests the data are recorded simultaneously.

Fig. 7.38 shows an example of the time of flight signal measured during a fatigue test experiment with a total strain amplitude $\varepsilon_{at} = 1\%$ at room temperature as a function of the load cycle number N. One can see that the deformation of the specimen during a fatigue cycle is reproduced in the TOF signal. The mean value of the time of

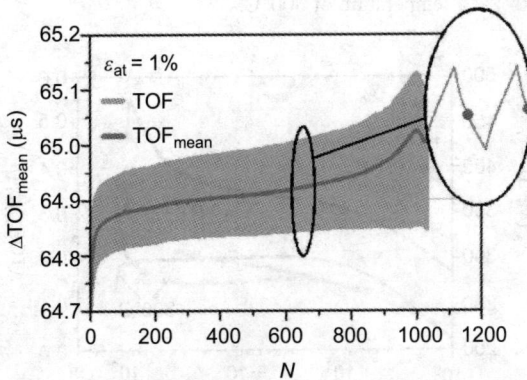

Figure 7.38 Ultrasonic time-of-flight (TOF) as a function of the load cycle number (N) for a fatigue experiment at room temperature.

flight signal (TOF_{mean}) over a fatigue cycle shows equivalent to the TOF signal changes depending on the load cycle number N during the fatigue experiment.

Due to an increase in dislocation density (Smaga et al., 2008), formation of deformation-induced α'-martensite, development of intrusions/extrusions at the specimen surface and finally formation of micro- and macrocracks, the change in the mean value of the time of flight ΔTOF_{mean} continuously increases from the beginning of the fatigue test up to specimen failure.

Results for the measuring value ΔTOF_{mean} recorded during a fatigue test at a temperature of 300°C and for different constant total strain amplitude loadings of 0.8, 1.0, 1.2 and 1.6% are shown in Fig. 7.39. For the graphs of the ΔTOF_{mean} curves as a function of the fatigue cycle number N, generally three sections over the fatigue life have been observed. First there is an initial increase correlated with cyclic hardening, followed by a decrease due to a cyclic softening and a secondary increase of the ΔTOF_{mean} values correlating with the development of intrusions and extrusions at the specimen surface, and finally the propagation of fatigue cracks.

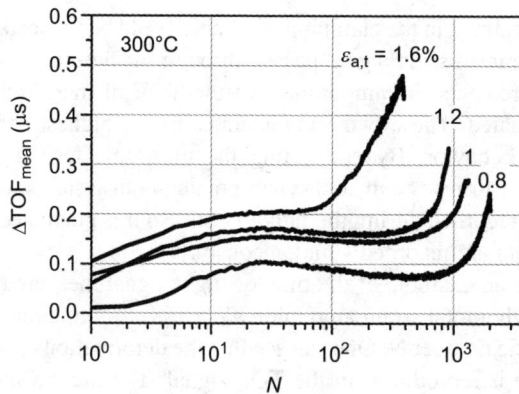

Figure 7.39 Changes of the mean value of the time of flight (ΔTOF_{mean}) as a function of the load cycle number (N) at a temperature of 300°C.

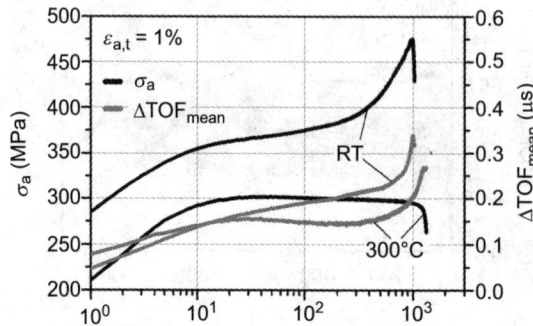

Figure 7.40 Development of stress amplitude σ_a and change in mean value of the time of flight (ΔTOF_{mean}) as a function of the load cycle number (N) at room temperature (RT) and at 300°C.

In Fig. 7.40 a comparison of the cyclic deformation behavior of AISI 347 at RT and $T = 300°C$ is presented as well for the mean changes in time of flight ΔTOF_{mean} as for the stress amplitudes σ_a. Here the high potential of the change in the mean value of time of flight to characterize the cyclic deformation behavior and the actual fatigue state is demonstrated.

Comparing the σ_a and the ΔTOF_{mean} curves at 300°C in the range of cycling number $N = 1 - 300$, it is obvious that both cyclic hardening and softening can be detected evaluating ΔTOF_{mean}. Furthermore, after cyclic softening or saturation typical for cyclical loaded austenic steels at elevated temperatures, a secondary increase of the ΔTOF_{mean} values for $N > 300$ can be directly correlated with the specimen surface and the development of microcracks. In contrast to the stress amplitude σ_a, which continuously decreases, ΔTOF_{mean} starts to increase as a response to the formation of the first microcracks. Therefore, the changes of the TOF of the ultrasonic signal of the radially polarized shear wave can be used as a reliable measuring value for the detection of fatigue processes.

7.4 Conclusions

After a short summary of the basics of the ultrasonic technique, selected examples of applications for materials characterization have been presented for the determination of mechanical and microstructural properties of metallic components such as grain size, texture, surface-hardening depth, hardness, fracture toughness, yield strength, stress states and fatigue.

As ultrasonic measuring parameters, scattering, sound velocity and attenuation are applied. Besides piezoelectric ultrasonic transducers, EMATs are used, especially for dynamic measurements, and in situations if a dry and contactless generation and reception of ultrasonic waves is necessary. In particular, the use of SH waves is only possible with EMATs.

With the advances in electronics and digital technology, ultrasonic measuring parameters, influenced by changes in material properties, can be measured with high accuracy.

For quantitative material characterization, in most cases empirical correlations and calibration must be established for each material.

For fatigue characterization at service temperatures in the range of 300°C, the ultrasonic measuring parameter of time of flight of shear waves excited and received by EMATs is a proper tool especially in the last third of the materials' lifetime.

7.5 Future trends

Up to now mainly only single ultrasonic measuring parameters have been evaluated for materials characterization. Combining two or more measuring parameters in a multi-parameter approach as in the case of the electromagnetic technique (see chapter: Hybrid methods for materials characterization) the measuring accuracy and reliability should be further improved. In the future, a combination of different techniques, eg, a combination of the ultrasonic and magnetic technique, should improve the determination of microstructural/mechanical properties of metallic materials.

References

Achenbach, J.D., 1973. Wave Propagation in Elastic Solids. Elsevier, Amsterdam.

Allen, D.R., Langmann, R., Sayers, C.M., 1985. Ultrasonic SH-wave velocity in textured aluminum plates. Ultrasonics 9, 215.

Altpeter, I., Dobmann, G., Hübschen, G., Fraunhofer-Gesellschaft, 2010. Verfahren zur spannungsunabhängigen Texturbestimmung eines ferromagnetischen Werkstückes (DE patent 10 2008 033 755). January 21, 2010.

Altpeter, I., Tschuncky, R., Hällen, K., Dobmann, G., Boller, C., Smaga, M., Sorich, A., Eifler, D., 2011a. Early detection of damage in thermo-cyclically loaded austenitic materials. In: ENDE 2011 Proceedings, ENDE 2011 Conference, March 10–12. ISO Press, Chennai.

Altpeter, I., Tschuncky, R., Hällen, K., Dobmann, G., Boller, C., 2011b. Early Detection of Damage in Thermo-Cyclically Loaded Austenitic Materials — Method Development for Monitoring within the Context of the Aging Management. IZFP Saarbrücken, WKK Kaiserslautern, Final Report No. 1501379. Reactor Safety Research, Germany.

Auld, B.A., 1973. Acoustic Fields and Waves in Solids, vol. 1. Wiley, New York.

Batthia, A.B., 1967. Ultrasonic Absorption. Clarendon Press, Oxford.

Batthia, A.B., Moore, R.A., 1959. Scattering of high frequency sound waves in polycrystalline materials II. Journal of Acoustical Society of America 31, 1140.

Beecham, D., 1966. Ultrasonic scatter in metals its properties and its application to grain size determination. Ultrasonics 4, 67–76.

Borsutzki, M., Thoma, C., Bleck, W., Theiner, W., 1993. On-line Bestimmung von Werkstoffeigenschaften an kaltgewalztem Feinblech. Stahl und Eisen 113 (Nr. 10), 93–98.

Badidi Bouda, A., Benchaala, A., Alem, K., 2000. Ultrasonic characterization of materials hardness. Ultrasonics 38 (1–8), 222–227.

Brown Jr., W.F., 1970. Review of Developments in Plane Strain Fracture Toughness Testing. ASTM STP-463. American Society for Testing and Materials, Philadelphia.

Briks, A.S., Green Jr., R.E., 1991. Nondestructive Testing Handbook. American Society for Nondestructive Testing.

Bunge, H.J., 1968. Kristall und Technik 3, 431–438.

Bunge, H.J., 1979. Textur und Anisotropie. Zeitschrift fuer Metallkunde 70, 411.

Bunge, H.J., 1982. Texture Analysis in Materials Science. Butterworth's, London.

Clark Jr., A.V., Govada, A., Thompson, R.B., Smith, J.F., Blessing, G.V., Delsanto, P.P., Mignogna, R.B., 1987a. The use of ultrasonics for texture monitoring in aluminium alloys. In: Thompson, D.O., Chimenti, D.E. (Eds.), Review of Progress in Qantitative NDE, vol. 6B. Plenum Press, New York, pp. 1515–1524.

Clark Jr., A.V., Moulder, J.C., Mignogna, R.B., Delsanto, P.P., 1987b. Ultrasonic determination of absolute stresses in aluminium and steel alloys. In: Macherauch, E., Hauck, V. (Eds.), Residual Stresses in Science and Technology, Oberursel, vol. 1, pp. 207–214.

Debbie, 2004. Ultrasonic Measurement of Hoop Stress in the Rim of Monobloc Railroad Wheel, User's Manual. Debro UMS, Warsaw.

Deputat, J., Kwaszczynska-Klimek, A., Szelazek, J., 1989. Monitoring of residual stress in railroad wheels with ultrasound. In: Boogard, J., van Dijk, G.M. (Eds.), Proceedings of 12th World Congress. Elsevier Science Publishers B.V., Amsterdam, pp. 974–976.

Deputat, J., 1993. Ultrasonic measurement of residual stress under industrial conditions. Acoustica 79, 161–169.

Eagle, D.M., Bray, D.E., 1976. Measurement of acoustoelastic and third order elastic constants for rail steel. Journal of Acoustical Society of America 60, 741–744.

Fay, B., 1973. Theoretische Betrachtungen zur Ultraschallrückstreuung. Acustica 28, 354–357.

Fukuoka, H., Hirao, M., Yamasaki, T., Ogi, H., Petersen, G.L., Fortunko, C.M., 1993. Ultrasonic resonance method with EMAT for stress measurement in thin plates. In: Thompson, D.O., Chimenti, D.E. (Eds.), Review of Progress in Quantitative Nondestructive Evaluation, vol. 12. Plenum Press, New York, pp. 2119–2136.

Goebbels, K., Höller, P., 1976. Quantitative determination of grain sizes by means of scattered ultrasound. In: WCNDT '76. 8th World Conference on Non-destructive Testing. Proceedings. World Conference on Non-Destructive Testing, Cannes.

Goebbels, K., 1980. Structural analysis by scattered ultrasonic radiation. In: Sharpe, R.S. (Ed.), Research Techniques in Non Destructive Testing, vol. 4. Academic Press, London, pp. 87–158.

Goebbels, K., Höller, P., 1980. Quantitative determination of grain size and detection of in homogeneities in steel by ultrasonic backscattering measurements. In: Berger, H., Linzer, M. (Eds.), Proceedings of the First International Symposium on Ultrasonic Materials Characterization Held at NBS, Gaithersburg, MD, June 7–9, 1978, pp. 67–74.

Goebbels, K., 1994. Materials Characterization for Process Control and Product Conformity. CRC Press, Boca Raton.

Hecht, A., Thiel, R., Neumann, R., Mundry, E., 1981. Nondestructive determination of grain size in austenitic sheet by ultrasonic backscattering. Materials Evaluation 39, 934–938.

Herzer, R., Schneider, E., 1989. Instrument for the automated ultrasonic time of flight measurement — a tool for materials characterization. In: Höller, P., Hauk, V., Dobmann, G., Ruud, C., Green, R. (Eds.), Nondestructive Characterization of Materials. Springer Verlag, Berlin, Heidelberg, pp. 673–680.

Heyman, J.S., Chern, E.J., 1982. Ultrasonic measurements of axial stress. Journal of Testing and Evaluation 10, 202–211.

Hirao, M., Fukuoka, H., Fujisawa, K., Murayama, R., 1994. On-line monitoring of steel sheet drawability using EMATS. In: Proceedings of the 1994 Far East Conference on NDT (FENDT '94) and ROCSNT Ninth Annual Conference. R.O.C., Taipei, China, pp. 63–70.

Hirao, M., Ogi, H., Suzuki, N., et al., 2000. Ultrasonic attenuation peak during fatigue of polycrystalline copper. Acta Metallica 48, 517–521.

Hübschen, G., 2012. Electromagnetic acoustic transducers. In: Nakamura, K. (Ed.), Ultrasonic transducers — materials and design for sensors and medical applications. Woodhead, Cambridge, pp. 36–69.

Hughes, D.S., Kelly, J.L., 1953. Second order elastic deformation of solids. Physics Reviews 92 (5), 1145–1149.

Hyoguchi, T., Kawashima, K., Nippon Steel Corporation, 1995. Methods for Measuring Properties of Cold Rolled Thin Steel Sheet and Apparatus Therefor (U.S. patent application 419, 968), April 10, 1995.

IZFP Rht Testing Device 3121. http://www.izfp.fraunhofer.de.

Kawashima, K., 1990. Nondestructive characterization of texture and plastic strain ratio of metal sheets with electromagnetic acoustic transducers. Journal of Acoustical Society of America 87 (2), 681−690.

Kino, G.S., Barnett, D.M., Gayeli, N., Herrmann, G., Hunter, J.B., Ilié, D.B., Johnson, G.C., King, R.B., Scott, M.P., Shyne, J.C., Steele, C.R., 1980. Acoustic measurements of stress fields and microstructure. Journal of Nondestructive Evaluation 1, 67−77.

Kopineck, H.J., Otten, H., Bunge, H.J., 1989. On-line determination of technological characteristics of cold and hot rolled steelbands by a fixed angle texture analyzer. In: Höller, P. (Ed.), Nondestructive Characterization of Materials III. Springer, Berlin/West.

Krautkrämer, J., Krautkrämer, H., 1977. Ultrasonic Testing of Materials, second ed. Springer Verlag, Berlin.

Krautkrämer USLT, 2000. http://www.GEInspectionTechnologies.com.

Langenberg, K.J., Marklein, R., Meyer, K., 2009. Theoretische Grundlagen der zerstörungsfreien Materialprüfung mit Ultraschall. Oldenburg Wissenschaftsverlag, München.

Lankford, W.T., Snyder, S.C., Bauscher, J.A., 1950. New criteria for predicting the performance of deep drawing sheets. Transactions ASM 42, 1197−1232.

Lukomski, T., Stepinski, T., 2010. Steel hardness evaluation based on ultrasonic velocity measurements. Insight 52 (10), 592−596.

Metalscan TRAPUS. http//www.metalscan.fr.

Murayama, R., Fujisawa, K., Fukuoka, H., Hirao, M., 1996. Development of an on-line evaluation system of formability in cold-rolled steel sheets using electromagnetic acoustic transducers (EMAT's). NDT & E International 29, 141−146.

Murnaghan, F.D., 1951. Finite Deformation of an Elastic Solid. Wiley, New York.

Musgrave, M.J.P., 1970. Crystal Acoustics. Holden-Day, San Francisco.

Nakamura, K., 2012. Ultrasonic Transducers-Materials and Design for Sensors, Actuators and Medical Applications. Woodhead Publishing, Cambridge.

Othani, T., Nishiama, K., Yoshikawa, S., et al., 2006. Ultrasonic attenuation and microstructural evolution throughout tension-compression fatigue of a low carbon steel. Materials Science and Engineering A442, 466−470.

Palanichamy, P., Jayakumar, T., Raj, B., 1995. Ultrasonic velocity measurements for estimation of grain size in austenitic stainless steel. NDE & E International 28 (3), 179−185.

Palanichamy, P., Mathew, M.D., Latha, S., Jayakumar, T., Bhanu, K., Rao, S.K., Mannan, S.L., Raj, B., 2001. Assessing microstructural changes in alloy 625 using ultrasonic waves and correlation with tensile properties. Scripta Materialia 45 (9), 1025−1030.

Papadakis, E.P., 1976. Ultrasonic velocity and attenuation measurement methods with scientific and industrial applications. In: Physical Acoustics: Principles and Methods, vol. XII. Academic Press, New York.

Pongratz, H.J., Willems, H., Arnold, W., 1983. Numerical smoothing of ultrasonic backscattered signals by convolution. Nondestructive Testing Communications 1, 19−25.

Roe, R.J., 1965. Description of crystalyte orientation in polycrystalline materials. Journal of Applied Physics 36, 2024−2031.

Rose, J.L., 1999. Ultrasonic Waves in Solid Media. University Press, Cambridge.

Schneider, E., Herzer, R., Bruche, D., 1989. Automatisierte Bestimmung Oberflächennaher Spannungszustände in Walzen Mittels Ultraschallverfahren. DGZfP Berichtsband, Berlin, pp. 419−426.

Schneider, E., Herzer, R., Bruche, D., Frotscher, H., 1994. Ultrasonic characterization of stress states in rims of railroad wheels. In: Green Jr., R.E., et al. (Eds.), Nondestructive Characterization of Materials. Plenum Press, New York, pp. 383−390.

Schneider, E., 1997. Ultrasonic techniques. In: Hauk, V. (Ed.), Structural and Residual Stress Analysis by Nondestructive Methods. Elsevier Science B.V., Amsterdam, pp. 522−533.

Schneider, E., Herzer, R., 1998. Ultrasonic evaluation of stress states in rims of railroad wheels. In: Proceedings of the 7th ECNDT, Copenhagen, pp. 1972−1979.

Schneider, E., 1999. Untersuchung der materialspezifischen Einflüsse und verfahrenstechnische Entwicklungen der Ultraschallverfahren zur Spannungsanalyse an Bauteilen (Ph.D. thesis, D 82). RWTH, Aachen.

Schneider, E., Stroh, M., Lejeune, I., 2007. Zerstörungsfreie Bestimmung der Oberflächenhärtetiefe mittels Ultraschallverfahren-Erfahrungen und notwendige Weiterentwicklungen. In: DGZfP-jahrestagung Fürth 2007, DGZfP-berichtsbände 104-CD, Vortrag 19.

Schneider, E., Herzer, R., Hübschen, G., Wildau, M., Steinhoff, K., 2009. Bestimmung des oberflächennahen Spannungszustandes von Walzen. In: DGZfP-jahrestagung, Poster 31, Münster.

Shankar, P., Palanichamy, P., Jayakumar, T., et al., 2001. Nitrogen redistribution, microstructure and elastic constant evaluation using ultrasonics in aged 316LN stainless steels. Metallurgical and Materials Transactions A 32, 2959−2968.

Sayers, C.M., Proudffood, G.G., 1986. Angular dependence of the ultrasonic SH-wave velocity in rolled metal sheets. Journal of the Mechanics and Physics of Solids 34, 579−592.

Smaga, M., Walter, F., Eifler, D., 2008. Deformation induced martensitic transformation in metastable austenitic steels. Materials Science and Engineering A 483, 394−397.

Spies, M., Schneider, E., 1989. Nondestructive analysis of the deep-drawing behavior of rolling sheets with ultrasonic techniques. In: Höller, P. (Ed.), Nondestructive Characterization of Materials III. Springer, Berlin/West, pp. 296−302.

Spies, M., Schneider, E., 1990. Nondestructive analysis of textures in rolled sheets by ultrasonic techniques. Textures and Microstructures 12, 219−231.

Stickels, C.A., Mould, P.R., 1970. The use of Young's modulus for predicting the plastic strain ratio of low-carbon steel sheets. Metallurgical and Materials Transactions 1, 1303−1311.

Thompson, R.B., 1979. Generation of horizontally polarized shear waves in ferromagnetic materials using magnetostrictive coupled meander-coil electromagnetic transducers. Applied Physics Letters 34, 175−179.

Thompson, R.B., Lee, S.S., Smith, J.F., 1985. Inference of stress and texture from the angular dependence of ultrasonic plate mode velocities. In: Wadley, H.N.G. (Ed.), NDE of Microstructure for Process Control. ASM.

Thompson, R.B., Lee, S.S., Smith, J.F., 1987. Relative anisotropy of plane waves and guided modes in thin orthorhombic plates: implication for texture characterization. Ultrasonics 25, 133−137.

Thompson, R.B., Smith, J.F., Lee, S.S., Johnson, G.C., 1989. A comparison of ultrasonic and X-ray determinations of texture in thin Cu and Al plates. Metallurgical Transactions A 20, 2431−2447.

Truell, R., Elbaum, C., Chick, B.B., 1969. Ultasonic Methods in Solid State Physics. Academic Press, New York.

Vary, A., 1976. Correlations among ultrasonic propagation factors and fracture toughness properties of metallic materials. In: NASA Technical Memorandum, NASA TM X-71889, Technical Paper Presented at Spring Conference of American Society for Nondestructive Testing, March 8−11, 1976, Los Angeles, California.

Vasudevan, M., Palanichamy, P., 2002. Characterization of microstructural changes during annealing of cold worked austenitic stainless steel using ultrasonic velocity measurements and correlation with mechanical properties. Journal of Materials Engineering and Performance 11, 169−179.

Vlad, C.M., 1972. Verfahren zur Ermittlung der Anisotropie-Kennzahlen für die Beurteilung des Tiefziehverhaltens kohlenstoffarmer Feinbleche. Materialprüfung 14, 179–182.

Viktorov, I.A., 1967. Rayleigh and Lamb Waves. Plenum Press, New York.

Willems, H., Goebbels, K., 1980. Gefügebeurteilung durch die Ultraschallrückstreuung von Oberflächenwellen. Materialprüfung 22 (9), 356–358.

Willems, H., Goebbels, K., 1981. Characterization of microstructure by backscattered ultrasonic waves. Metal Science 15, 549–554.

Willems, H., Hirsekorn, S., Repplinger, W., 1987. Weiterentwicklung der Ultraschall-Gefügeanalyse. In: Abschlussbericht zum Forschungsvertrag Nr. 7210-GA/123, Kommission der Europäischen Gemeinschaften, Technische Forschung Stahl, Brüssel-Luxemburg.

Willems, H., 1991. Nondestructive determination of hardening depth in induction hardened components by ultrasonic backscattering. In: Thompson, D.O., Chimenti, D.E. (Eds.), Review of Progress in Quantitative Nondestructive Evaluation, vol. 10B. Plenum Press, New York.

Willems, H., Neumann, K., Fraunhofer-Gesellschaft, 1996. Vorrrichtung zum zerstörungsfreien Messen der Dicke einer Härteschicht (DE patent application DE 1992–4239159 A).

Wormley, S.J., Forouraghi, K., Li, Y., Thompson, R.B., Papadakis, E.P., 1990. Application of a Fourier transform-phase-slope technique to the design of an instrument for the ultrasonic measurement of texture and stress. In: Thompson, D.O., Chimenti, D.E. (Eds.), Review of Progress in Quantitative Nondestructive Evaluation, vol. 9A, pp. 951–958. New York.

Zuev, L.B., Semukhin, B.S., Bushmelyova, K.I., 2002. Ultrasound velocity measurement of strain in metallic polycrystals. Material Research Innovations 5, 140–143.

Electromagnetic techniques for materials characterization

8

I. Altpeter[1], R. Tschuncky[2], K. Szielasko[2]
[1]Formerly Fraunhofer Institute for Nondestructive Testing (IZFP), Saarbrücken, Germany;
[2]Fraunhofer Institute for Nondestructive Testing (IZFP), Saarbrücken, Germany

8.1 Introduction

Most structural components are made of ferromagnetic steel. The mechanical properties of steel are determined by the microstructure, texture, and residual stress state. Microstructure state, residual stress state, and texture are influenced, for example, by a heat treatment or hardening process. Microstructural changes are changes in the dislocation density, the phase content, and grain size. These microstructural changes determine essentially the mechanical technological properties of steels, for instance, yield strength, hardness, toughness, and residual stresses. Knowledge of micro- and macroresidual stresses is necessary to evaluate the stress condition of a component. From the various kinds of residual stresses, residual stresses of the first kind can arise due to forming under applied mechanical stress or due to the different cooling rates of different cross-sections of homogeneous materials. Residual stresses of the second kind forms due to the different thermal expansion coefficients between precipitates and matrix in a two phase material. Residual stresses of a third kind appear when the lattice parameter of the second phase particles embedded coherently in the matrix and the lattice parameter of the matrix are different (Macherauch and Kloos, 1987). Destructive test procedures for determination of microstructure state and mechanical properties, ie, mechanical hardness, need a sample preparation. X-ray techniques are well known for residual stress and texture determination (see chapter: X-ray diffraction (XRD) techniques for materials characterization), but they are time consuming and therefore expensive.

With the understanding of ferromagnetic properties, especially micromagnetic theory, we are able to explain interactions between lattice imperfections and Bloch walls. Most of these studies were performed on single crystals and unalloyed polycrystalline materials (Seeger, 1966). Since these studies, micromagnetism has been a most successful tool to design new magnetic nondestructive techniques.

Since the early 1940s, there have been basic investigations on microstructure and residual stress analysis by magnetic techniques (Förster, 1940; Förster and Stambke, 1941). Since the 1950s and 1960s, when the theoretical bases of micromagnetism were laid, nondestructive methods have been derived from irreversible, magnetostrictive active Bloch wall movements.

Most micromagnetic nondestructive techniques started in the 1960s.

Materials Characterization Using Nondestructive Evaluation (NDE) Methods
http://dx.doi.org/10.1016/B978-0-08-100040-3.00008-0

At the end of 1960s, developments began on the evaluation of microstructure and stress states by means of the inductive Barkhausen noise effect (Leep and Pasley, 1969). Further investigations were carried out by Pawlowski and Rulka (1973) and Tiitto (1980). This research resulted in development of the first prototype testing devices and industrial applications (Tiitto, 1980).

Developments since 1980 use the micromagnetic multiparameter NDE-approach (3 MA: Micromagnetic Multiparameter Microstructure and stress-Analyze) for materials characterization (see chapter: Hybrid methods for materials characterization). This technique enables a nondestructive materials characterization even under rough environmental conditions, also in the industrial praxis.

Electromagnetic techniques are able to indicate nondestructively and quickly changes of residual stresses, texture, microstructure states, and mechanical properties, and are, therefore, very useful tools for materials characterization and damage assessment of in-service engineering components. Various magnetic methods for this task have been reviewed by Jiles (1998) and Altpeter et al. (2002). Out of all available methods, magnetic Barkhausen noise, incremental permeability, upper harmonics, and dynamic magnetostriction are very promising techniques for materials characterization.

8.2 Principles of electromagnetic techniques

8.2.1 Magnetic hysteresis

When a ferromagnetic material is exposed to an alternating magnetic field, a characteristic relationship is observed between B (magnetic induction) and H (magnetic field strength). It is difficult to completely describe the magnetic hysteresis loop using a few parameters, but the following magnetic hysteresis parameters are commonly used to characterize a ferromagnetic material: saturation magnetization, M_S; remanence, B_R; coercivity, H_C (Cullity,1972) (Fig. 8.1).

Figure 8.1 Magnetic hysteresis and derived measuring quantities.

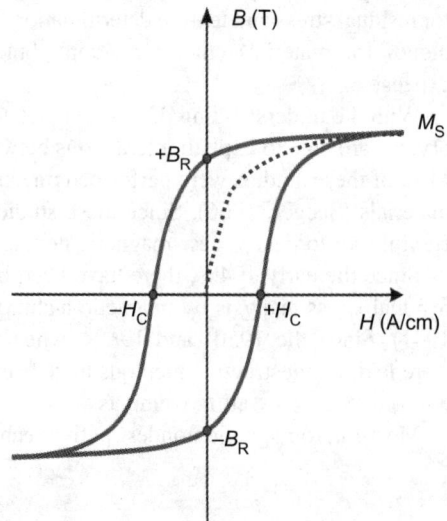

In physics, regular hysteresis measurements providing reproducible and reliable results can only be performed by using special measurement devices asking for specially shaped test specimens like spheres and cylinders with well-designed geometry combined with encircling coils to measure the magnetic induction. As a consequence, the technique cannot be applied to real components, eg, vessel shells, pipes, or high-speed running steel sheets in a cold rolling mill (Dobmann, 1999). All techniques based on magnetic approaches suffer from influence of liftoff on absolute value, subsequent value fluctuations, and shearing of the hysteresis curve (Jiles and Thoelke, 1989).

Only in the case of transformer steel sheets do standards exist based on destructive batch tests (ie, Epstein frame measurements). An Epstein frame is a standardized (IEC 60404-2: 1996 and BS EN 60404-2: 1998 + A1) measurement device for measuring the magnetic properties of soft magnetic materials, especially used for testing of electrical steels. An Epstein frame is comprised of a primary and a secondary winding. The sample under test should be prepared as a set of a number of strips cut from electrical steel. This technique cannot be applied to real components and is only used in the laboratory.

8.2.2 Magnetic Barkhausen noise

The magnetic Barkhausen noise was discovered by Heinrich Barkhausen (1919). The magnetic Barkhausen noise is a very sensitive method for characterization of the microstructure and, consequently, suitable to determine the mechanical technological properties and residual stress states.

As a result of dynamic magnetization of the material, Barkhausen events result from local abrupt changes of the magnetization caused by time-restricted Bloch wall pinning. These changes can only be seen in an enlarged cutout of the hysteresis curve (indicated by a circle, see Fig. 8.3). Micropulsed eddy currents are induced, which diffuse in the bulk volume. The diffusion length is mainly determined by the frequency spectrum of the pulses (eddy current damping), the electrical conductivity, and magnetic permeability of the material, the latter being a function of the magnetization state. Bloch walls are the real sensors in the material. In iron material, we differentiate between 180° and 90° Bloch walls. The potential of micromagnetic testing methods for detection of microstructure states is based on the analogy between the interaction of dislocations and of Bloch walls with microstructure states. Similar to the movement of dislocations under mechanical load, Bloch walls move under magnetization. Therefore, an analogy exists between technological values (hardness values) and magnetic values (coercivity force) (see Section 8.3.4).

The inductive Barkhausen noise generated by dynamic magnetization is a statistical signal and can be detected with an inductive sensor (pickup air coil, ferritic core coil, tape recorder head). The frequency content of the Barkhausen events extends from a few hundred Hz up to the MHz range. Barkhausen events excited at greater distances from the surface will result in voltage pulses with lower frequency content than near-surface events. By changing the analyzing frequency of a bandpass filter from low to high, a weighted characterization of the Barkhausen noise from near-surface

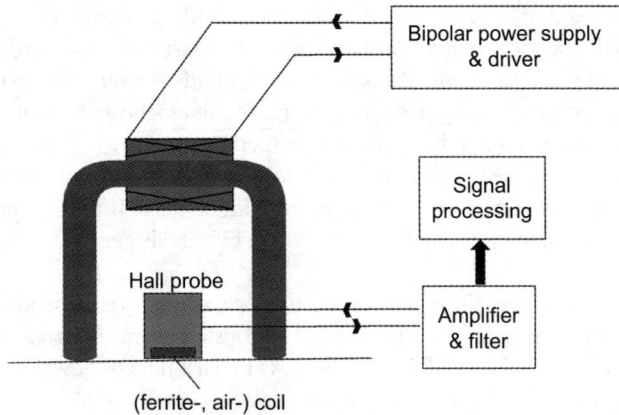

Figure 8.2 Experimental set up for Barkhausen noise measurements.

regions can be obtained (see Section 8.3.4). An electromagnet produces an alternating magnetic field, which can be chosen between 0.1 and several hundred Hz depending on the testing problem. The inductive sensor is located between the magnetic poles along with a Hall probe to measure the tangential magnetic field strength (see Fig. 8.2). The voltage pulses induced in the sensor are amplified at various stages between 60 and 100 dB. They are then filtered and rectified and their envelope called the inductive Barkhausen noise amplitude M is recorded and plotted against the tangential magnetic field strength H_t (see Fig. 8.3). Fig. 8.3 shows a typical Barkhausen noise profile curve. Typical measuring quantities derived from the magnetic Barkhausen noise are the maximum of the profile curve M_{MAX}, the H-field position of this, H_{CM}, and the half widths ΔH by 75%, 50%, and 25% of the maximum value (Fig. 8.3).

Different measuring quantities derived from the magnetic Barkhausen noise profile curve are still used by several working groups worldwide. For example, Stresstech (2015) from Finland and other working groups, eg, Moorthy et al. (2000) and Vengrinovich et al. (2006) use the Root Mean Square (RMS) voltage of the magnetic Barkhausen emission signal as a function of the current applied to the yoke. Working groups in the IZFP uses the Barkhausen noise profile curve $M(H)$ and derived measuring quantities, eg, maximum amplitude value M_{MAX}, the H-field position of this maximum H_{CM}, and the half widths ΔH by 75%, 50%, and 25% of the maximum amplitude value. A working group from Fraunhofer IKTS-MD uses the fractal dimension of the magnetic Barkhausen noise signal (Schreiber et al., 2008).

8.2.3 Incremental permeability

The incremental permeability μ_Δ is obtained by superimposing small magnetic field changes produced by the eddy current sensor and cyclic magnetic field tangential to the surface. In this case, the eddy current operating frequency has to be at least a factor of 100 higher than the magnetizing frequency. The total magnetization follows a hysteresis with at least 100 small inner loops as shown in Fig. 8.4 (Altpeter et al., 2002).

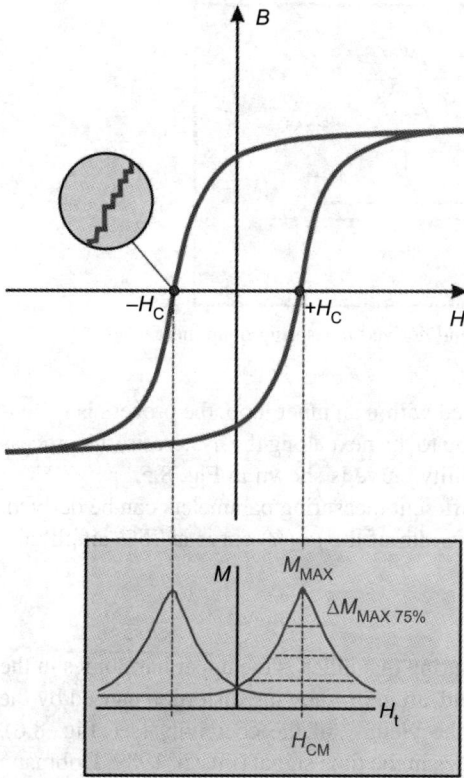

Figure 8.3 Magnetic Barkhausen noise curve and derived measuring quantities in comparison to hysteresis curve.

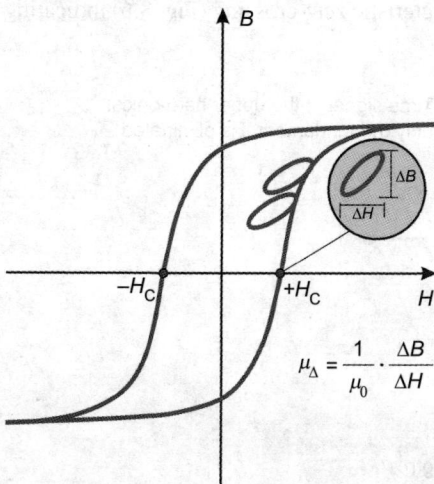

Figure 8.4 Incremental permeability.

$$\mu_\Delta = \frac{1}{\mu_0} \cdot \frac{\Delta B}{\Delta H}$$

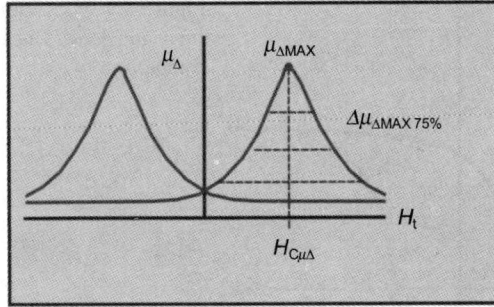

Figure 8.5 Incremental permeability curve and derived measuring quantities.

As long as the magnetization is changed within an inner loop, the process is reversible. In the time period from one inner loop to the next along the hysteresis, the process is irreversible. The incremental permeability curve is shown in Fig. 8.5.

From the incremental permeability, different measuring parameters can be derived, for example, $\Delta\mu_{\Delta MAX75\%}$ defined as the width of the $\mu_\Delta(H)$ curve at 75% amplitude.

8.2.4   Upper harmonics

One of the microstructure sensitive measuring quantities is the upper harmonics in the time signal of the tangential magnetic field strength, showing distortion caused by the nonlinearity of the hysteresis curve in the vicinity of the coercivity (see Fig. 8.6), which is the occurrence of upper harmonics in the time signal (Pitsch, 1989; Dobmann and Pitsch, 1988).

This distortion is influenced by changes in microstructure and residual stress state. A measuring quantity derived from the time signal of the tangential field strength is the coercivity H_C. The tangential field strength is sensed by a Hall probe. The time signal of the tangential field strength shows a characteristic zero crossing (Fig. 8.6) indicating

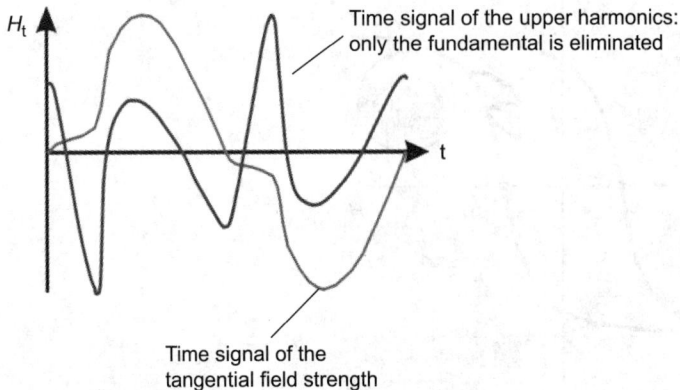

Figure 8.6 Time signal of the upper harmonics.

the time moment where the magnetic field meets the coercivity. Therefore, the coercivity of the hysteresis can be nondestructively picked up.

A further measuring quantity is the distortion factor K, which is given by:

$$K = 100\% \cdot \sqrt{\frac{A_3^2 + A_5^2 + \cdots + A_n^2}{A_1^2}}$$ [8.1]

A_1 is the amplitude of the fundamental wave, A_3, A_5,..., A_n are the amplitudes of the odd higher harmonics. In most cases, the evaluation up to the seventh harmonic is sufficient, especially when magnetization frequencies of several 10 or 100 Hz are used. High-order harmonics ($n > 31$) can be neglected in any case.

8.2.5 Dynamic magnetostriction

This nondestructive measuring parameter is influenced by the magnetostrictively active reversible and irreversible Bloch wall motions and rotation processes. A quasi-static magnetic field is superimposed by an alternating magnetic field with a frequency ≤ 3 MHz. In order to excite and to receive ultrasonic waves, we use a contactless Electromagnetic Acoustic Transducer (EMAT, see chapter: Ultrasonic techniques for materials characterization). The excited amplitude of the ultrasonic wave E_λ is a function of the longitudinal magnetostriction λ_L and therefore directionally dependent. Fig. 8.7 shows the excited amplitude of the ultrasonic wave E_λ as function of the magnetic field strength (Altpeter and Dobmann, 1994).

8.2.6 Barkhausen noise microscopy

Barkhausen noise microscopy (Altpeter and Theiner, 1994) offers the possibility for a materials characterization with high spatial resolution. A very efficient manipulation system precisely guides the miniaturized probe (video recorder head, see Fig. 8.8) with a local accuracy of 1 μm across the test surface. The measurement time including

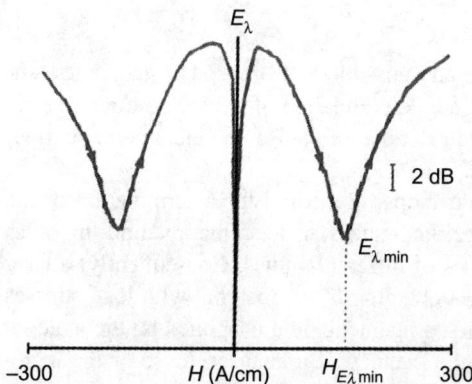

Figure 8.7 Dynamic magnetostriction curve: E_λ as function of magnetic field strength H.

Figure 8.8 Barkhausen noise microscopy: manipulation system together with the miniaturized probe (in the white circle).

positioning is approximately 1 s per point of measurement. Fig. 8.8 shows the Barkhausen noise microscope (Bender, 1997).

8.3 Applications

8.3.1 Microstructure state

A microstructure characterization is generally possible by using electromagnetic techniques. A typical measuring quantity for the determination of the microstructure state is the coercivity derived from the magnetic Barkhausen noise, the incremental permeability, or the dynamic magnetostriction.

Altpeter and Szielasko (2014) have developed a method for determination of the microstructure state independent of residual stress state. This method involves charging a material with two load stresses of different values. Consequently, a laser beam can be used. This laser allows a fast charging of the material with load stresses in a small area of the sample. Additionally, a magnetic field is applied on the material and a measuring quantity derived from the magnetic Barkhausen noise or the incremental permeability is received for each load stress (see Fig. 8.9).

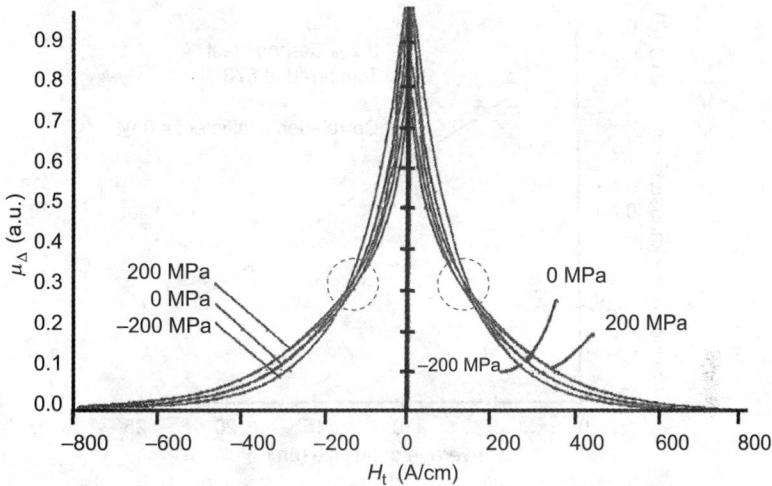

Figure 8.9 Determination of microstructure state independent of residual stress state.

The microstructure of the material is characterized by the characteristic magnetic field strength, for which the measuring quantity at two different load stresses has the same value (see the red circles in Fig. 8.9).

8.3.2 Grain size

The grain size influences the mechanical property of a material, especially the strength. Grain size is usually determined from light microscopy. This method often required cutting of samples from the material and is therefore time consuming. It is favorable to apply a nondestructive method for the determination of grain size. Besides ultrasonic methods (see chapter: Ultrasonic techniques for materials characterization) electromagnetic methods have a potential for determination of grain size.

There is a correlation between coercivity and grain size (Cullity, 1972):

$$H_C \sim 1/d \qquad\qquad\qquad\qquad [8.2]$$

H_C is the coercivity and d is the diameter of grain. The coercivity H_C derived from the magnetic Barkhausen noise is a very sensitive measuring quantity for determination of grain size. H_C is defined as the H field position of the Barkhausen noise maximum.

In a two-phase material, eg, tempered ferritic steels, very often a double peak is observed in the magnetic Barkhausen noise curve (see Section 8.3.3.1; Fig. 8.11). The lower H field position of the first maximum (peak 1) can be allocated to the softer phase (ferrite) and the higher field position of the second maximum (peak 2) can be allocated to the harder phase (cementite).

Moorthy et al. (2000) found a correlation between the position of peak 1 and average grain size for different tempered carbon steel samples (see Fig. 8.10). Moorthy used the current applied to the yoke instead of the magnetic field.

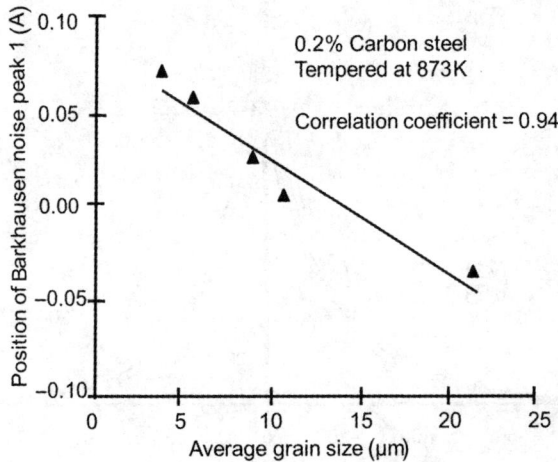

Figure 8.10 Correlation between the position of Barkhausen noise peak 1 and grain size for different tempered 0.2% carbon steel samples.

8.3.3 Phase content

8.3.3.1 Ferromagnetic phase content

Technical materials in general have a microscopic heterogeneous structure, ie, they contain two or more phases with different properties. The experimental investigation of such phase mixtures is of increasing interest and is required for the development of components with optimized material properties. Cementite, for example, plays an important role in steel with several percent carbon content C. Cementite is present as a second phase as in the tempered martensite states, in the ferritic/pearlitic structures, and in white cast iron. The volume fraction and morphology of cementite predicts the strength and the toughness of these materials. Cementite is a ferromagnetic phase and, therefore, it has its own domain structure and stray fields. Batista et al. (2014) have shown that by using a magnetic microscope with a local resolution of 50 nm, the stray fields of inclusions may interact with the domain walls within the iron matrix. The other source of interaction of the cementite precipitates with the domain walls in the ferrite matrix are the tensile residual stresses of the second kind, which are built up due to different expansion coefficients of the two phases below the Curie temperature of cementite (Altpeter, 1990, 1996) (see also, Section 8.3.6.2).

Based on this domain structure, cementite produces its own magnetic Barkhausen noise. Barkhausen noise profile curves of ferrite—pearlite microstructure states support this thesis. At room temperature, RT in magnetic Barkhausen noise profile curves of ferrite—pearlite microstructure states, a double peak structure can be observed (Altpeter, 1996; Moorthy et al., 2000; Batista et al., 2014). The first peak corresponds to the ferrite and the second peak is related to the presence of the second-phase cementite.

With increasing cementite content, the second peak increases (Batista et al., 2014). Another observation that supports the assumption that the stronger field peak is a signature of the cementite phase is provided by measurement of the temperature dependence on the magnetic Barkhausen noise signal. For example, in the reactor pressure vessel steel DIN 22 NiMoCr3-7 (ASTM A 508 Grade 2) containing globular and rod-shaped cementite at room temperature, a double peak was observed in the magnetic Barkhausen noise signal (Fig. 8.11). After heating the sample up to the Curie temperature of cementite, the stronger field peak M_{MAX2} decreases (Altpeter, 1996).

A comparison of both curves at room temperature and Curie temperature demonstrates that the widths of the Barkhausen noise curve are reduced. The reason for that is the decrease of the Barkhausen noise activity in the cementite phase above the Curie temperature.

Furthermore, Altpeter (1996) demonstrated on compact cementite samples and unalloyed white cast iron samples that cementite actively produces its own magnetic Barkhausen noise signal.

Altpeter (1996) observed a Barkhausen noise amplitude of the compact cementite specimen, which decreases with increasing temperature and disappears at the Curie temperature of cementite. The Curie temperature of the compact cementite sample is caused by the Mn content of 3% about 168°C. Fig. 8.12 shows the ratio M_{MAX} (T)/M_{MAX} (RT) for soft iron and cementite. In the temperature range RT - 240°C this ratio decreases for soft iron about 10% and for cementite about 100%.

Furthermore, the magnetic Barkhausen noise amplitude of white cast iron shows qualitatively the same temperature behavior. In addition, the magnetic Barkhausen noise signal of white cast iron decreases more strongly with increasing amount of cementite (see Fig. 8.13).

The influence of different cementite contents and cementite modifications on the magnetic Barkhausen noise for steel with ferritic, pearlitic, martensitic annealed, and tempered martensitic microstructure states, and for white cast iron, was determined. The Curie point T_C for cementite lies between 150 and 210°C depending on the alloying elements, whereas the Curie point for iron lies at 769°C. Therefore, Barkhausen noise measurements at room temperature and at Curie temperature are

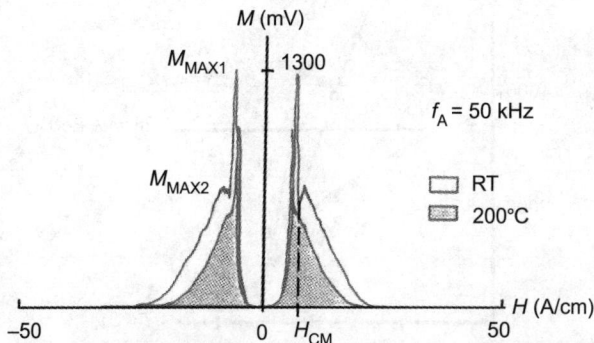

Figure 8.11 Magnetic Barkhausen noise profile curve for a ferrite–pearlite microstructure state (22 NiMoCr3 -7) at room temperature and 200°C.
Ph.D. thesis Altpeter (in German).

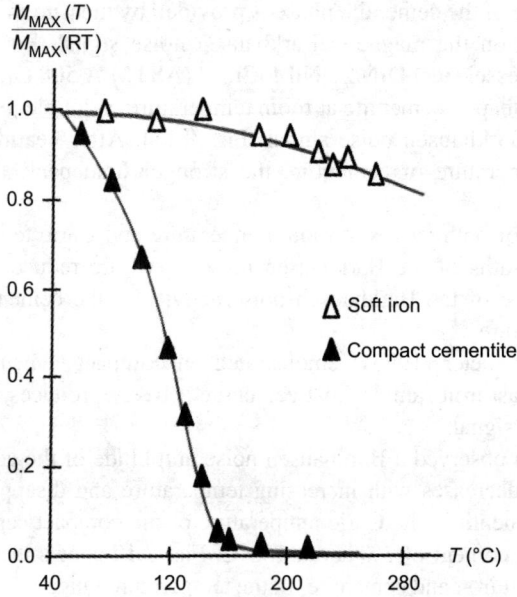

Figure 8.12 Ratio $M_{MAX}(T)/M_{MAX}(RT)$ for compact cementite in comparison to soft iron. Ph.D. thesis Altpeter (in German).

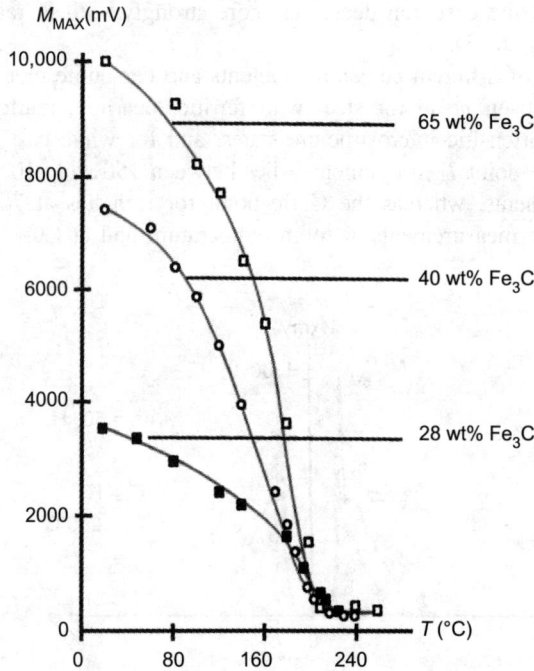

Figure 8.13 Magnetic Barkhausen noise amplitude M_{MAX} as function of temperature for white cast iron samples with different cementite content. Ph.D. thesis Altpeter (in German).

necessary for a determination of cementite content. Altpeter (1996) has found that the parameter M^* given by $M^* = (M_{MAX} (RT) - M_{MAX} (T \geq T_{Curie}))/M_{MAX} (RT)$ can be used to estimate the cementite content in steel with different microstructure states and in white cast iron. Depending on the microstructure state, different calibration curves exist. For example, in tempered martensitic structures, a calibration curve has been generated that shows the variation in M^* over the cementite content (Fig. 8.14). This calibration curve has been applied to estimate the cementite content in two specimens (black triangle) of 22 NiMoCr3-7 steel, which was not part of the calibration set. Between the cementite content estimated from the calibration curve (1.3 ± 0.13 wt% and 2.8 ± 0.28 wt%) and the content obtained by transmission electron microscopy (1.2 wt% and 2.9 wt%), a good correlation was found (Altpeter, 1996; see Fig. 8.14).

A quantitative determination of cementite content is possible by using the magnetic Barkhausen noise at room temperature and Curie temperature of cementite.

In addition, a qualitative determination of cementite is possible by using the analysis of upper harmonics. With the help of this method, the tendency to chilling in cast iron products can be determined. Tendency to chilling means the undesired occurrence of cementite or ledeburite in graphitic cast iron products. The potential of upper harmonics for detection of tendency to chilling was demonstrated on several hundred cast iron products (Kröning et al., 1996; Maisl et al., 2000).

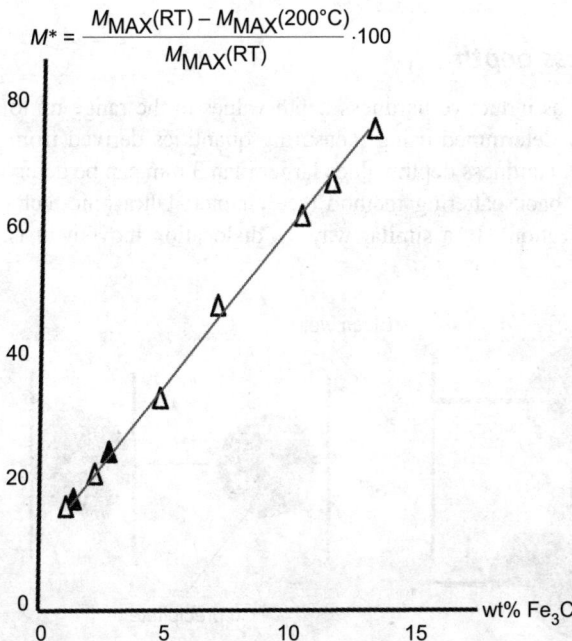

$$M^* = \frac{M_{MAX}(RT) - M_{MAX}(200°C)}{M_{MAX}(RT)} \cdot 100$$

Figure 8.14 Measuring quantity M^* as function of weight percent cementite for tempered martensitic structures of steel grade 22 NiMoCr3-7.

8.3.3.2   Nonferromagnetic phase content

Besides ferromagnetic phases, nonferromagnetic phases play an important role in steel. An example of such a nonferromagnetic phase is copper. Copper precipitates improve the hardness and reduce the toughness of a material, which can be characterized by a reduction of the Charpy energy and a shift in the ductile to brittle transition temperature Δdbt to higher temperatures (Dobmann, 2006). Copper precipitates in WB36 (15 NiCuMoNb-5) impede the Bloch wall movement in a similar way as the dislocation movement (see Fig. 8.15) and can lead to a thermally induced embrittlement (see Section 8.3.7).

Kersten's (1944) and Dijkstra and Wert's (1950) theories describe the correlation between coercivity and volume of small and coherent precipitates. All these theories show an increase in coercivity with rising volume in precipitates. Ayere (1988) showed that the increase of coercivity with rising precipitate diameter reaches a maximum when the precipitate diameter is equal to the Bloch wall thickness. The coercivity decreases for precipitates with diameter greater than the Bloch wall thickness. Pirlog (2005) also showed (in her doctoral thesis) an increase of coercivity with rising copper content similar to the behavior of mechanical hardness.

An increase in copper precipitate content causes also an increase of residual stresses of the third kind (coherent residual stresses) in the α-iron matrix. Besides the Barkhausen noise amplitude, the half width of Barkhausen noise profile curve at different heights from the amplitude are sensitive quantities for determination of copper precipitate content. Fig. 8.16 shows the width of Barkhausen noise profile curve at 25% of the amplitude value M_{MAX} versus copper precipitate content.

8.3.4   Hardness, hardness depth

Near-surface hardness, as well as inductive hardness depth values in the range up to 3 mm, can be nondestructively determined using measuring quantities derived from the magnetic Barkhausen noise. Hardness depth values larger than 3 mm can be determined by using the ultrasonic backscattering method (see chapter: Ultrasonic techniques for materials characterization). In a similar way, as dislocation movement is

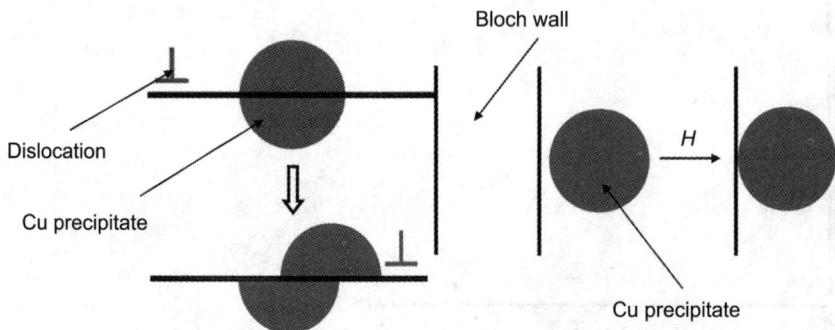

Figure 8.15 Dislocation and Bloch wall motion across a precipitate.

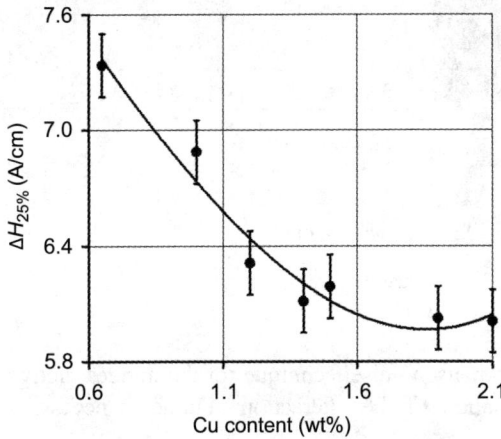

Figure 8.16 Width of the Barkhausen noise profile curve at 25% of M_{MAX} as a function of copper content. Ph.D. thesis Pirlog (in German).

impeded by the lattice defects, these defects impede also the movement of Bloch walls under the forces of magnetic loads during continuous or cyclic magnetization. Bloch walls are temporarily pinned by the lattice defects. A further increase in the magnetic field strength is needed to overcome the pinning effect and to increase the magnetization. The overall magnetization behavior is, therefore, strongly influenced by the density of the pinning points, eg, the density of the lattice defects. This can be documented in micromagnetic properties like coercivity derived from the Barkhausen noise (Altpeter et al., 2002). Magnetic coercivity is discussed in many textbooks as magnetic hardness. Very often similarities between magnetic hardness and mechanical hardness can be observed.

In Fig. 8.17 such an example of a nearly 1:1 correlation between the coercivity and the Rockwell hardness HRC is documented (Altpeter et al., 2002).

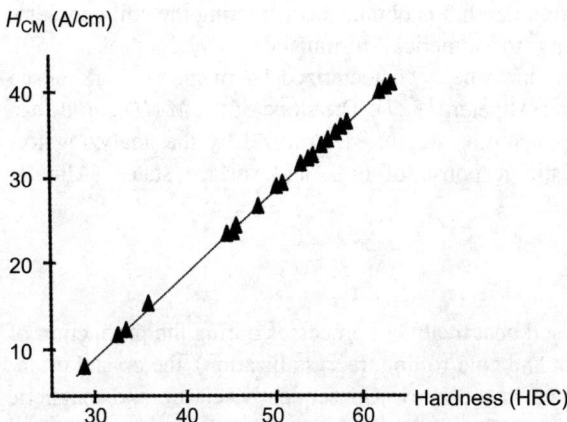

Figure 8.17 Correlation between coercivity and Rockwell hardness HRC for CK 45.

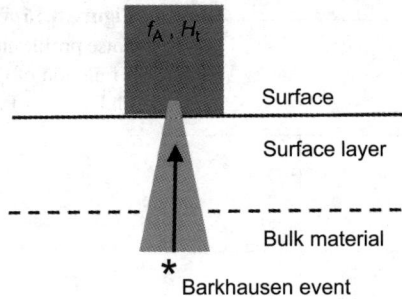

Figure 8.18 Interaction volume.

In order to apply the magnetic Barkhausen noise technique for the nondestructive evaluation of hardness depth, a variation of the interaction volume is necessary (Fig. 8.18).

The basic mechanism adjusting the depth of interaction is characterized by the law of the eddy current damping. Local magnetic events, eg, Bloch wall jumps and rotations of the magnetization directions in domains, dissipate energy by inducting pulsed eddy currents in their vicinity. These pulses are relatively broadband with bandwidth from dc up to some MHz, and they are damped following an exponential law. The higher the frequency in the spectrum, the higher the damping is. The damping constant for each frequency component f is $1/\delta$. Therefore, the higher-frequency part in the received noise is generated only in near-surface zones of a magnetized component (Skin effect).

Applying the theorem of reciprocity, the standard eddy current penetration depth is defined as:

$$\delta = \frac{1}{\sqrt{\pi f_A \sigma \mu}} \qquad [8.3]$$

(σ, electrical conductivity; μ, permeability; f_A, analyzing frequency). The factor f_A is selected to be the center frequency and frequency range of the pickup system. A more exact calculation of the penetration depth δ is obtained considering the coil geometry and sample geometry according to numerical formulation (Dodd et al., 1969). Laser-hardened Ck45 specimens have been characterized by magnetic Barkhausen noise measurements (Theiner and Altpeter, 1987). The slope of the $M(H)$ curves can be explained by the different penetration depths, determined by the analyzing frequency f_A, and the characteristic response of bulk and surface states (Altpeter et al.,1987) (see Fig. 8.19).

8.3.5   Texture

Texture is generated by rolling and heat treatment processes during the production of sheets. After such processes (hot and cold rolling, recrystallization), the crystal orientations are not equally distributed in the alloy. Consequently, elastic and magnetic properties of the sheets are anisotropic. There is a correlation between elastic and

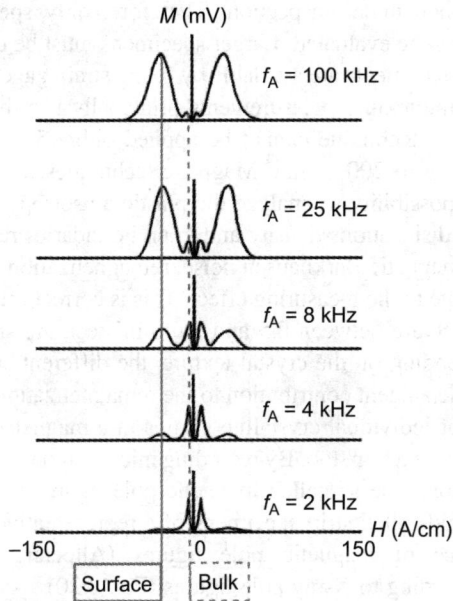

Figure 8.19 Magnetic Barkhausen noise curve as function of analyzing frequency f_A.

plastic anisotropy. The elastic anisotropy is determined by the crystal texture, whereas the plastic anisotropy is determined by the microstructural parameters like dislocation density and grain boundary. Accordingly, the techniques usually applied to determine the texture are different. For instance, differences in plastic deformation can be detected by the determination of the average Lankford plastic strain ratio r_m, elastic anisotropy by the determination of the Young's modulus and X-ray texture analysis (see chapter: X-ray diffraction (XRD) techniques for materials characterization). The X-ray measurements that can be applied under practical conditions are state of the art. In addition to the X-ray measurements, ultrasonic and magnetic techniques offer a possibility of fast and simple texture analysis. The measurement of ultrasonic velocities in different directions is an alternative approach to the X-ray measurement and has been developed for practical applications (Spies and Schneider, 1989), see chapter: Ultrasonic techniques for materials characterization. Worldwide activities in research and development have been observed in recent years in the applications of ultrasonic phase velocity measurements in order to determine the coefficients of the orientation distribution function, ie, C_{411} and C_{413} correlating with r_m- and Δr-values, respectively.

In the body-centered cubic (bcc) iron single crystal, the magnetization is directionally dependent. The reason for this is the crystal anisotropy energy. By magnetization parallel to the crystal directions (100), (110), (111), different magnetization curves can be obtained (Seeger, 1966). The measurements of hysteresis cycles need quasi-static magnetization, ie, magnetizing frequencies lower or equal to 50 mHz with sensor coils encircling the specimen under inspection. Therefore, only specimens with special shapes (cylindrical) can be evaluated. Larger specimens must be destructed by cutting. Magnetostriction measurements are available by fixing strain gages to the surface under inspection. Because the texture measurements principally use the data acquisition in different directions, the technique cannot be applied online in a steel mill on a strip running with speeds up to 200 m/min. Magnetic techniques in addition to ultrasonic techniques offer the possibility to analyze the plastic anisotropy, because microstructural parameters like dislocations density and grain boundaries restrict the mobility of Bloch walls. By the magnetic Barkhausen noise, remagnetization processes in the individual grains contribute to the measuring effect. This is correct if the distances between the Bloch walls, which are between 0.5 and 5 μm in steel, are smaller in comparison with grain size. Depending on the crystal texture, the different crystallite groups produce a directionally dependent contribution to the remagnetization process. Depending on the participation of individual crystallite groups in a magnetization direction noise activity changes in the direction, too. By recording micromagnetic quantities in different magnetization directions, the so-called magnetic pole figures are obtained. Measurements on different cold-rolled strip specimens of a representative number have documented the existence of magnetic pole figures (Altpeter, 1991; Altpeter and Dobmann, 1994) according to X-ray pole figures (Fig. 8.20).

A comparison of magnetic pole figures before and after a stress relieves annealing reveals the additional influence of residual stresses on the magnetic pole figures. The change of the pole figure under load stress illustrates the influence of a superimposing of texture and stress state on the measuring quantity M_{MAX}.

To solve the multiparameter inspection task of texture characterization independent of the influence of mechanical stress states, a combination of several electromagnetic and ultrasonic measuring quantities is required (see chapter: Hybrid methods for materials characterization). A further approach for a stress-independent texture measurement is given by dynamic magnetostriction measurements with SH waves (horizontally polarized shear waves) EMATs (Altpeter et al., 2010). These SH waves can be generated in sheets or plates by using EMATs, which do not require couplant between the transducers and sheets. The received ultrasonic signal as function of a magnetic field strength is shown in Fig. 8.7.

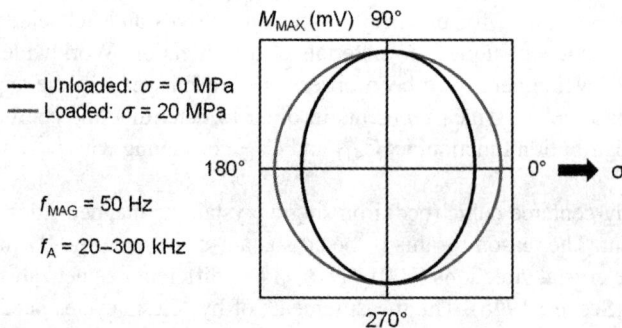

Figure 8.20 Influence of residual stresses on magnetic pole figures.

The dynamic magnetostriction curve has a second minimum $H_{E\lambda min}$. The position of this minimum allows a determination of texture independent of residual stress state (see chapter: Ultrasonic techniques for materials characterization).

8.3.6 Residual stresses

The basic requirement for a micromagnetic stress measurement is a magnetostrictively active ferromagnetic material, eg, steel and white cast iron. In contrast to the X-ray stress measurement and the ultrasonic methods, which are based on physically understood interactions, the micromagnetic stress measurement always has to be based on calibration curves because the dependence of electric and magnetic measuring quantities on stress is not yet quantitatively understood. Nevertheless, micromagnetic stress determination has found wide application in practice.

8.3.6.1 Residual stresses of the first kind

Ferromagnetic materials change their magnetic domain structure under the influence of mechanical stresses (Kneller, 1962; Cullity, 1972). These micromagnetic changes, caused by Bloch wall movements and rotation processes, are the reason for the well-known hysteresis shearing under residual stresses (see Fig. 8.21).

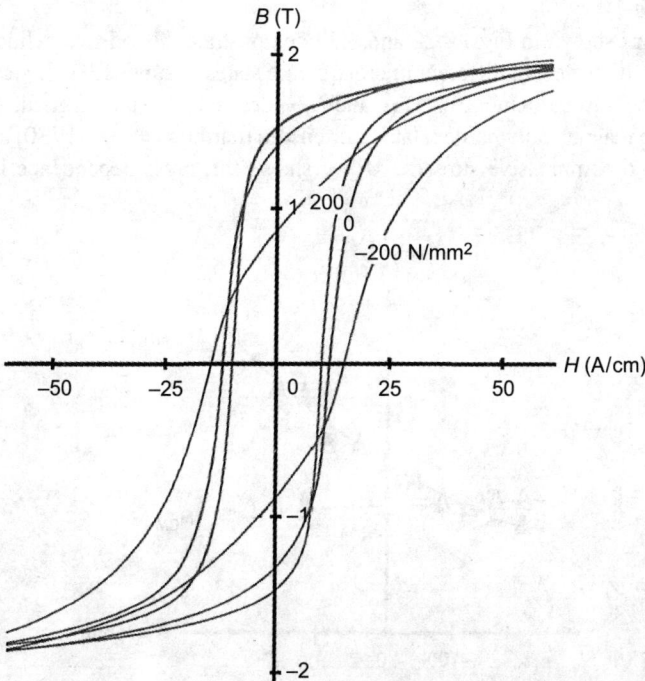

Figure 8.21 Hysteresis shearing under tensile and compressive residual stresses.

In magnetostrictive positive materials, tensile stresses cause an increase of the differential susceptibility X_{diff}, and in the region of the coercive force H_C, an H_C-shift to smaller values. Compressive stresses cause a decrease of X_{diff} in the region of the coercive force and an H_C shift to greater magnetic field values (see Fig. 8.21). However, the dependence of H_C and X_{diff} on tensile and compressive stresses cannot be used as a direct nondestructive measuring quantity for residual stress determination because it is not possible to measure the magnetic flux density B absolutely in the setup technique. In order to overcome this restriction, it is necessary to use electromagnetic measuring quantities that are sensitive to reversible and irreversible Bloch wall movements (Kneller, 1962; Seeger, 1966).

In ferromagnetic materials, the magnetostrictively active (100)-90° and (111)-90° Bloch walls and the rotation processes interact directly with stresses. All measuring quantities that have their origin in these remagnetization processes are stress sensitive like the dynamic magnetostriction (see chapter: Ultrasonic techniques for materials characterization) and different quantities derived from the incremental permeability. Because of the coupling of 90° and 180° Bloch walls, measuring quantities that use mainly the interactions of 180° Bloch walls are also stress sensitive but in an indirect manner like the magnetic Barkhausen noise. All ferromagnetic nondestructive testing (NDT) methods are more or less sensitive to mechanical stress and microstructure states of the tested material.

For a residual stress measurement independent of microstructure state, we need at least two measuring quantities derived from an electromagnetic method (Theiner and Altpeter, 1987).

The results shown in Figs. 8.22 and 8.23 demonstrate this on two cylindrical specimens (8 mm diameter) of different microstructure states of super 13% Cr steel. Fig. 8.22 shows the measuring quantities M_{MAX} and H_{CM} derived from the magnetic Barkhausen noise for the magnetically harder state (martensite) (hardness = 527HV30) as a function of tensile and compressive stresses. M_{MAX} shows the stress dependence in the stress

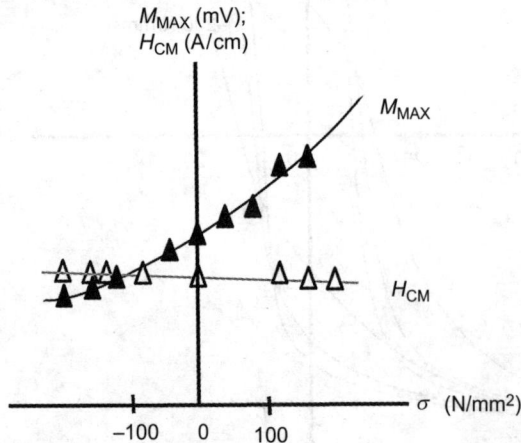

Figure 8.22 Measuring quantities M_{MAX} and H_{CM} as function of load stresses for a magnetically hard microstructure state (martensite).

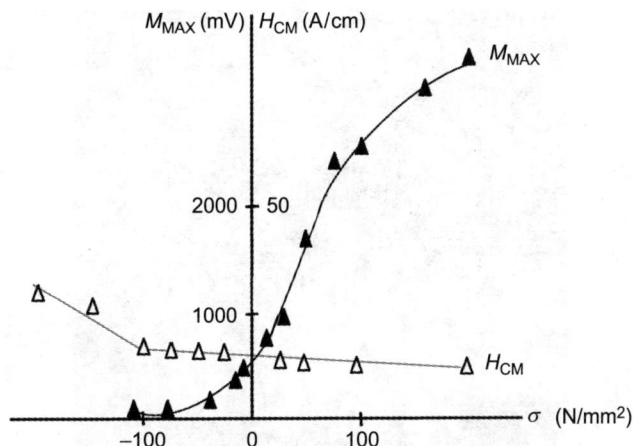

Figure 8.23 Measuring quantities M_{MAX} and H_{CM} as function of load stresses for a magnetically soft microstructure state (annealed martensite).

region between +200 and −200 N/mm². The measuring quantity H_{CM} shows a nearly constant value in the tensile and compressive region.

Fig. 8.23 shows the two measuring quantities for the magnetically softer material annealed martensite (250HV30). Both measuring quantities show a greater stress dynamic than for the harder material. In the magnetically harder material, the lower stress dependency is caused by the higher dislocation density, which pins all magnetostrictively active 90° Bloch walls. By measuring H_{CM}, it is possible to separate the two microstructure states of this steel independent of the stress state. This example shows the necessity of two measuring quantities for a stress measurement independent from microstructure state.

For a stress measurement independent from microstructure state, texture, and other influences, further electromagnetic methods such as the incremental permeability and the upper harmonics are necessary (see chapter: Hybrid methods for materials characterization).

For a quantitative residual stress measurement, a calibration of the magnetic measuring quantities with X-ray residual stress values is necessary. The reason is that in a physico–mathematical description is not possible because the interaction mechanism between microstructure and measuring quantities is too complex (Altpeter et al., 2002).

For a multiaxial residual stress measurement, a miniaturized electromagnetic probe was developed within the framework of a research project (Altpeter et al., 2009). This so-called rotating field probe (Fig. 8.24) was integrated in a deep drawing tool, the plunger. Magnetic pole figures were measured during the deep drawing process (see Fig. 8.25).

The shape of the pole figures, which is representative for the residual stress state of deep drawn sheets, allows inferences about the critical load stress, which can lead to tearings.

Figure 8.24 Rotating field probe (a prototype) for an online multiaxial process control.

Fig. 8.25 shows the maximum amplitude M_{MAX1}, derived from the magnetic pole figures, as function from punch position for different blank holder forces F-BH (Altpeter et al., 2009).

With these results, first basics for an online multiaxial process control have been created.

8.3.6.2 Residual stresses of the second kind (thermally induced residual stresses)

Thermally induced residual stresses of the second kind play an important role in the fracture mechanical analysis of thermally cycled materials and thus in lifetime analysis of such affected components. For example, forming rollers in a rolling plant made of white cast iron can develop cracking due to stress risers caused by thermally induced residual stresses. These residual stresses of the second kind are caused by the difference between the thermal expansion coefficient of matrix material (iron) and the second phase (cementite).

By heating higher than the Curie temperature of cementite, only the cementite content can be determined; by applying an additionally tensile load, the thermally induced residual stress state of the second kind caused by cementite can be determined (Altpeter et al., 1999). A testing method to determine thermally induced residual stresses was developed, tested, and patented (Altpeter, 1997a, EP patent) within the

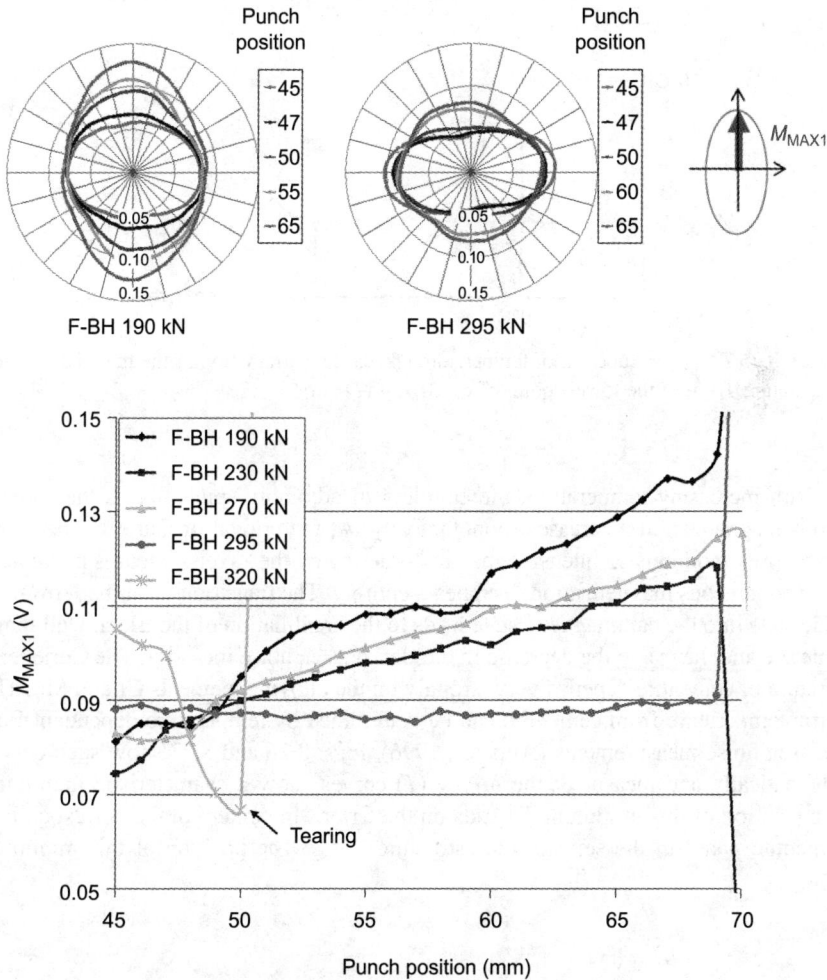

Figure 8.25 Maximum amplitude M_{MAX1}, derived from the magnetic pole figures, as function from punch position for different blank holder forces F-BH.

framework of the German Research Society project "Micromagnetic determination of thermally induced residual stresses in steel and white cast iron."

The developed testing method permits a quantitative determination of residual stresses without the need of calibration using a reference method such as the X-ray residual stress measurement. Therefore, this testing technique opens up a wide range of possible industrial applications. The developed testing method is an integral temperature and load stress-dependent Barkhausen noise technique. The test concept consists of three steps presented in Figs. 8.26 and 8.27, as follows:

First step: Determination of the materials Curie temperature T_C from the lowest point of the M_{MAX} (T) curve (Figs. 8.26 and 8.27):

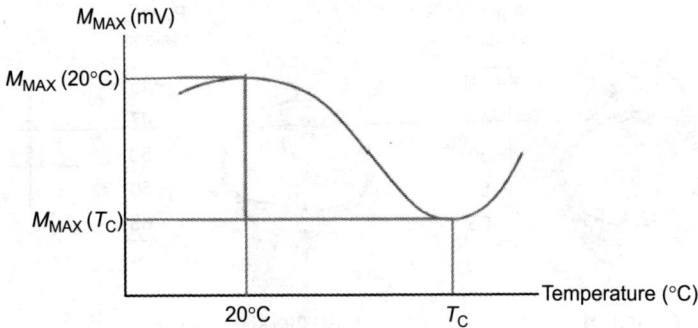

Figure 8.26 M_{MAX} as function of temperature (sketch); determination of the materials Curie temperature T_C from the lowest point of the M_{MAX} (T)-curve.

With increasing temperature, the coupling of adjoining spins, that is the spontaneous magnetization, decreases monotonically. At the critical or Curie temperature, the thermal motion is so intensive that any ordering by the "Weiss" field is prevented. The spontaneous magnetization disappears entirely. The transition from the ferromagnetic state into the paramagnetic state leads to the annihilation of the Bloch wall range structure and, hence, to the decrease in the Barkhausen noise intensity. The Curie temperature of cementite depends very strongly on the alloying elements Cr and Mn. The Curie temperature from cementite can be determined by temperature-dependent Barkhausen noise measurements (Altpeter, 1996). Figs. 8.26 and 8.27 show such curves schematically and measured; the M_{MAX} (T) curves show a characteristic minimum. The position of this minimum depends on the Cr or Mn content of the corresponding cementite phase in the several microstructure states. The position of this minimum

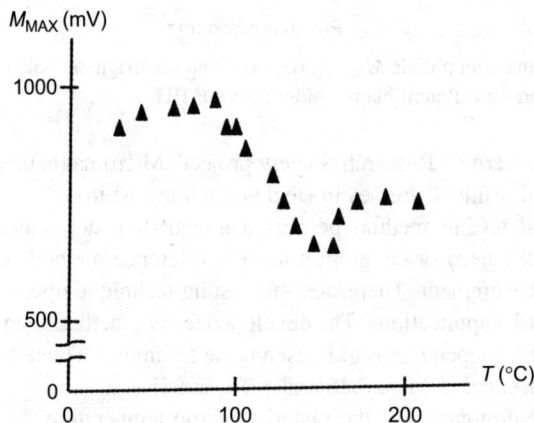

Figure 8.27 M_{MAX} as function of temperature (measured values); determination of the materials Curie temperature T_C from the lowest point of the M_{MAX} (T)-curve.

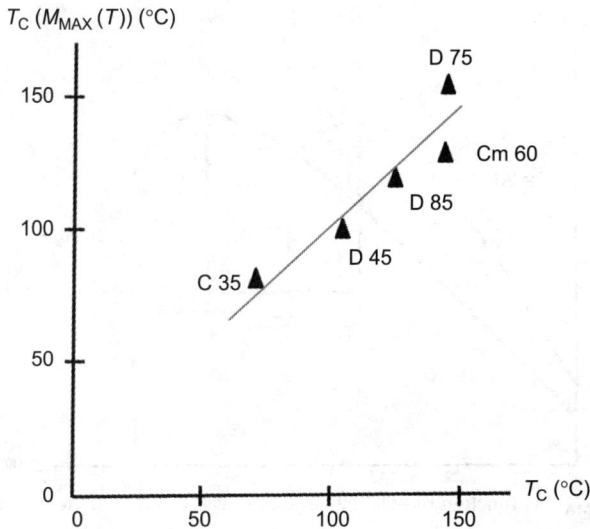

Figure 8.28 Correlation between T_C determined by Mössbauer spectroscopy and the temperature-dependent Barkhausen noise for steel samples with different carbon content (C 35: 0.35% C; D45: 0.45% C; D85: 0.85% C; Cm60: 0.6% C; D75: 0.75% C). Ph.D. thesis Altpeter (in German).

correlates very well with the Curie temperature from cementite determined by Mössbauer spectroscopy (Altpeter, 1996). The Curie temperature can be determined by using both these methods with an accuracy of $\pm 12°C$ (Altpeter et al., 1997a) (see Fig. 8.28).

Second step: Determination of the Barkhausen noise amplitude M_{MAX} as a function of tensile stresses at room temperature and at the determined Curie temperature (Fig. 8.29).

Third step: Determination of the distance $\Delta\sigma$ between the peaks of both $M_{MAX}(\sigma)$ curves. (Fig. 8.29)

Essential for this test concept are the reversibility and reproducibility of the established $M_{MAX}(T)$ curves during the sample heat-up and cool-down cycles. The reversibility of the $M_{MAX}(\sigma)$ curves, loaded and unloaded at a fixed temperature, and the reproducibility of the $M_{MAX}(\sigma)$ curves at repeatedly applied loads are documented in (Altpeter et al., 1999) (Fig. 8.30).

This concept was tested on several hypereutectoid steel samples with 1.2 and 1.85 wt% C (Fig. 8.31.), 1.95 and 2.18 wt% C (Fig. 8.32). $M_{MAX}(\sigma)$ curves were measured at room temperature and at the Curie temperature of cementite phase, both curves passing through a maximum. The stress difference between both maxima corresponds with the thermally induced tensile residual stresses of the second kind (Table 8.1).

Using the load stress-related Barkhausen noise measurement method at room temperature and Curie temperature, thermally induced stresses of the second kind can be determined directly and with sufficient accuracy. The micromagnetic determined values are in the scatter band of the X-ray values.

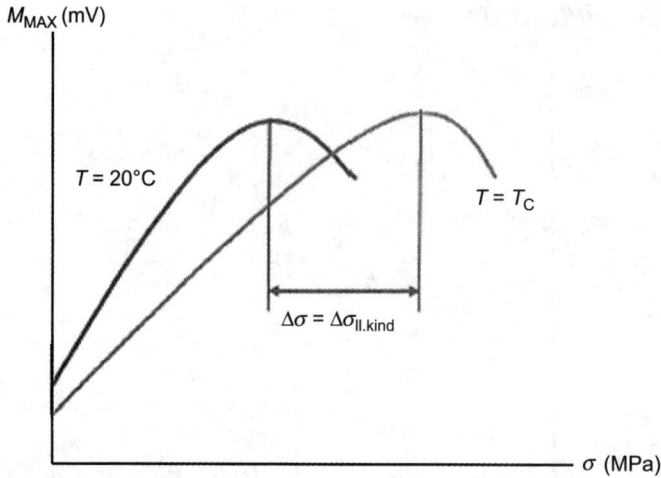

Figure 8.29 M_{MAX} as function of residual stresses σ at RT and Curie temperature T_C.

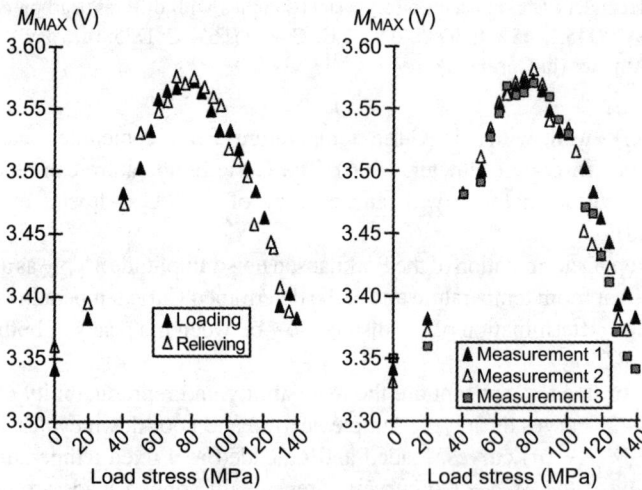

Figure 8.30 Reversibility (left-hand side) and reproducibility (right-hand side) of the $M_{MAX}(\sigma)$ curve at 20°C for a hypereutectoid steel sample with 1.85 wt% C.

8.3.6.3 Determination of residual stresses of second kind with high spatial resolution

With the X-ray technique, the measuring time increases with the lateral resolution, so that the legitimate measuring time leads to a limitation of the lateral resolution. Barkhausen noise microscopy offers the possibility to measure residual stresses with high spatial resolution. This method is not as time consuming and therefore not as expensive as the

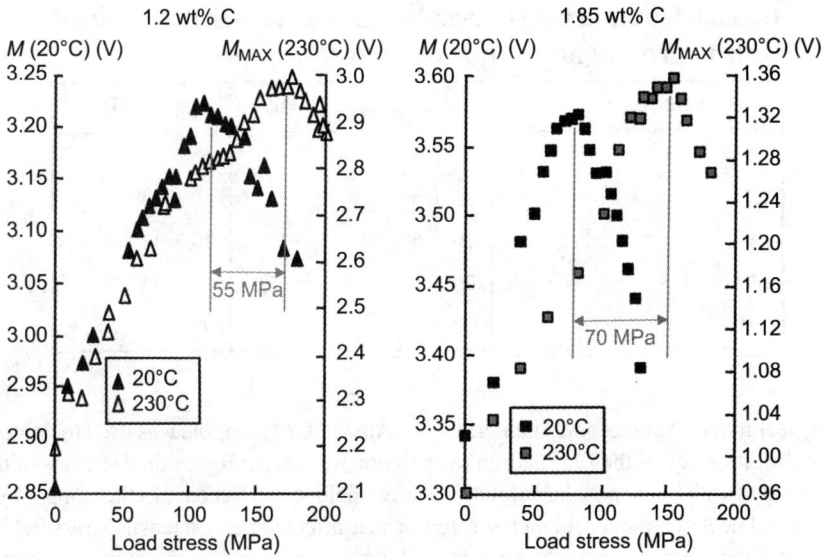

Figure 8.31 $M_{MAX}(\sigma)$ curves measured on hypereutectoid steel samples with 1.2 and 1.85 wt% C at room temperature and the Curie temperature of cementite phase.

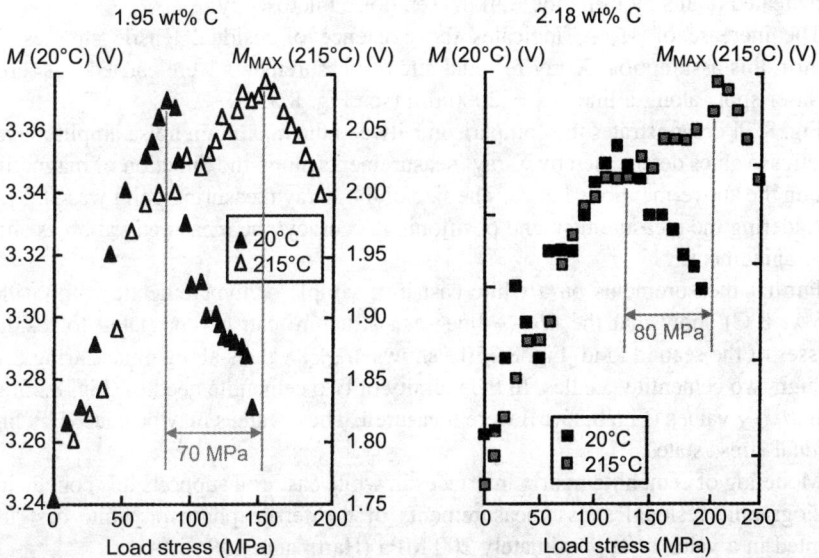

Figure 8.32 $M_{MAX}(\sigma)$ curves measured on hypereutectoid steel samples with 1.95 and 2.18 wt% C at room temperature and Curie temperature of cementite phase.

Table 8.1 **Comparison of residual stress values determined with micromagnetic and X-ray techniques**

Carbon content (wt% C)	$\sigma_{\parallel micro\ magnetic}$ (MPa)	$\sigma_{\parallel\ X\ ray}$ (MPa)
1.1	50 ± 5	40 ± 11
1.2	55 ± 5	60 ± 11
1.85	70 ± 5	70 ± 11
1.95	70 ± 5	70 ± 11
2.18	80 ± 5	90 ± 11

X-ray technique (Altpeter and Theiner, 1994). An X20Cr 13 sample was used to demonstrate the efficiency of the Barkhausen microscope for measuring residual stresses of the second kind with high spatial resolution. A stress field was induced into the sample by a point focal heat treatment using a laser. In the heat-affected zone, a tensile stress field is induced if the temperature in the laser-treated zone remains below the martensite transformation point. This is achieved by local heating and subsequent cooling through removing the heat into the neighboring material. The value of the residual tensile stresses results from the difference between the yield point of material in the hot and in the cold state. The surrounding cold material acts as a fixing, so that the thermal expansion after CO_2 laser irradiation cannot diminish and thus tensile residual stresses are generated. Fig. 8.33 presents the maximum Barkhausen noise amplitude measured at two laser-treated points by using the Barkhausen noise microscopy.

The increase of M_{MAX} indicates the existence of residual tensile stresses. To confirm this assumption, X-ray residual stress measurements were carried out across the laser spots along a line at $x = 2000$ μm (see Fig. 8.33).

Fig. 8.34 demonstrates the comparison between the maximum noise amplitude and the stress values determined by X-ray measurements along the direction of magnetization on the above-mentioned trace. The size of the X-ray measuring point was 500 μm. Considering the measurement and positioning accuracy, the measured variables show good agreement.

Further measurements on a white cast iron sample of hypereutectic composition (4.6 wt% C) show that the M_{MAX} line scan structure can be correlated to residual stresses of the second kind. Fig. 8.35(b) shows M_{MAX} values along a measuring trace through two cementite needles. In the vicinity of two cementite needles (Fig. 8.35(a)), high M_{MAX} values (Fig. 8.35(b)) were measured. These values may be caused by high residual stress states.

Modeling of cementite/matrix interfaces in white cast iron supports this conclusion. Radiographic residual stress measurements of the ferrite phase of white cast iron resulted in a value of approximately 200 MPa (Hartmann, 1994).

Within the frame of a thesis (Kühn, 1999), an algorithm was developed that allows the quantitative determination of residual stresses of the second kind independent of grain orientation. Therefore, different measuring quantities derived from the magnetic

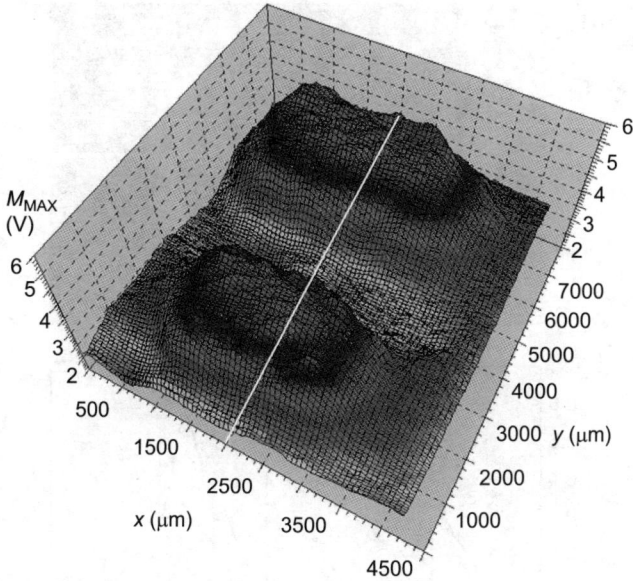

Figure 8.33 Maximum Barkhausen noise amplitude measured at two laser-treated points by using the Barkhausen noise microscopy.

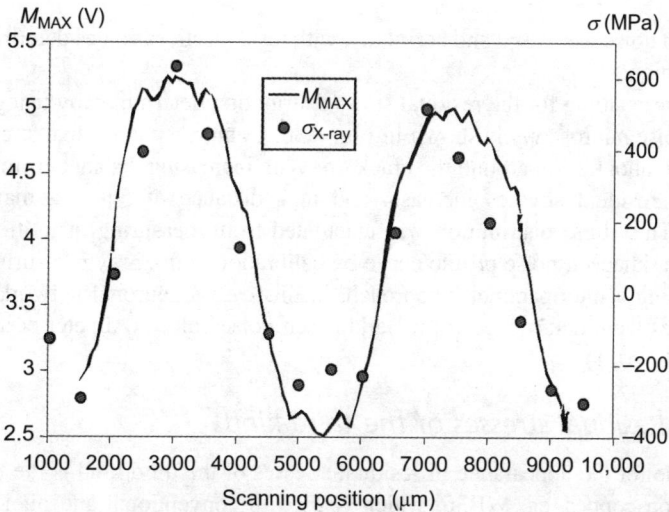

Figure 8.34 M_{MAX} values in comparison to residual stress values determined by X-ray method along the measuring trace (white line at $x = 2000$ μm in Fig. 8.33).

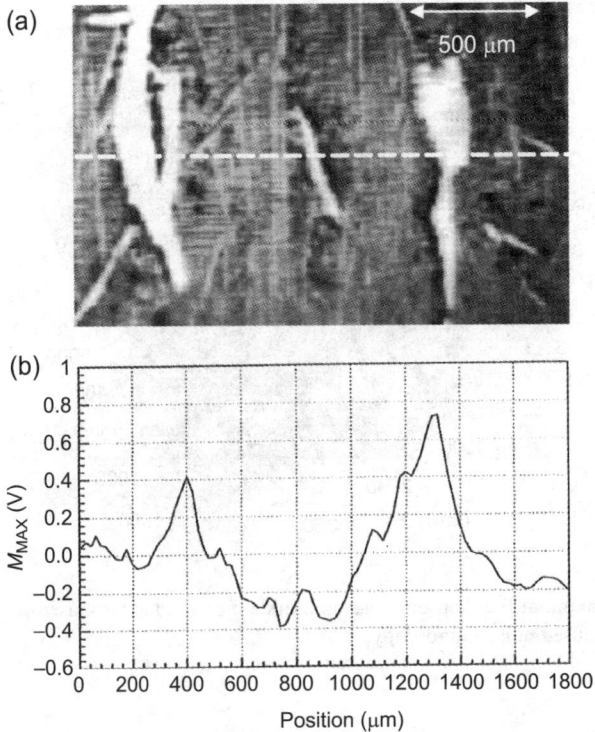

Figure 8.35 (a) Cementite needles embedded in Ledeburitic matrix. (b) Barkhausen noise line scan along a measuring trace through both cementite needles.

Barkhausen noise were used and correlated with residual stress values determined with X-ray techniques.

A further example for the residual stress distribution determined by using the Barkhausen noise microscopy is shown in Fig. 8.36. Compressive residual stresses were measured to about 2 μm around the crack tip. With increasing distance from the crack tip, tensile residual stresses increase, and in a distance of 5 μm, a maximum is observed. This stress distribution was calculated from measuring quantities derived from the Barkhausen noise profile curve by calibration with X-ray measuring quantities. By using a multiparameter approach, in this case a neuronal network, residual stress values were determined from Barkhausen noise values (Altpeter et al., 1997b; Boller et al., 2011).

8.3.6.4 Residual stresses of the third kind

An example for the appearance of residual stresses of the third kind is the two-phase material iron copper, eg, WB36, which is used in conventional and nuclear power plants (see Section 8.3.7).

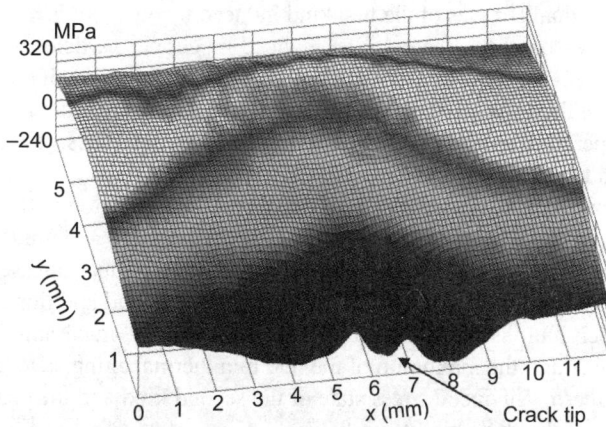

Figure 8.36 Residual stress field in front of a crack tip.

The residual stresses of the third kind increase during nucleation and formation of the coherent copper particles, and decrease during the transformation of the coherent copper particles into incoherent copper particles. To determine residual stresses of the third kind, a process is being used that was developed and patented within the framework of a German Research Society (DFG) project for steel containing cementite (Altpeter et al., 1997a). Altpeter et al. (2013) showed that this approach could generally be extended to the determination of high-degree internal stress for copper precipitates. In the frame of a DFG research project, it was shown that micromagnetic measurement techniques based on the tensile loading-dependent maximum, Barkhausen noise amplitude really can be used for the analysis of residual stresses of the third kind (Rabung et al., 2012). The usability of this procedure was demonstrated on binary Fe—Cu alloys. In order to obtain a precipitation-hardened alloy, a typical heat treatment was performed: solution annealing in the monophase area at 850°C for 2 h in a vacuum furnace quenching into water and thermal aging in the two-phase area at 500°C for 390 and 1500 min (Rabung et al., 2012). During the thermal aging, copper precipitates nucleate, grow, and change their microstructure from body center cubic (bcc) (coherent with the iron matrix) into face centered cubic (fcc) (incoherent with the iron matrix). Small and coherent copper precipitates cause the increase of the residual stresses of the third kind, whereas incoherent precipitates cause the decrease of the residual stresses of the third kind. Additional to the residual tensile stresses of the third kind, the precipitation of copper causes thermal-induced compressive residual stresses of the second kind. These residual stresses of the second kind are caused by the difference between the thermal expansion coefficient of iron and copper. The fact that the thermal expansion coefficient of copper is larger than that of iron causes compressive residual stresses of the second kind in the α-iron matrix. These residual stresses of the second kind do not change during the thermal aging.

Four steps are necessary in order to detect the residual stresses induced by the precipitation of the coherent copper particles, to eliminate the influence of the compressive residual stresses of the first kind induced by quenching, and the compressive residual stresses of the second kind induced by copper particles:

Step 1: Measurement of the $M_{MAX}(\sigma)$ curve and determination of the position A of the maximum in the $M_{MAX}(\sigma)$ curve after quenching into water after the solution heat treatment temperature at 500°C (Fig. 8.37). In that state, compressive residual stresses of first kind induced by quenching exist.

Step 2: Measurement of the $M_{MAX}(\sigma)$ curve and determination of the position B of the maximum in the $M_{MAX}(\sigma)$ curve after a thermal aging (390 min/500°C), when coherent Cu particles form. In that state compressive residual stresses of the first kind induced by quenching and residual stresses of the second and third kind induced by copper precipitations exist. The difference between the maximum values of the quenched state A and the maximum of the 390 min thermal aging state B corresponds to the sum of thermal-induced stress state of the second kind and the coherence residual stress state of the third kind (Fig. 8.37).

Step 3: After the thermal aging (1500 min/500°C), when coherent copper particles transform into incoherent copper particles, and therefore the residual stress of the third kind disappears, the residual stress of the second kind remains in the α-iron matrix. The stress difference between the maxima of the quenched state A and the maximum of the 1500 min thermal aging state C corresponds with the thermal-induced residual stress state of the second kind (Fig. 8.37).

Step 4: The measured shift between the $M_{MAX}(\sigma)$ curve during the early stage of copper precipitation (390 min/500°C) and the $M_{MAX}(\sigma)$ curve corresponding to the overaged thermal state (1500 min/500°C) represents the residual stress state of the third kind (Fig. 8.37).

The developed test method allows to quantitatively determine residual stresses of the third kind independent of residual stresses of first and second kind without the need for a reference method such as X-ray residual stress measurement. The developed testing method is patented (Altpeter et al., 2013).

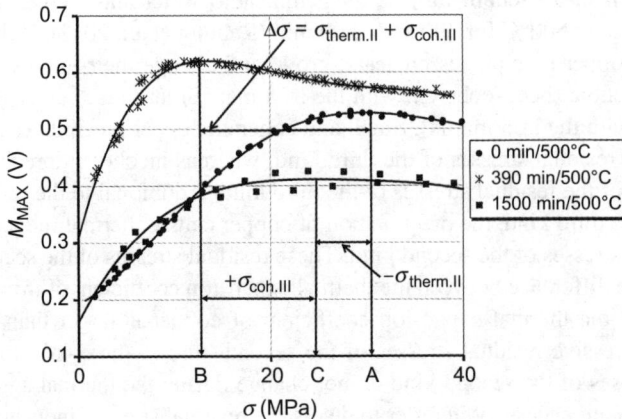

Figure 8.37 Concept for determination of residual stresses of the third kind (coherent residual stresses).

8.3.7 Aging (thermally induced embrittlement)

The potential of micromagnetic testing methods for the evaluation of thermally induced embrittlement was demonstrated on the Cu-rich, low-alloy, heat-resistant martensitic bainitic structural steel 15 NiCuMoNb-5 (WB36, material number 1.6368). This steel is used in conventional power plants at operating temperatures of up to 450°C, whereas German nuclear power plants use the material mainly for pipes at operating temperatures below 300°C, and in some rare cases in pressure vessels up to 340°C.

This material, when exposed at elevated temperatures between 280 and 350°C, shows the effect of precipitation hardening by Cu-rich precipitates. In the typical as-delivered state of WB36, half of the contained Cu (1.65 wt%) is already precipitated, while the other half remains in solid solution. After long-term service exposure above 320°C (thermal aging), damage was observed due to further precipitation of copper (Fig. 8.38).

An increase in yield strength $\Delta\sigma = +150$ MPa and a shift of the ductile to brittle transition temperature Δdbt $= +70°$C could be measured. Small-angle neutron-scattering measurements revealed the fact that the mechanical property changes are caused by copper precipitates ranging from 1 to 3 nm in size (Willer et al., 2001). The copper particles are coherent, have a bcc structure that induces a high level of compressive residual stress in their vicinity, balanced by tensile stresses in the environmental matrix.

The precipitation hardening can be characterized by Vickers hardness measurements in the laboratory; the effect is in the range of +40 HV 10 units. However, hardness measurements cannot be applied in large scale at a component in service, therefore there is demand for the development of an NDT technique.

By exposing a WB36 cylindrical specimen under shielding gas in an oven at 400°C, the thermal aging, ie, the precipitation of copper, was simulated. The hardness

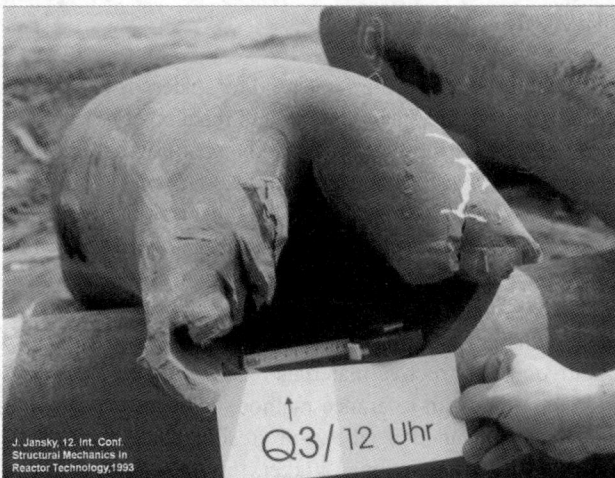

Figure 8.38 Damage of a pipe of WB36.
Structure and Mechanics in Reactor Technology, 1993.

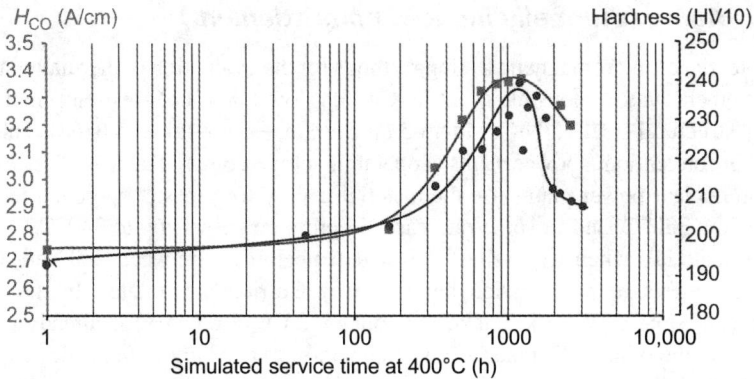

Figure 8.39 Magnetic coercivity in comparison with the mechanical hardness as a function of simulated service time at 400°C for WB36.

maximum was reached at 1000 h, followed by a coarsening of the precipitates combined with a hardness decrease (Altpeter and Dobmann, 2003; Altpeter et al., 2006). Fig. 8.39 shows the magnetic hardness, ie, the coercivity H_{CO} in comparison with the mechanical hardness, HV10, as function of length-of-service simulation. The analogy between mechanical and magnetic hardness is shown in the behavior of both hardness values as a function of simulated service time at 400°C.

By taking into account additional micromagnetic measuring quantities in order to define a regression approach, an NDT hardness prediction can be obtained with accuracy comparable to those of destructive tests (see chapter: Hybrid methods for materials characterization).

8.4   Conclusions

Electromagnetic techniques have a high potential for materials characterization, eg, for determination of microstructure state, texture, and residual stress state. A well-known electromagnetic technique is the magnetic Barkhausen noise. Many working groups worldwide use this technique for materials characterization. For several applications this technique alone is sufficient, for example, the determination of cementite content or residual stresses of the second and third kind. But an additional parameter, ie, temperature or load stress, is necessary: to determine cementite, the heating over the Curie temperature is needed; for the determination of residual stresses of the second or third kind, an additional tensile load stress is needed.

For a quantitative nondestructive materials characterization, a calibration of magnetic measuring quantities with reference methods is necessary. For example, in the case of residual stress measurement, the X-ray residual stress method; in the case of hardness measurement, a conventional Vickers hardness measurement is required. Only in the case of determination of residual stresses of the second or third kind, such a calibration is not required.

8.5 Future trends

For most industrial applications we have an overlapping of several influences as texture, residual stress, and microstructure state. One measuring quantity is not enough to solve the task. In these cases, two or more electromagnetic quantities are needed, for example, a combination of magnetic Barkhausen noise, incremental permeability, and dynamic magnetostriction. Such a multiparameter approach called 3MA (Micromagnetic Multiparameter Microstructure and stress Analysis) is described in chapter: Hybrid methods for materials characterization. Furthermore, a combination of magnetic and ultrasonic techniques can be used for a better texture analysis (see chapter: Hybrid methods for materials characterization).

References

Altpeter, I., 1990. Spannungsmessung und Zementitgehaltsbestimmung in Eisenwerkstoffen mittels dynamischer magnetischer und magnetoelastischer Messgrößen (PH.D thesis). Saarland University, Saarbrücken (in German).

Altpeter, I., 1991. Texture analysis with 3MA-techniques. In: Ruud, C., Bussiere, J.F., Green, R. (Eds.), Proceedings of the 4th International Symposium on Nondestructive Characterization of Materials. Plenum Press, New York, pp. 501–509.

Altpeter, I., Kern, R., Theiner, W., 1987. Determination of sub-surface microstructure states by micromagnetic NDT. In: Bussière, J.F., Green, R.E., Monchalin, J-P, Ruud, C. (Eds.), Nondestructive Characterization of Materials II. Plenum Press, New York.

Altpeter, I., Theiner, W.A., Fraunhofer-Gesellschaft, 1994. Vorrichtung zum ortsaufgelösten, zerstörungsfreien Untersuchen des magnetischen Barkhausenrauschens (DE patent 42 35 387 C1). März 24, 1994.

Altpeter, I., Dobmann, G., 1994. Nondestructive characterization of textures in cold-rolled steel products using the magnetic technique. In: Green, R.E., Kozaczek, K.J., Ruud, C. (Eds.), Nondestructive Characterization of Materials. Plenum Press, New York, pp. 807–816.

Altpeter, I., 1996. Nondestructive evaluation of cementite content in steel and white cast iron using inductive Barkhausen noise. Journal of Nondestructive Evaluation 15 (2), 45–60.

Altpeter, I., Dobmann, G., Kern, R., Theiner, W., Fraunhofer-Gesellschaft, 1997a. Verfahren zur zerstörungsfreien Werkstoffprüfung (EP patent 0 683 393 B1). October 1, 1997.

Altpeter, I., Dobmann, G., Meyendorf, N., Blumenauer, H., Horn, D., Krempe, M., 1997b. Residual stress measurement in front of a crack tip with high spatial resolution by using Barkhausen microscopy. In: Proceedings of the International Symposium on Nondestructive Characterization of Materials, vol. 8, pp. 659–664 (Boulder, Colorado).

Altpeter, I., Becking, R., Kern, R., Kröning, M., Hartmann, S., 1999. Mikromagnetische Ermittlung von thermisch induzierten Eigenspannungen in Stählen und weißem Gusseisen. In: Deutsche Forschungsgemeinschaft, Eigenspannungen und Verzug durch Wärmeeinwirkung, Forschungsbericht. Wiley-VCH, pp. 407–426.

Altpeter, I., Becker, R., Dobmann, G., Kern, R., Theiner, W., Yashan, A., 2002. Robust solutions of inverse problems in electromagnetic nondestructive evaluation. Inverse Problems 18, 1907–1921.

Altpeter, I., Dobmann, G., Chimenti, D.E., Thompson, O., 2003. NDE of material degradation by embrittlement and fatigue. In: Review of Progress in Quantitative Nondestructive Evaluation, vol. 22. American Institute of Physics (AIP), Melville, New York, pp. 15−21 (Conference proceedings).

Altpeter, I., Dobmann, G., Szielasko, K., 2006. Detection of copper precipitates in 15NiCu-MoNb-5 (WB36). Electromagnetic Nondestructive Evaluation (VII) 26, 239−246.

Altpeter, I., Sklarczyk, Ch, Kopp, M., Kröning, M., Hübner, S., Behrens, B.A., 2009. Nondestructive characterizing stress states in conventional deep drawing processes by means of electromagnetic methods. Electromagnetic Nondestructive Evaluation (XII) 131−139.

Altpeter, I., Dobmann, G., Hübschen, G., Fraunhofer-Gesellschaft, January 21, 2010. Verfahren zur spannungsunabhängigen Texturbestimmung eines ferromagnetischen Werkstückes (DE patent 10 2008 033 755).

Altpeter, I., Rabung, M., Szielasko, K., Schmauder, S., Binkele, P., Fraunhofer-Gesellschaft, March 11, 2013. Method for Non-destructive Quantitative Determination of the Micro-residual Stress II th and III th Type (EP patent 2 647 989 B1).

Altpeter, I., Szielasko, K., Fraunhofer-Gesellschaft, October 6, 2014. Determination of Micro-structure State Independent of Residual Stress State (EP patent 2063 266).

Ayere, Q., 1988. Analyse von Mikrogefügezuständen ausscheidungshärtender Modell - Legierungen mit transmissionselektronenmikroskopischen Verfahren und mikromagnetischen zerstörungsfreien Messgrößen (Saarbrücken, Doctoral thesis).

Barkhausen, H., 1919. Zwei mit Hilfe der neuen Verstärker entdeckte Erscheinungen. Physikalische Zeitschrift XX, 401−403.

Batista, L., Rabe, U., Altpeter, I., Hirsekorn, S., Dobmann, G., 2014. On the mechanism of nondestructive evaluation of cementite content in steels using a combination of magnetic Barkhausen noise and magnetic force microscopy techniques. Journal of Magnetism and Magnetic Materials 354, 248−256.

Bender, J., 1997. Barkhausen noise and eddy current microscopy (BEMI): microscope configuration, probes and imaging characteristics. In: Thompson, O., Chimenti, D.E. (Eds.), Review of Progress in Quantitative Nondestructive Evaluation. Plenum Press, New York, pp. 2121−2128.

Boller, C., Altpeter, I., Dobmann, G., Rabung, M., Schreiber, J., Szielasko, K., Tschuncky, R., 2011. Electromagnetism as means for understanding materials mechanics phenomena in magnetic materials. Materialwissenschaft und Werkstofftechnik 42 (4), 269−278.

Cullity, B.D., 1972. Introduction to Magnetic Materials. Addison-Wesley, London.

Dijkstra, L.J., Wert, C., 1950. Physical Review 79, 479.

Dobmann, G., Pitsch, H., Fraunhofer-Gesellschaft, 1988. Verfahren zum zerstörungsfreien Messen magnetischer Eigenschaften eines Prüfkörpers sowie Vorrichtung zum zerstörungsfreien Messen magnetischer Eigenschaften eines Prüfkörpers (DE patent 3813739 A).

Dobmann, G., 1999. State of the art to NDT for characterizing properties in the production of flat steel products. Joint Seminar of the German Iron and Steel Institute(VDEh) and the German Society for NDT(DGZfP). In: Düsseldorf to 'Material Properties Determination', Proceedings (in German) DGZfP Berlin.

Dobmann, G., 2006. NDE for material characterization of ageing due to thermal embrittlement, fatigue and neutron degradation. International Journal of Materials & Product Technology 26, 122−139.

Dodd, C.V., Deeds, W.E., Luquire, J.W., 1969. Integral solutions to some eddy current problems. International Journal of Nondestructive Testing 1, 29−90.

Förster, F., 1940. Messgerät zur schnellen Bestimmung magnetischer Größen. Zeitschrift fuer Metallkunde 32, 184.

Förster, F., Stambke, K., 1941. Magnetische Untersuchungen innerer Spannungen. Eigenspannungen beim Recken von Nickeldraht. Zeitschrift fuer Metallkunde 33 (3), 97—114.

Hartmann, S., 1994. Thermoelastische Eigenspannungen in weißem Gußeisen, Ph.D.thesis, University Saarbrücken, Germany.

IEC 60404-2: 1996 and BS EN 60404-2: 1998 + A1. Magnetic Materials — Part 2: Methods of Measurement of the Magnetic Properties of Electrical Steel Strip and Sheet by Means of an Epstein Frame.

Jiles, D.C., Thoelke, J.B., 1989. Theory of ferromagnetic hysteresis: determination of model parameters from experimental hysteresis loops. IEEE Transactions on Magnetics 25 (5), 3928—3930.

Jiles, D., 1998. Introduction to Magnetism and Magnetic Materials. CRC-Press, London, New York, Tokyo, Melbourne, Madras.

Kersten, M., 1944. Grundlagen einer Theorie der ferromagnetischen Hysterese und der Koerzitivfeldkraft (Leipzig).

Kneller, E., 1962. Ferromagnetismus. Springer, Berlin.

Kröning, M., Altpeter, I., Laub, U., 1996. Detecting the tendency to chilling in series — manufactured cast iron components using micromagnetic test procedures. Materials Science Forum 210—213, 55—62 (Switzerland, Transtec Publications).

Kühn, S., 1999. Eigenspannungsermittlung im Mikrometerbereich mittels mikromagnetischer Prüfverfahren. Diplomarbeit. IZFP, Saarbrücken.

Leep, R.W., Pasley, R.L., February 18, 1969. Method and System for Investigating the Stress Condition of Magnetic Materials (U.S. patent 3.427.872).

Macherauch, E., Kloos, K.H., 1987. Origin, measurement and evaluation of residual stresses. In: Residual Stresses in Science and Technology. DGM Informationsgesellschaft-Verlag.

Maisl, U., Frauendorfer, R., Kopp, M., Altpeter, I., 2000. Zerstörungsfreies Prüfverfahren zur Bestimmung von Werkstoffeigenschaften von Gußeisen. In: Sonderdruck aus Gießerei-Praxis, Heft 3,113—121. Fachverlag Schiele & Schön GmbH, Berlin.

Moorthy, V., Vaidyanathan, S., Jayakumar, T., Raj, B., 2000. Microstructural characterization of tempered and deformed ferritic steels using magnetic Barkhausen emission technique. Journal of Nondestructive Evaluation 20, 33—39.

Pawlowski, Z., Rulka, R., 1973. Measurement of internal stress using Barkhausen effect. In: VII ICNdT, Warschau, Vortrag J-13.

Pirlog, M., 2005. Mikromagnetischer Nachweis der Wekstoffalterung infolge von kohärenten Kupferausscheidungen (Doctoral thesis), Universität Saarbrücken.

Pitsch, H., 1989. Die Entwicklung und Erprobung der Oberwellenanalyse im Zeitsignal der magnetischen Tangentialfeldstärke als neues Modul des 3 MA - Ansatzes, Universität des Saarlandes (Doctoral thesis), IZFP - Bericht 900107-TW.

Rabung, M., Altpeter, I., Dobmann, G., Szielasko, K., 2012. Micromagnetic evaluation of microresidual stresses of the II and III orders. Welding in the World 56 (5/6), 29—34.

Schreiber, J., Kröning, M., Panin, V., Fraunhofer-Gesellschaft, 2008. Method for Determining the Remaining Service Life and/or Fatigue State of Components (EP patent 2 108 112 B1).

Seeger, A., 1966. Moderne Probleme der Metallphysik (zweiter Band). Springer-Verlag, Berlin, Heidelberg, New York.

Spies, M., Schneider, E., 1989. Nondestructive analysis of the deep- drawing behaviour of rolled sheets with ultrasonic techniques. In: Höller, P., Hauk, V., Dobmann, G., Ruud, C., Green, R. (Eds.), Proceedings of the 3rd International Symposium on Nondestructive Characterization of Materials. Springer-Verlag, Berlin.

Stresstech, 2015. Produktbeschreibung und Broschüren unter. www.stresstech.fi.

Theiner, W.A., Altpeter, I., 1987. Determination of Residual Stresses Using Micromagnetic Parameters, New Procedures in Nondestructive Testing, pp. 575–585.

Tiitto, S., 1980. Über die zerstörungsfreie Ermittlung der Eigenspannungen in ferromagnetischen Stählen. In: Berichte eines Symposiums in Bad Nauheim, Deutsche Gesellschaft für Metallkunde, pp. 261–270.

Vengrinovich, V., Tsukerman, V., Denkevich, Y., Bryantsev, D., 2006. New parameters to characterize internal stresses via Barkhausen noise. In: Proceedings of the 9th European Conference on Nondestructive Testing.

Willer, D., Zies, G., Kuppler, D., Föhl, J., Katerbau, K.-H., 2001. Betriebsbedingte Eigenschaftsänderungen kupferhaltiger ferritischer Behälter- und Rohrleitungsbaustähle. Reaktorsicherheitsforschung. Förderkennzeichen 150 1087, final report MPA Stuttgart.

Hybrid methods for materials characterization

9

R. Tschuncky[1], K. Szielasko[1], I. Altpeter[2]
[1]Fraunhofer Institute for Nondestructive Testing (IZFP), Saarbrücken, Germany;
[2]Formerly Fraunhofer Institute for Nondestructive Testing (IZFP), Saarbrücken, Germany

9.1 Introduction

In nondestructive testing (NDT) a hybrid method is understood as an approach which combines two or more NDT methods. Due to the variety of NDT techniques, many combinations are possible. In the nondestructive evaluation area of fault detection, for example, several combinations of NDT methods (hybrid methods) are in use (Willems et al., 2010a,b). Several combined methods are possible in order to achieve optimized materials characterization. This section will give insight into combinations of micromagnetic, eddy current, and ultrasonic methods. Combinations are presented where hybrid sensors can be used. Several NDT can therefore be applied with the help of such hybrid sensors. Due to the combination of several NDT methods, multiple measuring quantities are always used in materials characterization applications. Therefore, many hybrid methods use computational algorithms, eg, regression analyses or pattern recognition algorithms, for a connection between measuring quantities with different physical information content and reference values of the tested materials. These are relevant quality features like residual stresses, hardness, hardening depth, yield strength, etc.

This section shows the potential of hybrid approaches on several industrial applications and documents the benefits for the industrial user, which are mainly in the context of safety, quality, and lifetime. This section also documents some signal processing aspects for the use of eddy current techniques and micromagnetic hybrid methods (see Section 9.2) and gives insight into possible calibration procedures (see Section 9.3).

A hybrid method uses the combination between micromagnetic and ultrasonic methods and has potential for texture determination in steel sheets (see Section 9.4.2). A known micromagnetic hybrid method is the micromagnetic, multiparameter, microstructure and stress analysis, or 3MA, approach. This approach combines different micro- and electromagnetic measuring quantities and uses regression analyses or pattern recognition algorithms for the quantitative determination of material properties for ferromagnetic materials (see Section 9.2.1.3). An advantage of some hybrid approaches is that different material features such as residual stresses, hardness, and hardening depths are predicted simultaneously (see Section 9.4.7).

Materials Characterization Using Nondestructive Evaluation (NDE) Methods
http://dx.doi.org/10.1016/B978-0-08-100040-3.00009-2

9.2   Signal processing and analysis on the way toward software-defined nondestructive testing devices

Today, most of the electromagnetic NDT equipment supports multiparametric evaluation to a certain degree, which is particularly important to the field of materials characterization, eg, with eddy current approaches or micromagnetic techniques. The challenge in this field is the presence of superimposed disturbances especially in the industrial praxis, so that the target quantities (eg, hardness, hardening depth, residual stress [of the first kind] or applied stress) usually cannot be represented by a single nondestructive measuring quantity in a satisfactory way, with exception of several applications (eg, determination of residual stresses of second and third kinds and determination of cementite content; see section: Electromagnetic techniques for materials characterization). Disturbances such as surface condition, lot-related variation and tolerance of the material composition can be dealt with if several (ideally independent) measuring quantities are determined and their information combined. On the software side, a tradeoff has to be found between the powerful mathematical approaches (regression analysis, pattern recognition, neural networks) on the one hand, and a user interface that makes the multiparametric data space and the calibration easy to understand and to manage. On the hardware side, sensor and device complexity should be kept on a low level in order to reduce manufacturing costs and increase the reproducibility of characteristics.

This contribution illustrates how signal generation and evaluation affect the material information obtained with essentially the same probe design. Today, we are in the situation that mostly the software defines the testing method, so that with a common (or at least very similar) probe, different methods can be performed by means of software development. This circumstance enables the development of simplified multimethod (hybrid) and multiparameter NDT systems. Since there is a thin line between eddy current and micromagnetic techniques, they will be summarized under the term "electromagnetic."

9.2.1   Design and applications of inductive probes

Most electromagnetic NDT methods are based on or contain inductive probes. If we leave out the details such as coil shape and special coil arrangements, inductive probes contain at least one transmitter coil that generates a magnetic field in proximity to the material under test. They may also (optionally) contain a magnetically conductive core and/or set of receiver coils or probes that pick up the magnetic field. Common features of inductive probe designs; their applications are schematically represented in Fig. 9.1.

The coils inside an inductive probe are most frequently used for the following purposes:

- Excitation of small-amplitude magnetic fields (no irreversible domain wall motion)
- Excitation of large-amplitude magnetic fields (with irreversible domain wall motion)
- Pick-up of magnetic field strength variation
- Pick-up of magnetic flux variation (usually encircling coil)
- Pick-up of magnetic Barkhausen noise (series of domain wall jump events)

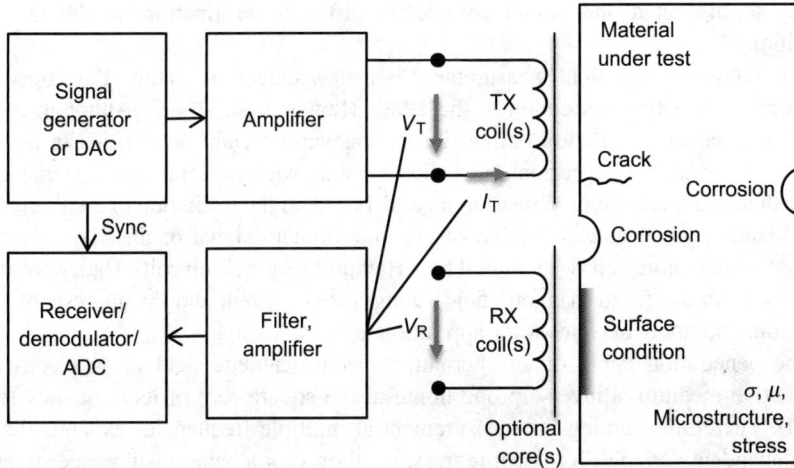

Figure 9.1 Simplified schematic design of inductive probes and their control circuits; voltage (V_T, V_R) measurement options in transmitter (TX) and receiver (RX), and current measurement option (I_T) in transmitter. Some of the relevant material conditions and flaws are indicated. Some probes contain magnetic cores.

Based on the signals obtained, the applications range from materials characterization (hardness, hardening depth, stress, strength, etc.) to corrosion detection, wall thickness estimation, and crack detection.

9.2.1.1 Signal processing for multifrequency eddy current techniques

As an example of inductive probes, eddy current probes are widely used for materials characterization, flaw detection, thickness measurement, and proximity detection, where magnetic and/or electrically conductive materials are involved (Udpa and Moore, 2004). Eddy current inspection is based on the impedance of a coil setup in close proximity to the material under test. The impedance is modulated by the material's apparent (ie, not only intrinsic, but also shape- and flaw-affected) permeability and conductivity.

On the evaluation side, multiparameter analysis can increase the robustness of the discrimination between different conditions (Bartels and Fisher, 1995). In the simplest scenario, real and imaginary parts (or magnitude and phase) of the probe impedance can be used as independent variables in a multiple linear regression approach. The condition qualifier (OK/NOK) or the value of a mechanical characteristic such as hardness usually is the dependent variable (target). A polynomial expression (mix), which approximates the target quantities or qualifier, is the result of the regression.

Under field conditions, disturbances and material variations may raise the need for additional or different information. Several approaches are applied in order to obtain other magnetic quantities as input for the regression analysis without having to modify the probe, sometimes even without the need for different hardware. Some of these approaches are described in the following sections in order to point out that, whereas the probes are of similar or same construction, signal generation and analysis can be highly sophisticated and obtain completely different information on the material condition.

Alternating current field measurement is a flaw detection method developed for offshore construction inspection in the 1980s (Lewis et al., 1988). Although sometimes referred to as an eddy current inspection variant (Jain, 2104), it can also be understood as alternating current flux leakage testing with a special way of signal analysis. Alternating current field measurement is based on the excitation of an alternating field (using a yoke) and evaluation of the time domain signal of the magnetic field strength in two dimensions (measured by a Hall probe or pick-up coil). Under presence of a crack, an increased magnetic field strength is observed, and the analysis of both directional components allows for approximate crack sizing.

The penetration depth of an alternating electromagnetic field in an electrically conductive medium is inverse proportional to the square root of its frequency (skin effect). Therefore, running the measurement at multiple frequencies expands the set of independent variables for linear regression, allows for a better disturbance suppression and delivers information on depth-dependent material conditions. In order to deal with surface-near material gradients or residual stress, or to explicitly obtain surface or depth properties, the eddy current inspection frequency is often varied. Most devices offer a frequency-multiplexing option, ie, automatic frequency alternation, measuring with only one frequency at a time. As a downside, multiplexing reduces the measuring speed, but state-of-the-art eddy current systems are very fast, so the remaining speed after multiplexing can still reach several thousand measurements per second.

Superimposed generations of several testing frequencies can be considered when the highest measuring speed is required, but it significantly reduces signal dynamics, because all frequencies share the same line in the signal processing chain. Therefore, this technique is less frequently used and has increased demands to the hardware, such as narrowband filters and/or high-resolution analog–digital converters.

Pulsed eddy current techniques represent a variant of multifrequency eddy current inspection (Giguère et al., 2001). The transmitter coil is driven with a pulse waveform instead of a continuous sine wave. The pulse waveform contains a wide-stretched frequency spectrum. As opposed to conventional eddy current inspection, the drive amplitudes are higher. In combination with the low-frequency content of the signal, this leads to high penetration depth and high lift-off capability, which is important when measuring through thick (mostly nonconductive) cover layers such as insulation material, brick, or concrete. One application of pulsed eddy current testing is corrosion detection (eg, pipe wall thickness measurement). The analysis procedures for pulsed eddy current signals differ from conventional eddy current analysis. Pulsed eddy current analysis is mostly performed in the time domain, revealing the distance to a defect by earlier or later occurrence of an indication in the received signal.

9.2.1.2 Signal processing for micromagnetic techniques

Quite often, there is only a thin line between different electromagnetic nondestructive evaluation techniques. When a ferromagnetic test object is magnetized periodically with sinusoidal coil voltage of sufficiently high voltage (ie, when domain walls move irreversibly), the time domain signal of the magnetic field strength contains characteristic harmonics, which can be measured inductively or with Hall probes. Amplitudes and phases of the low-order odd harmonics, the harmonic distortion factor and characteristic intersection points that depend on the coercivity may be used for materials characterization (Dobmann and Pitsch, 1989; see section: Electromagnetic techniques for materials characterization). Depending on the evaluation algorithms applied, harmonics analysis can also be considered a variant of multifrequency eddy current impedance analysis (Heutling et al., 2004). Both ways of processing the harmonics are used for microstructure and stress analysis by several kinds of electro-magnetic hybrid testing equipment today (Laub and Altpeter, 1998; Buschur, 2015). Besides the use of naturally occurring harmonics, a variation of the excitation frequency may be applied in order to reach different penetration depths. This approach was successfully applied to surface-hardened material (Szielasko, 2009). Plots of the harmonic distortion factor K as a function of frequency depended on the surface-hardening depth (Fig. 9.2).

Magnetic Barkhausen noise analysis, a method of micromagnetic materials characterization where domain wall jumps are detected with an inductive probe, has been used in NDT for a while (Matzkanin et al., 1979). The analyzed frequency range of the signal received is of crucial importance to the information content of the noise envelope (Altpeter, 1990; Batista et al., 2014). Depending on the frequency band analyzed, a different response to microstructure and stress is obtained (Fig. 9.3).

By analysis of the Barkhausen noise frequency spectrum, different material deformation conditions can be identified (Fig. 9.4). In modern testing equipment,

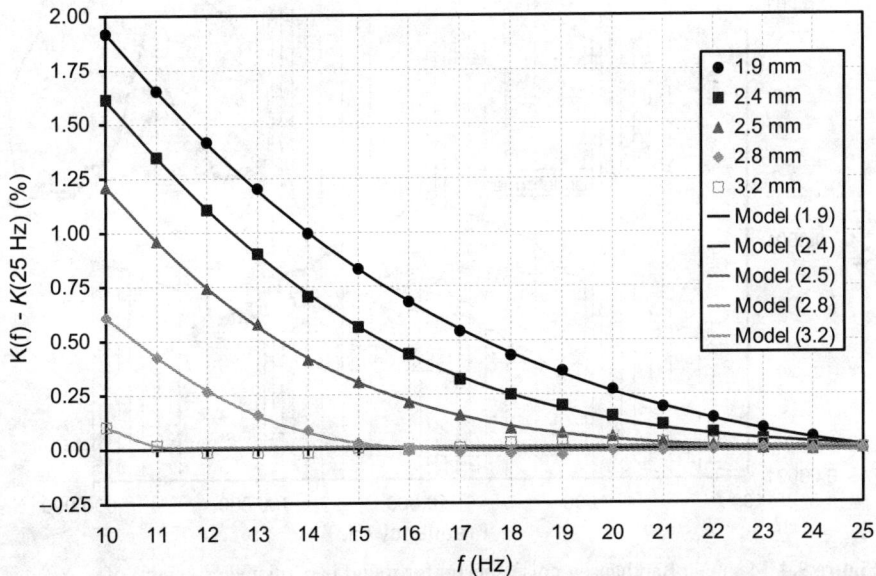

Figure 9.2 Differential change of harmonic distortion factor K with reference to its value at 25 Hz for samples of different surface-hardening depth from 1.9 up to 3.2 mm. Ph.D. thesis Szielasko (in German).

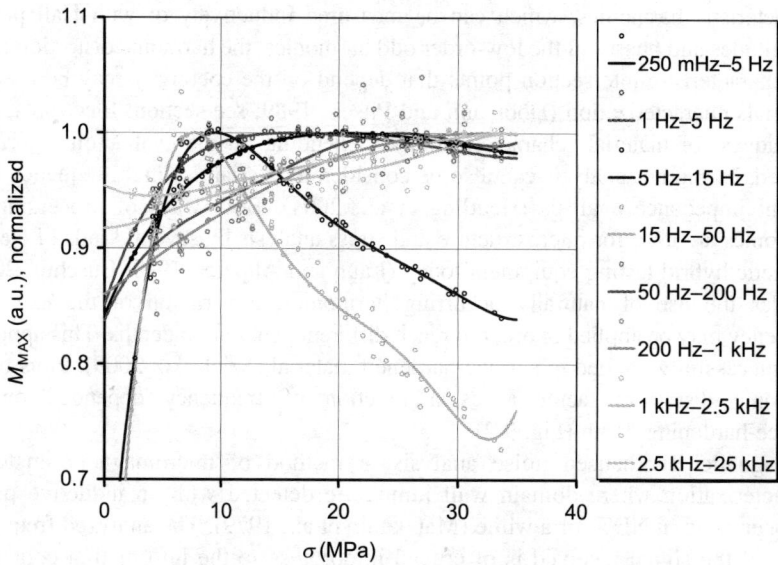

Figure 9.3 Magnetic Barkhausen noise peak amplitude M_{MAX} as a function of applied tensile stress in a Fe-0.65% Cu model alloy shows different responses in different frequency bands (excitations frequency: 0.05 Hz).
Ph.D. thesis Szielasko (in German).

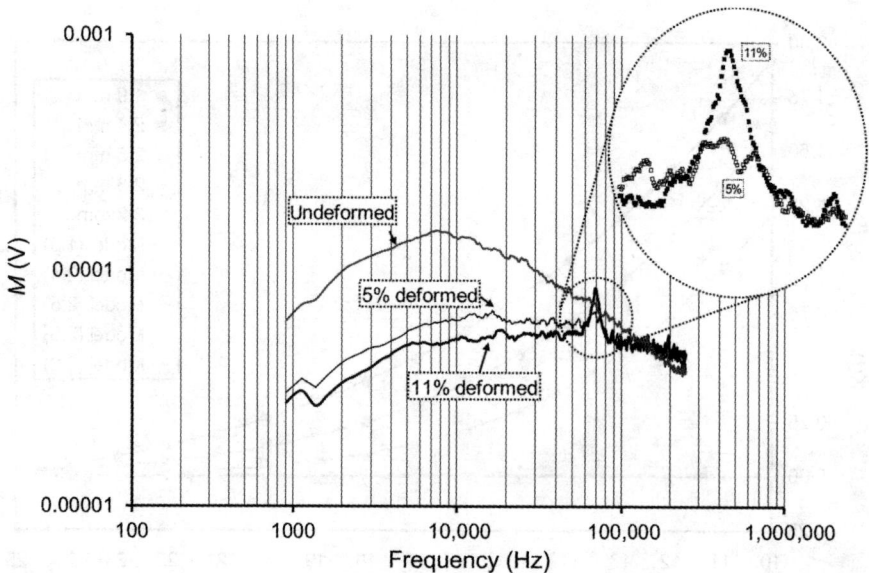

Figure 9.4 Magnetic Barkhausen noise spectra for round bars (diameter: 6 mm) of 15NiCuMoNb5 in undeformed, 5% plastically deformed, and 11% plastically deformed condition. Excitation frequency: 1 Hz.
Ph.D. thesis Szielasko (in German).

the choice of frequency can be made after the signal acquisition, since the Barkhausen noise signal is digitized and mostly software processed. There also exist approaches to filter the signal for a separation of deterministic and stochastic contents (Szielasko, 2009; Szielasko et al., 2009).

9.2.1.3 The 3MA approach

Developments since 1980 use the micromagnetic multiparameter nondestructive 3MA evaluation- and signal-processing approach. The 3MA approach uses application-dependent electromagnetic NDT techniques, ie, analysis of the magnetic Barkhausen noise, analysis of the upper harmonics in the time signal of the tangential magnetic field strength, incremental permeability analysis, and multifrequency eddy current techniques. (Detailed information and the physical basics for these techniques are described in section: Electromagnetic techniques for materials characterization.) The reason for using more than one measuring quantity for materials characterization is the increased robustness against disturbing influences. Disturbance means that additional parameters like microstructure, texture, material variations, and surface treatment, for example, are influencing stress measurements.

For this approach a multicoil probe is needed. Because of the necessary periodic magnetization of the material electromagnets are integrated in the multicoil probe. In the center of the electromagnet a Hall probe and air coil(s) are placed. The time signal of the tangential magnetic field at the surface of the component during the magnetization is detected by means of the Hall probe. By a coil the Barkhausen noise events are detected. But it can also serve as an eddy current transceiver that comes with the eddy current technique and incremental permeability analysis is used. A schematic drawing of such a multicoil probe is shown in Fig. 9.5.

The 3MA approach obtains a "magnetic fingerprint" of the material which consists of approximately 40 measuring quantities derived from measured signals. The 3MA approach solves the inverse problem of target quantity prediction from a set of calibration data for the quantitative determination of industry-relevant target quantities such as hardness, hardening depth, yield strength, residual stress, retained austenite as well as the detection of grinding burn in ferromagnetic materials (Dobmann and Pitsch, 1989; Dobmann and Höller, 1990; Altpeter et al., 2002; Szielasko, 2009; Tschuncky, 2011). Four electromagnetic methods are supported in 3MA-II testing devices

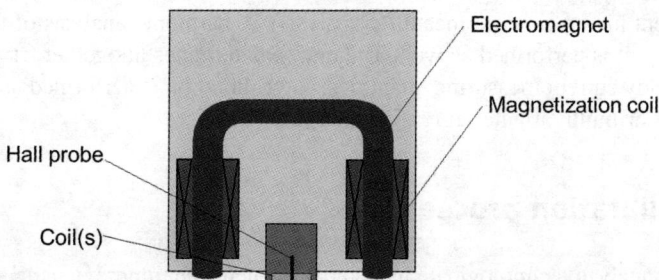

Figure 9.5 Schematic drawing of a multicoil probe for the 3MA approach. Ph.D. thesis Tschuncky (in German).

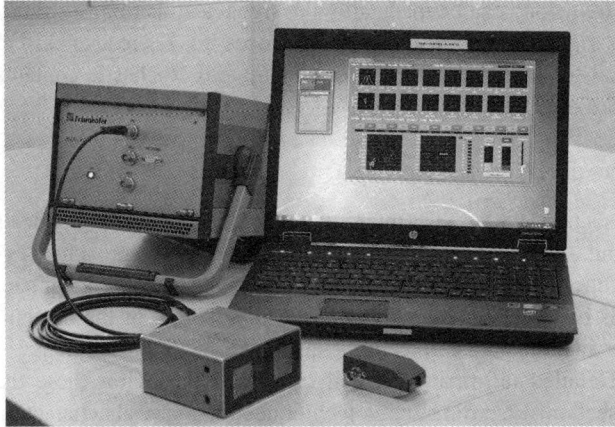

Figure 9.6 Micromagnetic testing device 3MA-X8 (three-channel version).

(analysis of the magnetic Barkhausen noise, analysis of the upper harmonics in the time signal of the tangential magnetic field strength, incremental permeability analysis, and multifrequency eddy current techniques) (Boller et al., 2011).

9.2.2 Method combination in software-defined measuring systems

Since high-dynamics signal processing hardware is available at a reasonable cost today, electromagnetic testing equipment can be highly simplified (Szielasko et al., 2013). Where in earlier systems a complex multicoil probe was required for eddy current impedance and incremental permeability analysis, today single-coil parametric systems with temperature-stabilized measurement can be implemented (Szielasko et al., 2014a,b). A recent development in this direction is the 3MA-X8 testing device, which requires only a single electromagnet with a single coil as transducer for eddy current impedance analysis, incremental permeability analysis, and harmonics analysis. Fig. 9.6 sows a three-channel version of the 3MA-X8 testing device.

The single-coil transducer generates a mix of magnetization and eddy current testing frequency, and based on current and voltage signal, eddy current analyses are performed in several magnetization states throughout the magnetic hysteresis curve (eddy current incremental permeability analysis). A harmonic analysis of the magnetization current is performed as well. The device determines altogether 21 micromagnetic and eddy current measuring quantities, and calibration is performed using pattern recognition or multivariante regression analysis.

9.3 Calibration procedure

In order to analyze technological characteristics such as hardness, tensile strength, or residual stresses mechanically, so-called target quantities, a calibration procedure of hybrid systems is required. During such calibration procedures the measuring quantities are assigned to the corresponding target quantities. This requires sample sets with conventionally determined reference data that can be used as target quantities in the calibration procedure. All measuring quantities derived from the hybrid test method must be determined for the selected target quantities. Several methods are available as mathematical tools for the determination of correlations between the measuring quantities and the material property values. On the one hand, there is the regression analysis, which analyzes the calibration data by means of the regression calculation in terms of the significance of the measuring quantities. Here the measured variables in different potencies are taken into account in the calculations as reciprocal and in the form of selected products and quotients (Albers and Skiera, 1999). On the other hand, pattern recognition approaches and neural networks can be used. One example is a pattern recognition approach that determinates the target values of samples to be characterized in terms of subjects on the basis of similarity to the measuring variables of the calibration samples and calculated the target quantity values by a review of the target quantity values from the calibration samples (Stork, 2001). In Fig. 9.7, the process of a calibration procedure of hybrid test systems is shown schematically. It can be seen that regardless of the calibration method the space of the measuring quantities derived from the calibration samples is always the basis for the solution of the multiparameter problem (calibration).

Such a calibration procedure needs calibration specimens. This calibration sample set has to represent the material states to be characterized. Therefore, reference investigations with destructive tests have to be performed and each calibration specimen has to represent a relevant material state.

The calibrated parameter settings of micromagnetic hybrid NDT methods like the 3MA approach can depend on both test sample properties and instrument, ie, probe characteristics and electronics. Any change of instrument may cause measurement errors or a decline of the result quality. A solution for this problem was researched in Tschuncky (2011). This thesis deals with the determination, analysis, and

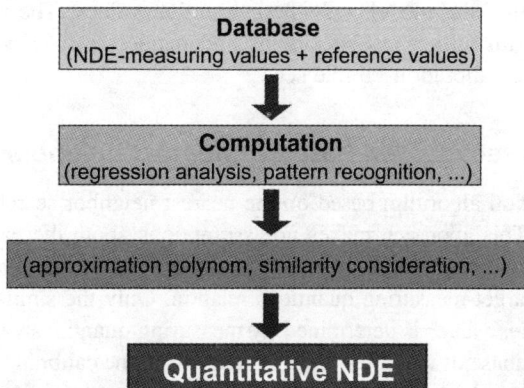

Database
(NDE-measuring values + reference values)

↓

Computation
(regression analysis, pattern recognition, ...)

↓

(approximation polynom, similarity consideration, ...)

↓

Quantitative NDE

Figure 9.7 Schematic illustration of the calibration procedure for a quantitative nondestructive evaluation (NDE) for hybrid systems.
Ph.D. thesis Tschuncky (in German).

implementation of calibration procedures for hybrid micromagnetic NDT methods that allows replacement of sensors and electronics.

9.3.1 Regression analysis

One option for the calibration procedure of nondestructive hybrid techniques is the use of a regression analysis. A regression analysis is a statistical analysis method, which aims to determine relationships between a dependent and one or more independent variables (Schach and Schäfer, 1978; Blume, 1974a,b). The required data matrix is composed from the measuring variables and interpreted as multiple independent variables. The corresponding target values vector is interpreted as the dependent variables, so that the regression analysis is to find correlations between the testing values and the associated result vectors.

The regression analysis models the correlation by a polynomial, which approximates the measuring-target-quantities relation. The approximate optimal coefficients of the regression polynomial are determined by applying the regression analysis based on the data of calibration, which are analyzed with regard to the significance of the calibration measuring quantities (Albers and Skiera, 1999). The calibration measurements are taken on a calibration sample set which must cover the target quantity value range.

When using the linear regression analysis for calibration of electromagnetic testing methods a polynomial is parameterized, this approximates the expected target quantity values from the measuring quantity values afterward. A regression analysis based on calibration polynomial can be written as:

$$y = a_0 + a_1 x_1 + a_2 x_2 + \cdots + a_N x_N$$

where y is the target quantity, a_i are the coefficients, x_i are the measuring quantities.

For online applications, robust solutions are required. The robustness is documented by a high correlation coefficient and a low error bandwidth (standard deviation) comparing the model with the calibration values. The regression model has to be tested with further test specimens, independently selected, and they must not be a part of the calibration sample set.

9.3.2 Pattern recognition based of nearest neighbor searches

A pattern recognition algorithm based on the nearest neighbor searches is a nonparametric approach. This approach makes no assumptions about the parametric form of the underlying distribution of the measuring quantities and it is not based on a model to determine the target-measuring quantities relation. Only the similarity between the measuring quantities values is determined by measuring quantities values contained in the calibration database. It identifies those samples from the calibration database which the least measured sample deviates from. With these nearest neighbors the target quantities values of the measured sample can be determined. Therefore, a weighting process of the target quantities values of the nearest neighbors can be used. A known nearest neighbor search variant, for example, is the k-nearest neighbor search. In this variant k, similar nearest neighbors are always selected from the calibration database. So less similar neighbors may also be included in the results analysis. This can produce some nonoptimal results. An alternative for this variant is described in the following section.

9.3.2.1 Nearest neighbor search with Euclidian distances

The first step of nearest neighbor searches with Euclidian distances is a normalization process for all measuring quantities. Since each measuring quantity is not necessarily in the same value range, normalization leads all measuring quantities in equivalent value ranges, thereby making an evaluation of the distances between the measuring quantities possible. After the normalization an assessment of the measuring quantities values of the sample is to be classified in terms of the Euclidian distance to the corresponding entries in the calibration database instead. There is a difference between each normalized measuring quantity of a calibration data set and the corresponding measuring quantity of the current sample. These differences are associated to the corresponding calibration data and added to a normalized total distance. After formation of the total distance the records from the calibration database are called nearest neighbors of the current sample, if their Euclidean distance is smaller than a barrier selected by the user (maximum Euclidian distance). These nearest neighbors are used to generate the result of the pattern recognition with the help of a weighting process with the corresponding target quantity values. For weighting processes the maximum Euclidian distances and the normal derivation function or other derivation functions can be used, for example (Tschuncky, 2011).

9.4   Applications

9.4.1   Residual stresses

The micromagnetic 3MA hybrid techniques have become increasingly important for stress measurements since 1976. This can be deduced from papers published between 1976 and 2015, for example (Altpeter et al., 1986, 2005; Altpeter and Theiner, 1989; Dobmann and Pitsch, 1989; Dobmann and Höller, 1990; Szielasko et al., 2013, 2014a,b; Wolter et al., 2003; Wolter and Dobmann, 2006; Theiner et al., 1999; Kern et al., 1996, etc.).

The monitoring of regions of high tensile residual stresses of hardened components like steering shafts, cam rings, crankshafts, camshafts, and gear wheels is an important objective in the industrial practice. Inspection costs are a main part in quality costs and are especially high, because most of the tests are destructive in nature and very time consuming. Hybrid 3MA-II testing devices can measure these characteristics in real time or in fast postprocess units. An example is presented in Fig. 9.8.

Fig. 9.8 demonstrates a good correlation between residual stresses on grinding states determined by micromagnetic hybrid method 3MA and residual stresses on grinding states determined by X-ray technique (Theiner et al., 1999).

A further example which documents the potential of micromagnetic hybrid methods for residual stress measurement is shown in Fig. 9.9.

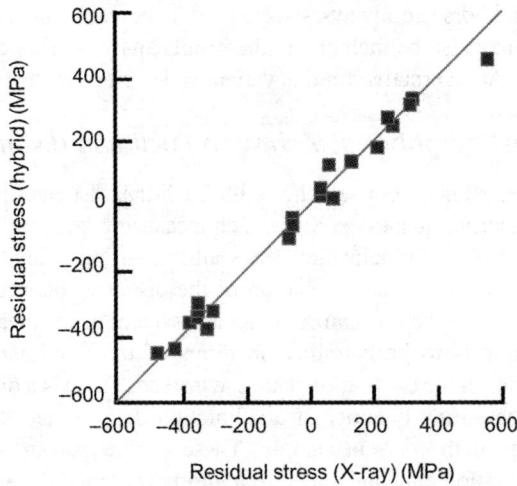

Figure 9.8 Residual stresses measured on grinding states: 3MA hybrid testing device results versus X-ray reference values.

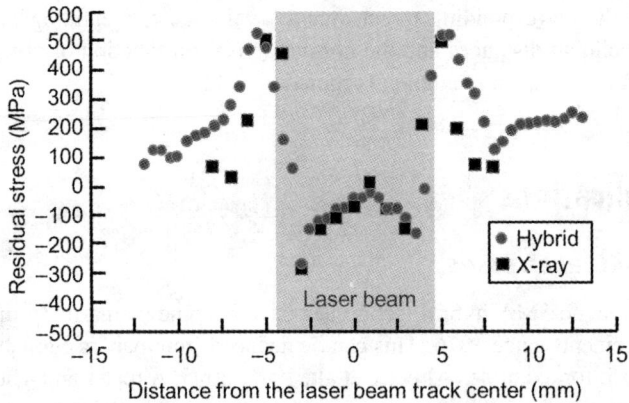

Figure 9.9 Residual stresses across a laser hardened track determined with micromagnetic hybrid method and X-ray.

After a calibration step the residual stresses across the laser hardened track can be well compared with the residual stresses measured by X-ray. Both results do not fit exactly because of different penetration depth and because of two-dimensional stress states which influence X-ray and micromagnetic measurements in different ways. The used micromagnetic hybrid method combines analysis of the magnetic Barkhausen noise, analysis of the upper harmonics in the time signal of the tangential magnetic field strength, and incremental permeability analysis (Kern et al., 1996; Altpeter et al., 2005).

9.4.2 Texture

As mentioned in section: Electromagnetic techniques for materials characterization, a combination of several micromagnetic and ultrasonic measuring quantities is required to solve the multiparameter inspection task of texture characterization independent of the influence of mechanical stress states.

In the framework of an European Community of Coal and Steel (ECCS) project (1988−91), ultrasonic and micromagnetic measurements on cold rolled strips (interstitial free steel grades with extremely small carbon content) have been carried out with the aim to develop a nondestructive hybrid method, which is fast and simple and has the potential to determine deep drawability values r_m and Δr, planar and vertical anisotropy parameters, online in a steel mill (Altpeter et al., 1994).

To solve the multiple parametric inspection task of texture characterization independent of the influence of mechanical stress state, a hybrid method was developed which combines the ultrasonic and micromagnetic approaches. The micromagnetic approach is distinguished from the ultrasonic approach in two specific ways:

- For the micromagnetic approach, so far no theory exists which allows a description of the coefficients of texture distribution function, C_4^{11} and C_4^{13}, in terms of micromagnetic measuring quantities. Therefore, this approach has its basis only in the empirical correlation with mechanical technological values r_m and Δr determined in a calibration procedure. But activities have been observed worldwide that the determination of the coefficients of texture distribution function is possible by using ultrasonic phase velocity.
- All micromagnetic measuring quantities are sensitive to texture and residual stresses, whereby the measuring effect is comparable. The influence of residual stresses on ultrasonic phase velocities is lying in per mill range but the influence of texture in the percent range.

A combination of both approaches with different information content allows a texture determination independent of residual stresses.

A hybrid probe was developed which receives micromagnetic and ultrasonic measuring quantities. The probe or transducer is an electromagnetic acoustic transducer, which transmits and receives the so-called shear horizontal waves (see section: Ultrasonic techniques for materials characterization). For the micromagnetic measurements, a 3MA-II testing device was used, and for the ultrasonic measurements a device was used that automatically measured the ultrasonic time-of-flight.

A multiple regression analysis approach was performed in order to model the deep drawability values r_m and Δr. The coefficients of the regression analysis approach were determined on a calibration set of about 30 cold rolled strips with different thickness, steel composition, and steel grade (ST12, ST13, ST14 according to delivery conditions of the German Iron and Steel Institute) under load stresses.

A multiple regression analysis algorithm was performed in order to model r_m and Δr, whereby the coefficients C_4^{11} and C_4^{13} were integrated additionally to the micromagnetic and ultrasonic measuring quantity values (Altpeter et al., 2002). The results are shown in Fig. 9.10.

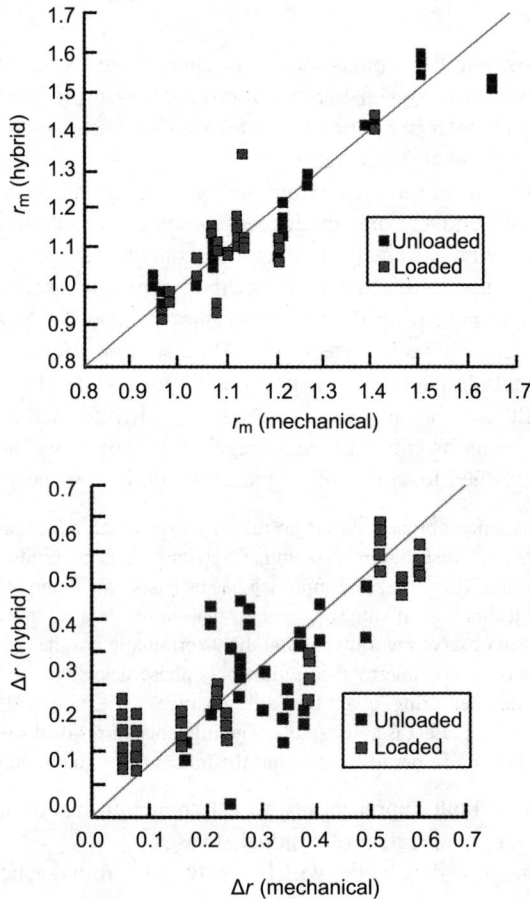

Figure 9.10 Deep drawability values r_m and Δr determined with nondestructive hybrid method (micromagnetic and ultrasonic methods) versus the mechanically determined reference values of samples with and without superimposed tensile loads.

9.4.3 Thermal aging

A typical example of the use of micromagnetic hybrid methods (like the 3MA approach) is the task of characterizing the lifetime of pressurized components in the nuclear power industry. In Germany, the low alloy, heat resistant steel 15 NiCuMoNb-5 (WB36) is used as piping and vessel material in boiling water reactors, pressurized water reactor nuclear power plants, and fossil fuel power plants as well.

After long-term service exposure at temperatures above 320°C, damage was observed during operation only in fossil fuel plants. The damage was related to service-induced decreases in toughness and increases in hardness of about 40 HV10 (Altpeter et al., 1999).

It has been shown that conventional Vickers hardness measurements are suitable to characterize embrittlement. However, the conventional Vickers hardness measurements are not repetitively applicable at the same locations and area-wide during service inspection. Therefore, the early nondestructive detection of the hardness increase is the most favorable solution of this problem.

In this way, micromagnetic hybrid surveillance of power plant components can inform the utility about ongoing aging processes.

The micromagnetic hybrid technique is very suitable to solve this task. By using 13 different micromagnetic measuring quantities derived from the analysis of the upper harmonics in the time signal of the tangential magnetic field strength and the analysis of Barkhausen noise, a hardness prediction of the service-simulated samples according to a micromagnetic hybrid model were possible (Altpeter et al., 2002). In the calibration step the micromagnetic measuring quantities were recorded at 10 different measuring positions on each sample.

First of all, the calibration was carried out on a calibration set, and second, the results were tested on independently selected samples with unknown hardness values. The quality of this regression analysis approach was also documented by proving independence of side effects such as plastic deformation (Fig. 9.11) and superimposed tensile loads (Fig. 9.12) in order to simulate residual stresses as disturbing influences. In each case the precipitation induced hardness can be predicted with high stochastic confidence. The Pearson correlation coefficient r^2 (Huber Carol et al., 2002) describes how good the correlation between the nondestructive determined hardness and the mechanical hardness is. The error band and the r^2 value describe the quality of the regression algorithm.

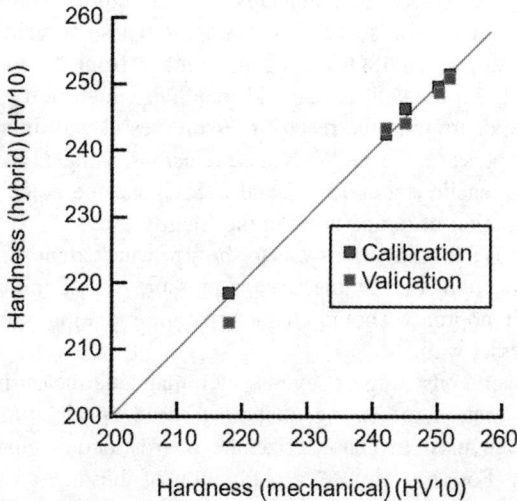

Figure 9.11 Hardness values determined with a nondestructive micromagnetic hybrid method versus the mechanically determined reference values of plastically deformed samples of WB36.

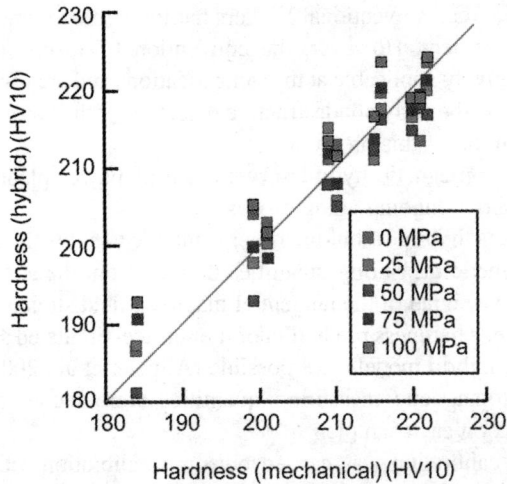

Figure 9.12 Hardness values determined with a nondestructive micromagnetic hybrid method versus the mechanically determined reference values of samples of WB36 with superimposed tensile loads.

9.4.4 Neutron induced embrittlement

In the use of nuclear power for energy production, some reactor core elements close to the reactor pressure vessel expose different high neutron fluxes. For this reason the installed materials will change in microstructure in form of embrittlement of the material due to aging time under operating conditions caused by the neutron irradiation. Usually, the plant safety regarding this microstructure changes is covered via destructive testing methods on surveillance samples. These surveillance samples are standard tensile samples and ISO-V specimens made from the exact same reactor pressure vessel material as well as its weld material, which hurry ahead in time at higher radiation exposure than the reactor pressure vessel wall in special irradiation channels of reactor pressure vessel. With these revisions, samples are taken and tested destructively in the tensile test at 150°C and 275°C, and the notch impact energy is determined as a function of temperature in the Charpy test.

The aim of the NDT investigations was to show to what extent some NDT methods can be used for the revisions of the reactor pressure vessel through an austenitic cladding. The NDT approach should characterize the radiation embrittlement of the reactor pressure vessel wall.

A hybrid approach consisting of several micromagnetic measuring quantities and dynamic magnetostriction measuring quantities based on electromagnetic acoustic transducers has been used to characterize the microstructure changes induced by neutron irradiation. For this purpose, a calibration of the target values, the shift in the ductile-to-brittle transition temperature measured at 41 J (ΔT41) and neutron fluence was carried out for both reactor pressure vessel steels and weld metals, which are located in western design nuclear power plants in use, as well as for reactor pressure vessel steels from nuclear power plants eastern designs. Therefore a demonstrator was used and built in the form of a nondestructive hybrid probe and measuring device for the combination of micromagnetic and magnetostriction measurements.

The hybrid method of micromagnetic (3MA: analysis of the magnetic Barkhausen noise, analysis of the upper harmonics in the time signal of the tangential magnetic field strength, and incremental permeability analysis) and dynamic magnetostriction has the potential for recurring revision of the reactor pressure vessel to characterize the neutron induced embrittlement through an austenitic cladding. For this purpose, each material (base material and inhomogeneous weld material) a pattern recognition-based calibration procedure was used. The calibration procedure was carried out with some surveillance samples from the base and the weld materials from nuclear power plants eastern and western designs. The hybrid measurements were acquired in a hot cell caused by the radiation of the samples. Fig. 9.13 ("Calibration") shows the calibration results for the base material.

The validation of the shift the ductile-to-brittle transition temperature measured at 41 J of the base material reached a correlation coefficient of 98.5% for the measurements, which could be characterized by means of pattern recognition (see Fig. 9.13, "Validation"). The validation of the neutron fluence obtained a correlation coefficient of 87.3% for the measurements, which could be characterized by means of pattern recognition-based calibration procedure. Caused by the microstructure changes during the welding process for the weld material a separate calibration procedure is needed. Therefore it is necessary to detect the location of the weld on the reactor pressure vessel through the austenitic cladding layer.

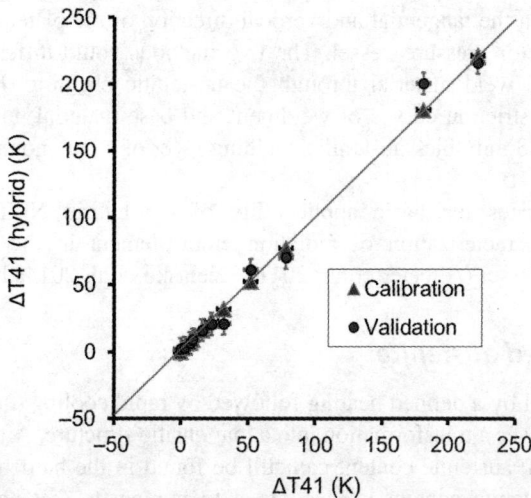

Figure 9.13 Correlation between nondestructively determined and reference values of the shift in the ductile−brittle transition temperature (ΔT41) of different reactor pressure vessel base materials.

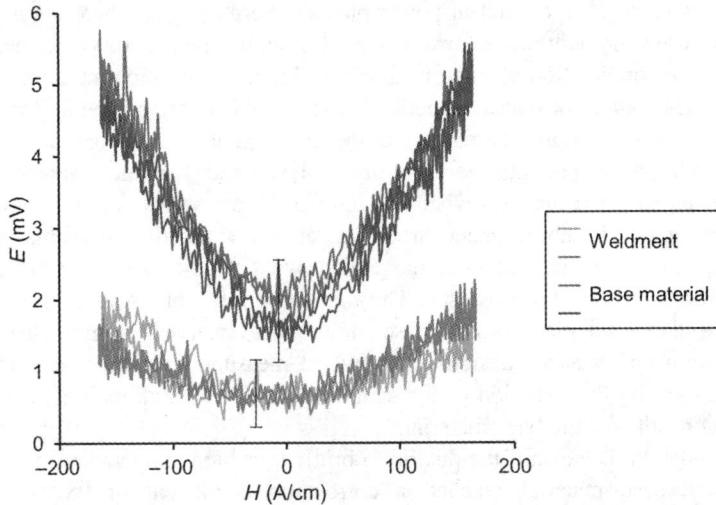

Figure 9.14 Dynamic magnetostriction curves of weld and base materials, measured through an approximately 8 mm thick austenitic cladding layer of a real, nonirradiated reactor pressure vessel (two plots for each material). The error bars indicate the standard deviation of the samples and were only shown once for improved visibility.

The potential of the nondestructive hybrid method for the recurring revision of reactor pressure vessels has been demonstrated in a nonirradiated reactor pressure vessel with an austenitic cladding. By the hybrid probe, scanning measurements were carried out in the tangential and vertical direction of the plated inside of a real, nonirradiated reactor pressure vessel. The two methods could differentiate between base material and weld material through the austenitic cladding. Fig. 9.14 shows dynamic magnetostriction curves of weldment and base material, measured through an approximately 8 mm thick austenitic cladding layer of a real, nonirradiated reactor pressure vessel.

This demonstrates the basic applicability of the hybrid NDT approach and nondestructive characterization of radiation embrittlement in the revision of the reactor pressure vessel (Altpeter et al., 2011; Szielasko et al, 2014a,b).

9.4.5 Retained austenite

Hardening of steel by a defined heating followed by rapid cooling (quenching) of the components achieves a transformation into a martensitic structure. After the hardening process a residual, austenite content can still be found in the hardened components. This residual austenite content is in contrast to martensite soft and nonmagnetic. The austenite can lead to undesirable changes in shape of the component, so the hardened component must be checked with regard to its residual austenite content.

The standard procedures for assessing residual austenite contents in heat treated components are the metallographic method and the X-ray diffraction. Both methods are destructive due to the necessary sample preparation and additionally connected to high time and cost. Therefore, the development of a NDT method for checking the residual austenite content is desirable. Electromagnetic hybrid measurement techniques allow nondestructive, rapid analysis of structural characteristics, and other material characteristics such as hardness, case depth, residual stresses, etc.

Therefore a calibration sample set with different residual austenite contents were produced with special heating and following deep freezing procedures. With these calibration sample set, the electromagnetic hybrid measuring quantity values were determined and correlated with the corresponding retained austenite content. But with a special probe adapted to the geometry of the anchor bolt, integral electromagnetic measuring quantity variables were recorded. Fig. 9.15 shows the hybrid probe prototype that the inspection task could be solved with.

The special hybrid probe is able to use the analysis of the magnetic Barkhausen noise, analysis of the upper harmonics in the time signal of the tangential magnetic

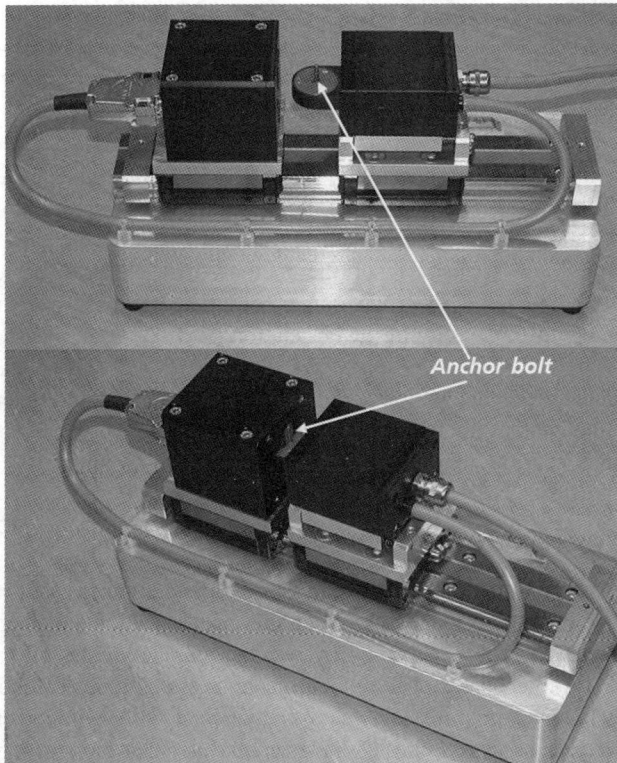

Figure 9.15 Special hybrid probe for anchor bolts with an inserted anchor bolt in the open state (top) and in the closed state (below). Hybrid measurements are possible in the closed state (connection between anchor bolt and special-shaped electromagnet on the left black part; coil(s) and Hall probe in the red part).
Ph.D. thesis Tschuncky (in German).

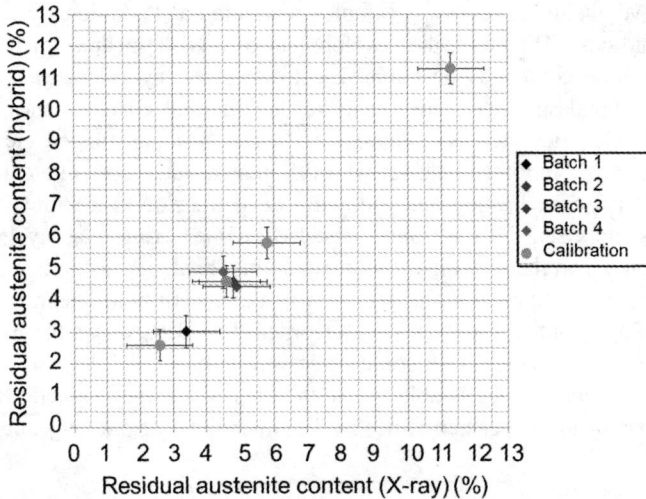

Figure 9.16 Correlation between the nondestructive (electromagnetic hybrid method based) residual austenite content values and the reference residual austenite content values determined with X-ray diffraction technique for the calibration sample set and samples from four different production batches.

field strength, incremental permeability analysis, and multifrequency eddy current techniques. The calibration procedures use methods of pattern recognition algorithms (described in Section 9.3.2.1), as shown in Fig. 9.7.

The suitability and robustness of the method was checked on the basis of components from the standard production of different hardness batches. Fig. 9.16 shows an example of the quality of the correlation (RMSE for all measured samples: 0.36%).

With electromagnetic hybrid methods, nondestructive and rapid determination of residual austenite content in anchor bolts can be achieved. The works form the basis of a process inspection system which can be integrated in the production.

9.4.6 Microstructure gradients

To reduce the weight of car combustion engines, cylinder crankcases of cast iron with vermicular graphite can be manufactured as this material allows the replacement of engines with greater cubic capacity by weaker engines with the same power (downsizing). However, the tool life when machining a cylinder crankcase made of cast iron with lamellar graphite are much greater than with cylinder crankcases of cast iron with vermicular graphite. Cylinder crankcases that have at key processing points (the cylinder inner surface) partially a microstructure gradient in the radial direction of lamellar graphite to vermicular graphite are a possible solution for this problem.

In order to be able to control the manufacturing process of such gradient cylinder crankcases nondestructive electromagnetic hybrid test methods can be used.

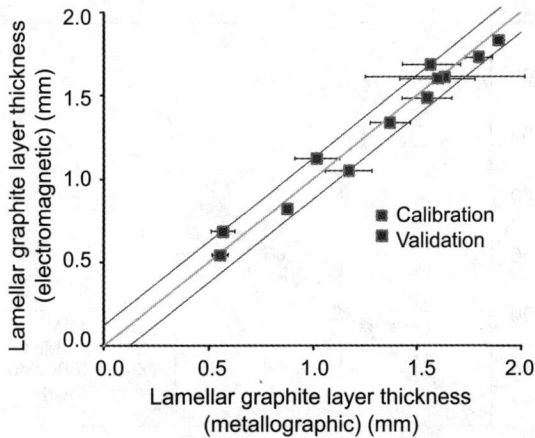

Figure 9.17 Electromagnetic-determined lamellar graphite layer thickness as a function of the metallographic-determined lamellar graphite layer thickness.

Electromagnetic hybrid methods can evaluate materials in different analyzing depths. Therefore, depth-dependent microstructure gradients can be analyzed by the combination of the methods.

A special hybrid probe adapted to the geometry of the cylinder to characterize the microstructure gradient has been developed. With this hybrid probe, surface scans were performed on the cylinder inner surface. The combined testing methods in this hybrid method are the analysis of the magnetic Barkhausen noise, analysis of the upper harmonics in the time signal of the tangential magnetic field strength, incremental permeability analysis, and multifrequency eddy current techniques.

For a quantitative nondestructive characterization of microstructure gradients a calibration procedure is needed. For this purpose, samples were taken from the gradient cylinder crankcases, and metallographic and optical analyses were used to determine the layer thickness of lamellar graphite. Based on the metallographic evaluation a calibration procedure by means of multiple regression analysis was carried out. After the calibration procedure the electromagnetic measuring quantities values can be used to determine the layer thickness of lamellar graphite. It was possible to achieve a good correlation between the electromagnetic and metallographic-determined layer thicknesses of lamellar graphite (correlation coefficient $R^2 = 0.93$; standard deviation = 0.06 mm; see Fig. 9.17). This calibration has successfully been validated (Abuhamad et al., 2007).

9.4.7 Simultaneous determination of different material properties

An advantage of nondestructive hybrid techniques is that different target quantities like residual stresses, hardness and hardening depth, deep drawability values, and yield strength can be predicted simultaneously. This can be used for the surveillance of industrial processes, eg, the laser hardening process.

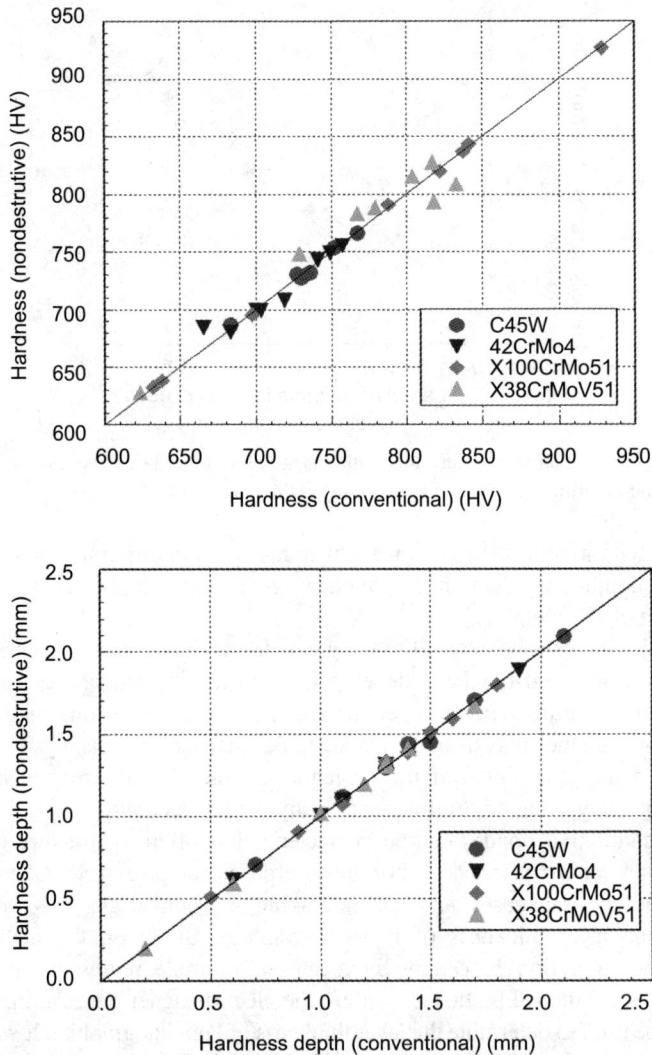

Figure 9.18 Correlations between nondestructive determined surface hardness (top) and hardness depth values (below) based on hybrid method and the conventional determined reference values.

One example is the surveillance of the laser hardening process by using nondestructive hybrid techniques. In this case, the micromagnetic hybrid probe can be integrated in a CO_2 laser-hardening system and measure the surface hardness and the case depth approximately 5 cm behind the laser spot (see Fig. 9.18).

A further application for micromagnetic hybrid techniques (like the 3MA approach) is the surveillance of grinding processes. The occurrence of grinding defects, eg, in gear wheels, has been a main problem for many years, which is caused by too much heat input during the grinding process. Modern grinding tools allow much higher grinding speed compared to former machines but on the other side, this can result in more defects. To get information on the quality of grinding microstructure states, the common method in industry is the nital etching technique. Grinding defects are indicated by the discoloration of the surface. This technique is effective as long as the production step defects are produced below the surface which are covered in the next production step by perfect finishing. Several examples of defective gear wheels investigated by hole drilling method and X-ray diffraction have shown that in a depth of surface, a perfect compressive state of several 100 MPa has been found. These hidden defects cannot be detected by nital etching.

Micromagnetic hybrid probes can detect the residual stresses at the surface and in a depth of approximately 100 μm at the same time. Fig. 9.19 displays the surface residual stresses determined by micromagnetic hybrid 3MA-II testing device versus the X-ray values and Fig. 9.20 displays the micromagnetic hybrid values in 100 μm depth versus the values determined by the hole drilling method (Wolter et al., 2003).

With such a micromagentic hybrid system, a testing speed of more than 10 Hz is available, and therefore the quality characteristics for the inspection of grinding states can be obtained in real time.

Experiences with micromagentic hybrid approaches and systems in the past two decades were the continuous mechanical property determination at steel strips, designed to produce car bodies, running with a speed of 300 m/min, for instance, in a continuous galvanizing and annealing line (Borsutzki, 1997; Borsutzki et al., 1999; Wolter and Dobmann, 2006).

Yield strength ($R_{p0,2}$), tensile strength (R_m), planar and vertical anisotropy parameters (r_m, Δr) are in the focus of quality assurance measures, all of them are defined by destructive test and cannot be measured continuously. Therefore a hybrid approach was used. The feasibility for practical use was demonstrated by performing online measurements in the steel plant.

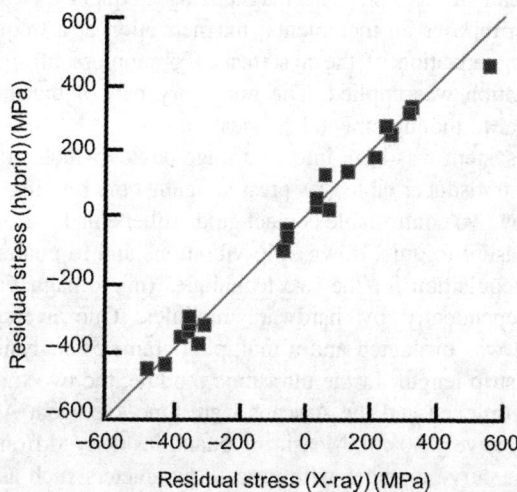

Figure 9.19 Residual stresses measured by 3MA-II hybrid testing device versus X-ray reference values on grinding states.

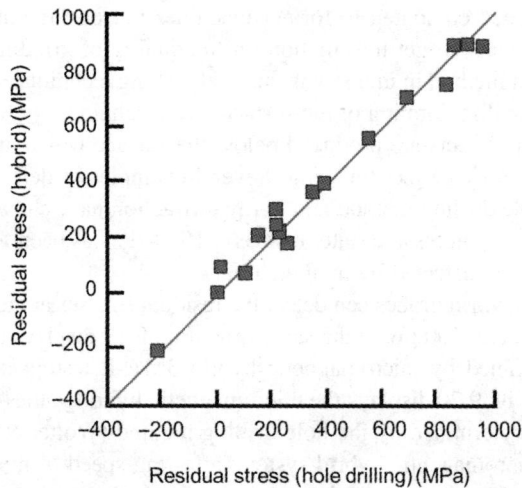

Figure 9.20 Residual stresses measured by 3MA-II hybrid testing device versus reference values measured by hole drilling method on grinding states.

Incremental permeability data were applied to predict the yield strength, ultrasonic time of flight data were performed in the rolling direction and at a 45 degree angle to it in order to determine the anisotropy effects. Two shear horizontal plate wave transmitter receiver combinations were applied to predict two coefficients of the orientation distribution function C_4^{11} and C_4^{13}. This transducer system was integrated into a sensor carrier. The incremental permeability probe is an electromagnetic yoke with 160 mm pole shoes distance from center to center. The eddy current coil was placed in a symmetric position to the pole shoes. A maximal magnetizing field strength of only 10 A/cm can be applied. The magnetizing frequency is 50 Hz. The eddy current probe superimposes an incremental magnetization at a frequency of 20 kHz to the dynamic magnetization of the hysteresis. To suppress lift-off changes of the probe, a phase rotation was applied. The imaginary part of the impedance is then directly proportional to the incremental permeability.

The transducer system was built into a carriage device which can be put in a roll table position. The transducer table was pressed against the running steel strip (speed 120−150 m/min) by two controllable contact guide rollers made of austenitic stainless steel. This is a measure to quiet down strip vibrations and to guarantee a controlled lift-off. The data acquisition for the two techniques (micromagnetic and ultrasonic) is performed independently by hardware modules. One averaged incremental permeability curve was evaluated and a multiple parameter analysis was performed at every 12 cm of strip length. In the ultrasonic module, the two transmitter receiver channels were multiplexed and the A scans were time averaged. A pair of (r_m, Δr) values was available every 80 cm. Measuring quantities derived from the incremental permeability profile curve together with process parameters such as sheet thickness, coiling temperature and carbon−nitrogen equivalent in the chemical composition were fed into a regression algorithm. The results are shown in Fig. 9.21.

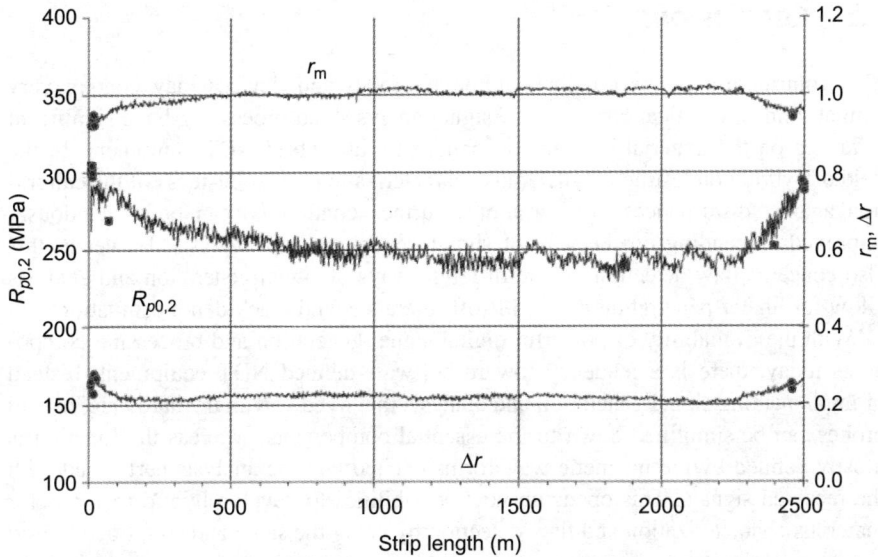

Figure 9.21 Nondestructively determined yield strength ($R_{p0,2}$) and deep drawing profiles (r_m, Δr) along 2500 m of a high strength strip steel (red dots: reference values determined after the nondestructive measurements).

Fig. 9.21 shows a yield strength profile along a coil of 2.5 km length (Borsutzki, 1997; Borsutzki et al., 1999). At the beginning and the end, an unacceptable increasing of strength is detected, higher than the upper acceptance level. The strength values are calculated by the 3 MA approach from measured micromagnetic data. The red dots indicate the selection of specimens taken to destructive verification tests after performing NDT. A good correlation between nondestructively (black lines) and destructively (red dots) obtained values was found. The residual standard errors found by validation are in the range 4−7% concerning the yield strength.

In several subsequent activities, micromagnetic/ultrasonic hybrid probes were developed and evaluated on line. The latest development in this series is a compact testing device and hybrid probe prototype called "MAGNUS," which performs the micromagnetic methods harmonics analysis in the magnetic tangential field strength, Barkhausen noise analysis, incremental permeability analysis, and eddy current impedance analysis together with ultrasonic time-of-flight analysis using guided SH (shear horizontal) and Lamb wave modes (Szielasko et al., 2014a,b). For determination of stresses time-of-flight measurements of SH-waves (see section: Ultrasonic techniques for materials characterization) and/or micromagnetic methods can be used. To determine texture coefficients and thus the anisotropy parameter the time-of-flight differences of several modes of the SH and Lamb wave (see section: Ultrasonic techniques for materials characterization) and/or micromagnetic methods can be evaluated. The micromagnetic and eddy current methods can determine several material properties like hardness, hardening depth, yield strength, tensile strength, etc.

9.5   Conclusions

This contribution points out that, although probes and devices may contain very similar, almost identical components, signal analysis and processing have significant influence on the material information obtained with hybrid NDT equipment. In the field of hybrid nondestructive materials characterization, the robustness of the calibration against disturbances (lot variations, surface conditions) can be tremendously improved, depending on the way of signal generation and analysis. However, this also concerns flaw detection, where different ways of signal generation and analysis allow for higher penetration depth, lift-off tolerance, and crack depth estimation.

With the availability of powerful digital signal generation and processing components today, there is a tendency toward software-defined NDT equipment. Instead of implementing signal generation and analysis in a fixed way in the hardware, hybrid probes can be simplified down to the essential components, whereas the function is mostly defined by the magnetic waveforms excited and the analysis performed with the received signals. This opens up new possibilities to develop hybrid methods for materials characterization and flaw detection by using the same and simplified hybrid probe and device components under different excitation signals and analysis procedures. This potential is already demonstrated today by the great bandwidth of the application range.

9.6   Future trends

Future applications of hybrid methods may also regard component testing, civil engineering, condition monitoring, and new applications in materials characterization since the hybrid probe design is scalable and can be adapted to many object shapes, sizes and NDT methods. In the future the number of effective and efficient combinations of NDT methods in hybrid approaches will grow rapidly and the use in industrial applications will also rise. Because of the advantages of hybrid methods opposite to disturbance influences, the wide application range and the possibility to determine several material properties at once, the profit of nondestructive hybrid methods for materials characterization is obvious.

References

Abuhamad, M., Altpeter, I., Dobmann, G., Kopp, M., 2007. Non-destructive characterization of cast iron gradient combustion engine cylinder crankcase by electromagnetic techniques. In: Proceedings of the DGZfP-Annual Assembly (2007), Fürth (in German).

Albers, S., Skiera, B., 1999. Regressionsanalyse. In: Hermann, A., Homburg, C. (Eds.), Marktforschung Grundlagen Methoden Anwendungen, Wiesbaden, pp. 205–236 (in German).

Altpeter, I., Theiner, W., Reimringer, B., 1986. Measurement of hardness and residual stresses using a nondestructive magnetic method. In: Hauk, V., Macherauch, E. (Eds.), Residual stresses, Oberursel, Deutsche Gesellschaft für Materialkunde (DGM), pp. 263–276.

Altpeter, I., Theiner, W., 1989. Spannungsmessung mit magnetischen Kräften, Handbuch für experimentelle Spannungsanalyse (in German), Düsseldorf, Rohrbach.

Altpeter, I., 1990. Spannungsmessung und Zementitgehaltsbestimmung in Eisenwerkstoffen mittels dynamischer magnetischer und magnetoelastischer Messgrößen (Ph.D. thesis). Saarland University, Saarbrücken (in German).

Altpeter, I., Dobmann, G., Meyendorf, N., Schneider, E., 1994. Erfassung von Eigenspannungen und Textur in Walzprodukten mit zerstörungsfreien Prüfverfahren (in German). In: Verein Deutscher Ingenieure (VDI): Blechbearbeitung'94. VDI Verlag, Düsseldorf, pp. 115−129 (VDI Berichte 1142).

Altpeter, I., Dobmann, G., Katerbau, K.-H., Schick, M., Binkele, P., Schmauder, S., 1999. Copper precipitates in steel 15NiCuMo Nb 5 (WB36) material properties and micro-structure, atomistic simulation, NDE by micromagnetic techniques. In: 25th MPA Seminar, Stuttgart, Germany.

Altpeter, I., Becker, R., Dobmann, G., Kern, R., Theiner, W., Yashan, A., 2002. Robust solu-tions of inverse problems in electromagnetic nondestructive evaluation. Inverse Problems 18, 1907−1921.

Altpeter, I., Dobmann, G., Kern, R., Schneider, E., Wolter, B., Spies, M., 2005. Industrial integration of residual stress measuring NDT-systems based on ultrasonics and micro-magnetics. In: Chimenti, D.E., Thompson, D.O. (Eds.), Review of Progress in Quantitative Nondestructive Evaluation, 24B. American Institute of Physics (AIP), Melville, New York, pp. 1387−1393 (AIP Conference Proceedings 760).

Altpeter, I., Dobmann, G., Hübschen, G., Kopp, M., Tschuncky, R., 2011. Nondestructive characterization of neutron induced embrittlement in nuclear pressure vessel steel micro-structure by using electromagnetic techniques. In: Chady, T., Gratkowski, S., Takagi, T., Udpa, S. (Eds.), Electromagnetic Nondestructive Evaluation (XIV): ENDE 2010, Studies in Applied Electromagnetics and Mechanics, vol. 35. IOS Press, Amsterdam, Washington, Tokyo, pp. 322−329.

Bartels, K.A., Fisher, J.L., 1995. Multifrequency eddy current image processing techniques for nondestructive evaluation. In: Proceedings of the International Conference on Image Processing, vol. 1. http://dx.doi.org/10.1109/ICIP.1995.529752.

Batista, L., Rabe, U., Altpeter, I., Hirsekorn, S., Dobmann, G., 2014. On the mechanism of nondestructive evaluation of cementite content in steels using a combination of magnetic Barkhausen noise and magnetic force microscopy techniques. Journal of Magnetism and Magnetic Materials 354, 248−256.

Blume, J., 1974a. Statistische Methoden für Ingenieure und Naturwissenschaftler. Band I: Grundlagen, Beurteilung von Stichproben, einfache lineare Regression, Korrelation. VDI Verlag, Düsseldorf, ISBN 3-18-403041-5 (in German).

Blume, J., 1974b. Statistische Methoden für Ingenieure und Naturwissenschaftler. Band II: Verteilungstests, einfache nichtlineare und mehrfache Regression, partielle Korrelation, Folgetests und Qualitätskontrolle. VDI Verlag, Düsseldorf, ISBN 3-18-403041-5 (in German).

Boller, C., Altpeter, I., Dobmann, G., Rabung, M., Schreiber, J., Szielasko, K., Tschuncky, R., 2011. Electromagnetism as a means for understanding materials mechanics phenomena in magnetic materials. Materialwissenschaft und Werkstofftechnik 42 (4), 269−278. http://dx.doi.org/10.1002/mawe.201100761.

Borsutzki, M., 1997. Process Integrated Determination of the Yield Strength and the Deep Drawability Properties R_m and Δr on Cold Rolled and Hot Dip Galvanized Steel Sheets (Ph.D. thesis). Saar University, Saarbrücken, Germany.

Borsutzki, M., Dobmann, G., Theiner, W.A., 1999. Online NDT characterization and mechanical property determination of cold rolled steel strips. Advanced sensors for metals processing. In: Proceedings of the International Symposium, 38th Annual Conference of Metallurgists of CIM.

Buschur, B., 2015. iSHA — ibg's Simultaneous Harmonic Analysis New Test Possibilities and Additional Test Reliability, vol. 21. TEST Patterns. http://www.ibgndt.com/issue21.php.

Dobmann, G., Pitsch, H., 1989. Magnetic tangential field-strength-inspection, a further NDT-tool for 3MA. In: Dobmann, Green, Hauk, Höller, Ruud (Eds.), Nondestructive Characterization of Materials III. Springer, Berlin, Heidelberg, New York, pp. 636−643.

Dobmann, G., Höller, P., 1990. Nondestructive determination of material properties and stresses. In: 10th Conference of NDE in the Nuclear and Pressure Vessels Industries, Glasgow, ASM International.

Giguère, S., Lepine, B.A., Dubois, J.M.S., 2001. Pulsed eddy current technology: characterizing material loss with gap and lift-off variations. Research in Nondestructive Evaluation 13 (3), 119−129. http://dx.doi.org/10.1080/09349840109409692.

Heutling, B., Reimche, W., Krys, A., Grube, L., Stock, M., Bach, F.-W., Kroos, J., Stolzenberg, M., Westkämper, G., 2004. Online NDE of mechanical-technological material characteristics of cold rolled steel strips by using the harmonic analysis of eddy current signals. International Journal of Applied Electromagnetics and Mechanics 19 (1−4), 445−451.

Huber Carol, C., Balakrishnan, N., Nikulin, M., Mesbah, M., 2002. Goodness of Fit, Tests and Model Validity Statistics for Industry and Technology. Birkhäuser, Boston.

Jain, N., 2014. The Rebirth of Eddy Current Nondestructive Testing. Quality Magazine. http://www.qualitymag.com/articles/92056-the-rebirth-of-eddy-current-nondestructive-testing.

Kern, R., Theiner, W., Valeske, B., Meyer, R., 1996. Process integrated nondestructive testing of laser-hardened components. In: Bartos, A.L., Green, R.E., Ruud, C. (Eds.), Nondestructive Characterization of Materials VII. Part 2. Trans Tech Publications, Zürich, pp. 687−694.

Laub, U., Altpeter, I., 1998. Cast iron inspection by means of micromagnetic procedures. In: 7th European Conference on Nondestructive Testing ECNDT, vol. 1, pp. 1094−1101.

Lewis, A.M., Michael, D.H., Lugg, M.C., Collins, R., 1988. Thin-skin electromagnetic fields around surface-breaking cracks in metals. Journal of Applied Physics 64 (8), 3777−3784.

Matzkanin, G.A., Beissner, R.E., Teller, C.M., 1979. The Barkhausen Effect and its Applications to Nondestructive Evaluation. NTIAC. Southwest Research Institute, San Antonio, Texas, pp. 79−82.

Schach, S., Schäfer, T., 1978. Regressions- und Varianzanalyse. Springer-Verlag, Berlin-Heidelberg, New York, ISBN 3-540-08727-3 (in German).

Szielasko, K., 2009. Development of Metrological Modules for Electromagnetic Multiparameter Materials Characterization and Testing (Ph.D. thesis). Saarland University, Saarbrücken (in German).

Szielasko, K., Rabung, M., Altpeter, I., 2009. Magnetic Barkhausen noise analysis in the audio frequency range. In: Proceedings of the 7th International Conference on Barkhausen Noise and Micromagnetic Testing.

Szielasko, K., Mironenko, I., Altpeter, I., Herrmann, H.-G., Boller, C., 2013. Minimalistic devices and sensors for micromagnetic materials characterization. IEEE Transactions on Magnetics 49 (1), 101−104.

Szielasko, K., Youssef, S., Niese, F., Weikert, M., Sourkov, A., Altpeter, I., Herrmann, H.-G., Dobmann, G., Boller, C., 2014a. Multi-method probe design for the electromagnetic characterization of advanced high strength steel. In: Capova, K., Udpa, L., Janousek, L., Rao, B.P.C. (Eds.), Electromagnetic Nondestructive Evaluation (XVII). IOS Press, Amsterdam, Washington, Tokyo, pp. 52−59.

Szielasko, K., Tschuncky, R., Rabung, M., Seiler, G., Altpeter, I., Dobmann, G., Herrmann, H.-G., Boller, C., 2014b. Early detection of critical material degradation by means of electromagnetic multi-parametric NDE. In: Chimenti, D.E., Thompson, D.O., Bond, L.J. (Eds.), Review of Progress in Quantitative Nondestructive Evaluation, 40th Annual Review of Progress in Quantitative Nondestructive Evaluation Incorporating the 10th International Conference on Barkhausen Noise and Micromagnetic Testing, vols. 33A and B. American Institute of Physics (AIP), Melville, New York, 8 pages (AIP Conference Proceedings 1581).

Stork, D., 2001. Pattern Classification. John Wiley & Sons, Inc.

Theiner, W., Kern, R., Graus, M., 1999. Process integrated nondestructive testing (PINT) for evaluation of hardness, case depth and grinding defects. In: Brusey, B.W., Bussière, J.F., Dubois, M., Moreau, A., Canadian Institute of Mining, Metallurgy and Petroleum (Eds.), Advanced Sensors for Metals Processing. Met Soc, Quebec, pp. 159–171.

Tschuncky, R., 2011. Sensor- und geräteunabhängige Kalibrierung elektromagnetischer zerstörungsfreier Prüfverfahren zur praxisorientierten Werkstoffcharakterisierung (Ph.D. thesis). Universität des Saarlandes, Saarbrücken, Germany (in German).

Udpa, S.S., Moore, P.O., 2004. Nondestructive Testing Handbook In: Electromagnetic Testing, third ed., vol. 5. American Society of Nondestructive Testing.

Willems, H., Jaskolla, B., Sickinger, T., Barbian, O.-A., Niese, F., 2010a. A new ILI tool for metal loss inspection of gas pipelines using a combination of ultrasound, eddy current and MFL. In: American Society of Mechanical Engineers (ASME): 8th International Pipeline Conference (IPC2010), pp. 557–564.

Willems, H., Jaskolla, B., Sickinger, T., Barbian, O.A., Niese, F., 2010b. Advanced possibilities for corrosion inspection of gas pipelines using EMAT-technology. In: European Federation for Non-destructive Testing (EFNDT): European Conference on Non-destructive Testing (10) ECNDT.

Wolter, B., Theiner, W., Kern, R., Becker, R., Rodner, Ch., Kreier, P., Ackeret, P., 2003. Detection and quantification of grinding damage by using EC and 3 MA techniques. In: Proceedings of the 4th International Conference on Barkhausen Noise.

Wolter, B., Dobmann, G., 2006. Micromagnetic testing for rolled steel. In: European Conference on Nondestructive Testing (9), pp. 25–29.

Index

图书在版编目（ＣＩＰ）数据

材料表征的无损检测方法：英文／（德）格哈德·惠布什等主编. --长沙：中南大学出版社，2017.9

ISBN 978－7－5487－2992－1

Ⅰ.①材… Ⅱ.①格… Ⅲ.①工程材料－无损检验－英文 Ⅳ.①TB302.5

中国版本图书馆 CIP 数据核字(2017)第 230378 号

材料表征的无损检测方法
CAILIAO BIAOZHENG DE WUSUN JIANCE FANGFA

Gerhard Hübschen Iris Altpeter

Ralf Tschuncky Hans-Georg Herrmann 主编

□责任编辑	胡　炜
□责任印制	易红卫
□出版发行	中南大学出版社
	社址：长沙市麓山南路　　　　邮编：410083
	发行科电话：0731－88876770　　传真：0731－88710482
□印　　装	长沙鸿和印务有限公司

□开　　本	720×1000　1/16　□印张 20.75　□字数 533 千字
□版　　次	2017 年 9 月第 1 版　□2017 年 9 月第 1 次印刷
□书　　号	ISBN 978－7－5487－2992－1
□定　　价	115.00 元